ENCYCLOPEDIA OF ORGANIC CHEMISTRY

ENCYCLOPEDIA OF ORGANIC CHEMISTRY

Vol. 3

By
Sananda Chatterjee
DAE, DCA, CMCNet
Scientific Consultant
Institute for Natural Sciences
Kolkata
(West Bengal)

DISCOVERY PUBLISHING HOUSE PVT. LTD.
NEW DELHI-110 002

Published by:
Tilak Wasan
DISCOVERY PUBLISHING HOUSE PVT. LTD.
4383/4B, Ansari Road, Darya Ganj
New Delhi-110 002 (India)
Phone : +91-11-23279245, 43596064-65
Fax : +91-11-23253475
E-mail : parul.wasan@gmail.com
discoverypublishinghouse@gmail.com
web : www.discoverypublishinggroup.com

***First Edition:* 2012**

ISBN: 978-81-8356-875-3 (Set)

Encyclopedia of Organic Chemistry

Printed at:
Shree Balaji Art Press
Delhi

Preface

Organic Chemistry is that branch of chemistry, which deals with the reactions related to the life form, or one of the source of the life form, carbon (we can say the end of it).

This carbon mainly conjugates with different elements such as nitrogen, hydrogen and oxygen sometimes metals and other non-metals also). Organic chemistry studies mainly the extraction, structure, properties, of these compounds.

In this book chapters deals with the different methods of purification, extraction of compounds, some processes were used and devised by Alchemists.

The extraction preservation of natural dyes are examples of alchemical work, these been discussed in the chapter "dyes indicators and pigments".

Dyes is the mal-pronunciation of the damathi word dyum meaning colour, it was, the ancient Egyptians who use different natural pigments to extract different colours.

In this book the organic compounds has been mainly classified into two broad groups, the aromatic and the aliphatic compounds. The extraction of alcohol is definitely an alchemical process'.

The destructive distillation of wood is one of the finest processes devised by the alchemists for the preparation of methanol is one of the important topics of this book.

Looking ahead in to the modem world, the world of future, the subject comes in vision is the biochemistry, the base of which is organic chemistry, the chapter containing carbohydrates, and the chapter, chemical compounds of biological and biochemical interest is the best element of focus for the students stepping foot in the advanced staircase of biochemistry.

The last chapter, chapter 41, discusses the different organic reactions and their suitably mechanisms, this will help those students who are very much afraid of organic formulas and reactions. This chapter also deals with the chirality one of the interest of Dr. Palash Gangopadhyay (author's friend).

Organic Chemistry is not a mere subject, it is the rhythm of life tuned in the essence of carbon, which looks black but carries the dazzleness of diamond in its heart.

SANANDA CHATTERJEE

Contents

CHAPTER

12

Aldehydes and Ketones

Introduction

Aldehydes and ketones are two important classes of organic compounds. They both contain the common divalent carbonyl [>C=O] group. In aldehyde one of two valences of this group is always satisfied by

Aldehydes or Alkanals

Formaldehyde, methaldehyde, methanal	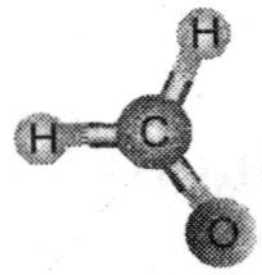
Acetaldehyde, ethaldehyde, or ethanal	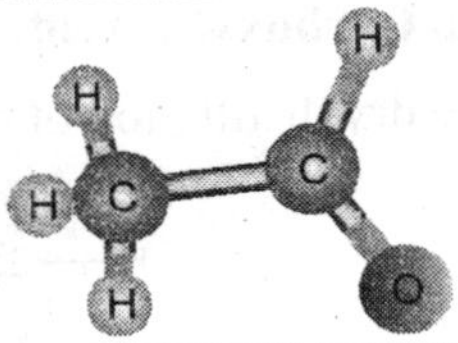
Propionaldehyde, propaldehyde, popanal	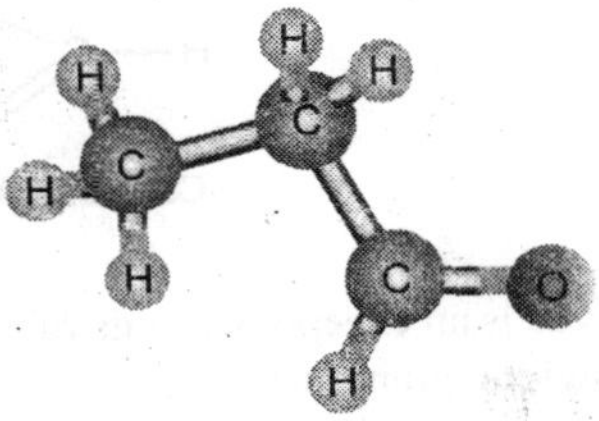

a hydrogen atom and other by an alkyl (R) group*. In ketones both the valences of the carbonyl group are satisfied by similar or different alkyl group e.g.

$$\begin{matrix} R \\ H \end{matrix}\!\!>C=O \qquad \begin{matrix} R \\ R \end{matrix}\!\!>C=O \qquad \begin{matrix} R \\ R_1 \end{matrix}\!\!>C=O$$

Aldehydes — Simple ketone both the alkyl are same — mixed ketones both the alkyl are different

Aldehydes and ketones are formed by the regulated oxidation of primary and secondary alcohols respectively. The aldehydes closely resemble the ketones as regards methods of preparation and also in properties.

General Methods of Preparation

Aldehydes may be preparation by the following general methods: -

1. From Primary Alcohols:

(a) By oxidation with potassium dichromate and sulphuric acid or air in the presence of platinum as catalyst.

(i) $CH_3OH \xrightarrow{O} HCHO + H_2O$

(ii) $C_2H_5OH \xrightarrow{O} CH_3CHO + H_2O$

(b) By catalytic decomposition of in the presence of copper at 300°C

(i) $\underset{\text{methanol}}{CH_3OH} \xrightarrow{Cu/300^\circ C} \underset{\text{formaldehyde}}{HCHO} + H_2$

(ii) $\underset{\text{ethanol}}{C_2H_5OH} \xrightarrow{Cu/300^\circ C} \underset{\text{acetaldehyde}}{CH_3CHO} + H_2$

2. From Mono Carboxylic Acid:

(a) By the dry distillation of calcium salt of the corresponding fatty acid with calcium formate.**

$$\underset{\text{Calcium formate}}{(HCOO)_2Ca} \longrightarrow CaCO_3 + \underset{\text{formaldehyde}}{H-CHO}$$

*Formaldehyde is an exception to this rule, it is the >C = O group carries 2-H atoms and no alkyl group and therefore atoms are unique properties.

**formaldehyde is mainly prepared by the dry distillation of calcium formate.

(b) ***By catalytic decomposition***: Mixed vapours of a fatty acid and formic acid are passed over heated catalyst Al_2O_3 (aluminium oxide) at 400°C.

(i) $HCOOH + CH_3COOH \rightarrow CH_3CHO + CO_2 + H_2O$
Formic acid, Acetic acid, Acetaldehyde

(ii) $HCOOH + HCOOH \rightarrow HCHO + CO_2 + H_2O$
Formic acid, Formic acid, Formaldehyde

3. From Acid Chloride:

By catalytic reduction with hydrogen in the presence of finely divided platinum or palladium precipitated on barium sulphate – Rosenmund's Reaction.

(i) $H\text{-}C(=O)Cl + 2H \xrightarrow{Pt} H\text{-}C(=O)H + HCl$
Formyl chloride, Formaldehyde

(ii) $H_3C\text{-}C(=O)OH + H\text{-}C(=O)OH \longrightarrow H_3C\text{-}CH(=O) + CO_2 + H_2O$
Acetic acid, Formic acid, Acetaldehyde

4. From Alkaledine Halides:

(With both the halogens attached to the same terminal carbon atom) by hydrolysis with aqueous alkalies or super heated steam.

$H\text{-}CH_2\text{-}Cl \xrightarrow{H_2O} H\text{-}CH(OH)\text{-}OH \longrightarrow H\text{-}CH{=}O + H_2O$
Chloromethane, Methanediol unstable, Formaldehyde

5. From Grignard's Reagent:

By the action of ethyl formate

$H_3C\text{-}Mg\text{-}I + H\text{-}C(=O)(OCH_2CH_3) \longrightarrow H\text{-}C(OH)(O\text{-}CH_2\text{-}CH_3)(CH_3) \xrightarrow{2H_2O} Mg(OH)I \longrightarrow C_2H_5OH + H_3C\text{-}CH{=}O$
Methyl magnesium Iodide, Ethyl formate, 1-ethoxy magnesium iodide, Magnesium hydroxy Iodide, Ethyl Alcohol, Acetaldehyde

6. From Acetylene Hydrocarbons:

By warming a solution of acetylene hydrocarbon in sulphuric with water in presence of mercuric sulphate.

$H\text{-}C\equiv C\text{-}H \xrightarrow[H_2SO_4,\ 50 - 60°C]{+H_2O} H_3C\text{-}CH{=}O$
Acetylene, Acetaldehyde

Chemical Properties

Aldehydes are very reactive compounds and show the following reactions: -

1. **Oxidation**: With $K_2Cr_2O_7$ and therefore they act as a strong reducing agents

$$\underset{\text{Acetaldehyde}}{CH_3CHO} + O \rightarrow \underset{\text{Acetic acid}}{CH_3COOH}$$

2. **Reduction**: (i) *The ammonical silver nitrate solution (Tollen's reagent) silver mirror experiment*

$$AgNO_3 + NH_4OH \rightarrow AgOH + NH_4NO_3$$
$$2AgOH \rightarrow Ag_2O + H_2O$$
$$CH_3CHO + Ag_2O \rightarrow CH_3COOH + 2Ag(\text{mirror})$$

(ii) *Fehlling's solution into red cuprous oxide:* When a solution of sodium hydroxide is added to a solution of copper sulphate, a blue precipitate of $Cu(OH)_2$ is formed. The precipitate dissolves in a solution of *Rochelle Salt,* sodium potassium tartrate

CHOHCOONa
|
CHOHCOOK

The blue solution thus prepared is called Fehlling's Solution. It oxidizes aldehydes to acids itself is reduced to red CuO

$$CuSO_4 + 2NaOH \rightarrow Cu(OH)_2 + Na_2SO_4$$
$$Cu(OH)_2 \rightarrow Cu_2O + H_2O$$
$$CH_3CHO + 2CuO \rightarrow CH_3COOH + Cu_2O$$

[Rochelle Salt simply helps in reduction].

The reduction of ammonical silver nitrate solution or Fehlling's solution can be used for the testing the aldehydes.

(iii) With sodium amalgam and dilute sulphuric acid or zinc dust and acetic acid.

(a) $$\underset{\text{Formaldehyde}}{HCHO} + 2H \rightarrow \underset{\text{Methyl alcohol}}{CH_3OH}$$

(b) $$\underset{\text{Acetaldehyde}}{CH_3CHO} + 2H \rightarrow \underset{\text{Ethyl alcohol}}{CH_3CH_2OH}$$

3. **Addition Reactions**: With sodium bi-sulphite, hydrogen cyanide ammonia and Grignard's reagent)

(i) ***With Sodium Sulphite Solution***: Sulphite compound precipitates are produced

$$\underset{\text{Acetaldehyde}}{H_3C{-}CH{=}O} + \underset{\text{Sodium Hydrogen Sulphite}}{NaHSO_3} \longrightarrow \underset{\text{Acetaldehyde bisulphite}}{H_3C{-}CH(OH){-}SO_2{-}ONa}$$

The Bisulphite compounds are crystalline and can be decomposed by warming with sodium carbonate are dilute hydrochloric acid. This reaction is used for the separation and identification of acetaldehydes.

(ii) ***With hydrocyanic acid***: Cyanohydrins are produced, the addition ethyl amides accelerates the reduction considerably

H—C(=O)—H + HCN ⟶ H—C(OH)(H)—C≡N

Acetaldehyde Hydrogen cyanide Acetaldehyde cynanohydrin

POISON 6

HIGHLY POISONOUS

The cyanohydrins are important starting substances in the in organic chemistry.

(iii) ***With Ammonia***: Aldehyde ammonia is formed

H_3C—C(=O)—H + NH_3 ⟶ H_3C—C(OH)(H)(NH_2)

Acetaldehyde Ammonia 1-aminoethanol ammonia aldehyde

Formaldehyde behaves in a different way and does not give addition compound. It is, however, gives hexa-methylene tetrammine or Urotropine (distinction from CH_3CHO)

H—C(H)=O + NH_3 ⟶ Urotropine

Formaldehyde Urotropine

Condensation Reaction

With Hydoxymine. (Reaction with NH_2OH, $NH_2 - NH_2$ $C_6H_5NHNH_2$ PCl_5 and alcohols)

H—C(H)=O + H_2N—OH ⟶ H—C(H)=NOH

Formaldehyde Hydroxylamine Formaldehyde oxime

H_3C—C(H)=O + H_2N—OH ⟶ H_3C—C(H)=NOH

Acetaldehyde Hydroxylamine Acetaldehyde oxime

The oximes are crystalline compounds and are decomposed by acids to give back the original aldehydes. Hence, the formation is used for the isolation and identification of aldehydes.

With Hydrazine: Hydrazines are formed

$$H{-}\overset{H}{\overset{|}{C}}{=}O + H_2N{-}NH_2 \longrightarrow H{-}\overset{H}{\overset{|}{C}}{=}NNH_2 + H_2O$$

Formaldehyde, Hydrazine, Formaldehyde hydrazone, Water

$$H_3C{-}\overset{H}{\overset{|}{C}}{=}O + H_2N{-}NH_2 \longrightarrow H_3C{-}\overset{H}{\overset{|}{C}}{=}NNH_2 + H_2O$$

Acetaldehyde, Hydrazine, Acetaldehyde hydrazone, Water

With Phenyl Hydrazine: Phenyl hydrazines are produced

$$H{-}\overset{H}{\overset{|}{C}}{=}O + C_6H_5{-}NH{-}NH_2 \longrightarrow H{-}\overset{H}{\overset{|}{C}}{=}N{-}\overset{H}{\overset{|}{N}}{-}C_6H_5 + H_2O$$

Formaldehyde, Phenylhydrazine, Formaldehyde phenylhydrazone, Water

$$H_3C{-}\overset{H}{\overset{|}{C}}{=}O + C_6H_5{-}NH{-}NH_2 \longrightarrow H_3C{-}\overset{H}{\overset{|}{C}}{=}N{-}\overset{H}{\overset{|}{N}}{-}C_6H_5 + H_2O$$

Acetaldehyde, Phenylhydrazine, Acetaldehyde phenylhydrazone, Water

The reaction may also employ for the separation and purification of aldehydes since the hydrazones regenerated aldehydes on warming with acid.

With Phosphorous Pentachloride: Alkaledine chlorides are formed

$$H{-}\overset{H}{\overset{|}{C}}{=}O + PCl_5 \longrightarrow H{-}\underset{Cl}{\underset{|}{\overset{H}{\overset{|}{C}}}}{-}Cl + POCl_3$$

Formaldehyde, Dichloromethane

$$H_3C{-}\overset{H}{\overset{|}{C}}{=}O + PCl_5 \longrightarrow H_3C{-}\underset{Cl}{\underset{|}{\overset{H}{\overset{|}{C}}}}{-}Cl + POCl_3$$

Acetaldehyde, 1,1-Dichloroethane

With alcohol: Acetals are produced

$$H{-}\overset{H}{\overset{|}{C}}{=}O + H_3C{-}OH \longrightarrow H{-}\underset{OCH_3}{\underset{|}{\overset{H}{\overset{|}{C}}}}{-}O{\cdot}CH_3$$

Formaldehyde, Methanol, Dimethoxymethane

$$H_3C{-}\overset{H}{\overset{|}{C}}{=}O + H_3CCH_2{\cdot}OH \longrightarrow H_3C{-}\underset{OCH_2-CH_3}{\underset{|}{\overset{H}{\overset{|}{C}}}}{-}O{\cdot}CH_2{-}CH_3$$

Acetaldehyde, Ethanol, 1,1-Diethoxyethane

A little sulphuric acid or ammonium chloride is found to act as a catalyst in the above reaction.

Other Reaction of Aldehydes

With hot Concentrated Caustic Alkalies: Dark brown complex compounds called resins are formed when acetaldehyde reacts with sodium hydroxide. The reaction is known as resinification and may also be employed as the testing of aldehydes.

Formaldehyde does not form resin but undergoes Cannizzaro's Reaction [distinction from acetaldehydes].

Cannizzaro's Reaction

In 1858, two years after Amadeo Avogadro's death, his fellow Italian **Stanislao Cannizzaro** (1826-1910) outlined a course in theoretical chemistry for students at the University of Genoa—where he had to teach without benefit of a laboratory. He used Avogadro's hypothesis as a pathway out of the confusion rampant among chemists about atomic weights and the fundamental structure of chemical compounds.

By all accounts Cannizzaro was much clearer in his explanations than Avogadro, and as an organic chemist he also showed how Avogadro's ideas could be applied to this branch of chemistry. In 1860 the First International Chemical Congress was held in Karlsruhe, Germany, to settle some of the contemporary chemical disputes—how to define "molecule" and "atom", what chemical nomenclature to use, how to determine atomic weights, and so on. After much discussion the chemists agreed to return home to decide for themselves how to proceed. However, many participants carried away a handout, a printed version of Cannizzaro's outline, that seemed convincing upon later reading. At this time Cannizzaro was in the midst of eventful chemical and political careers. He was born in Palermo, Sicily, where his father was a magistrate and the minister of police, and he later attended medical school there, which kindled an interest in chemistry. Despite his family's connections to the royal court in Naples, he joined the anti-monarchical 1848 revolution in Sicily. When it failed, he fled to Paris, where he resumed his chemical studies. After returning to Italy, he held academic appointments in Alessandria, where he worked out the "Cannizzaro reaction"—the self-oxidation and self-reduction of aldehydes—and Genoa, where he expounded Avogadro's hypothesis. He next supported Giuseppe Garibaldi's Sicilian revolt of 1860 and took part in the new government centered in Palermo. During this time he expanded the program of chemical studies at the university there. Upon Italian unification in 1871 he moved to Rome, where he continued his roles as a public figure and as a chemical scientist and educator.

The Definition

The reaction which certain aldehydes may undergo in concentrated alkali; one molecule of the aldehyde is reduced to the corresponding alcohol and another molecule is simultaneously oxidized to the salt of a carboxylic acid.

Aromatic aldehydes, and aliphatic ones with no α-hydrogen, give the Cannizzaro aldehyde disproportionation when treated with conc. NaOH or other strong bases. In this reaction one molecule of aldehyde oxidizes another to the acid and is itself reduced to the primary alcohol.

$$2\ RCHO \xrightarrow{OH^-} RCH_2OH + RCOO^-$$

$$\underset{\text{Formaldehyde}}{2HCHO} + NaOH \rightarrow \underset{\text{Methanol}}{HCH_2OH} + \underset{\text{Sodium formate}}{HCOONa}$$

, Aldehydes with an α-hydrogen do not give the reaction, because when these compounds are treated with base the aldol condensation is much faster. Normally, the best yield of acid or alcohol is 50% each, but this can be altered in certain cases.

When the aldehyde contains a hydroxide group in the ring, excess base oxidizes the alcohol formed and the acid can be prepared in high yield (the OH is reduced to H_2). On the other hand, high yields of alcohol can be obtained from almost any aldehyde by running the reaction in the presence of methanal. In this case the methanal reduces the aldehyde to alcohol and is itself oxidized into formic acid (which may be further oxidized into CO_2. In such a case, where the oxidant aldehyde differs from the reductant aldehyde, the reaction is called the crossed Cannizzaro reaction.

The mechanism of the Cannizzaro reaction involves a hydride shift. First OH^- adds to the C=O to give (a), which may lose a proton in the basic solution to give the diion (b)

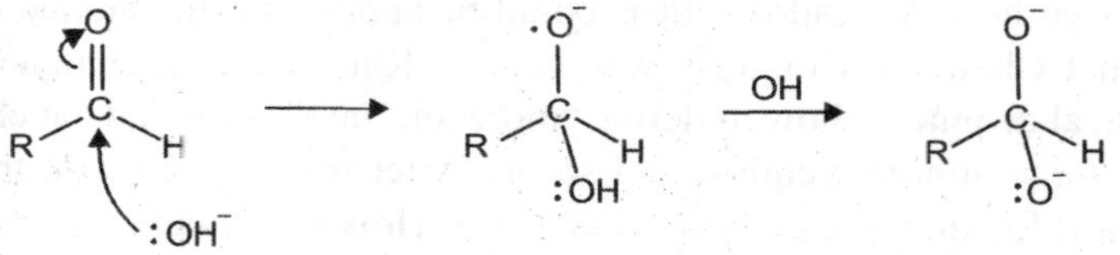

(a) (b)

The strong electron-donating character of O^- greatly facilitates the ability of the aldehydic hydrogen to leave with its electron pair. Of course, this effect is even strong in (b). When the hydride does leave, it attacks another molecule of aldehyde. The hydride can come from (a) or (b)

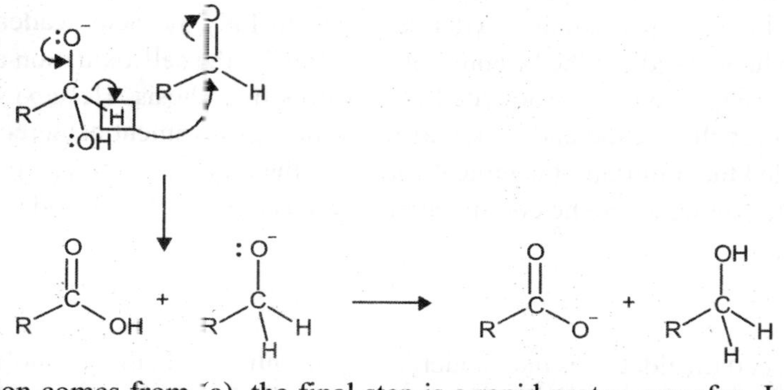

If the hydride ion comes from (a), the final step is a rapid proton reansfer. In the other case, the acid salt is formed directly, and the alkoxide ion acquires a proton from the solvent. Evidence for this mechanism is: (1) The reaction can be first order in base and second order in substrate (thus going through (a) or, at higher base concentrations, second order in each [going through (b)]; and (2) when the reaction was run in D_2O, the recovered alcohol contained no α-deuterium, indicating that the hydrogen comes from another mole of aldehyde and not from the medium.

Experimental Procedure for the Cannizzaro's Reaction

Dissolve 3.0 g of solid KOH (corrosive) in 3.0 mL of deionized (distilled) water by gently stirring the mixture in a 50 or 100 mL beaker. Cool the mixture to room temperature. Place 3.0 mL of benzaldehyde (toxic/flammable) (droppers should be marked) in the test tube provided and add the concentrated KOH you have just made to it. Cork the tube securely and shake the tube vigorously until an emulsion is formed. Place the test tube in the rack provided by the instructor. (Do not use separate racks). Allow the test tube to stand until the next lab period. (Or, at least, 4 days).

Add about 2.0 mL of water to the mixture, stopper and shake the test tube. If the entire solid has not dissolved then decant off the solution into a clean beaker and add more water to the test tube and repeat shaking of the test tube. If there are solid lumps, break them up with a stirring rod and add water and then shake the mixture. Repeat the procedure until all or most of the solid has dissolved. Pour the solution into a separatory funnel, and extract the solution 2 times with 8mL portions of diethyl ether (listed/ **Flammable**). Combine the ether extracts in one beaker. If a solid separates out of the solution when the ether is added, add water in small quantities until the solid has redissolved, and then proceed with the extraction. Be sure to avoid the formation of an emulsion by shaking the mixture gently at first, and more vigorously once you observe how long it takes the emulsion to breakup after shaking. The ethereal solution is dried using anhydrous $MgSO_4$. The dried ether solution is analyzed using the GC for the relative amounts of benzyl alcohol to benzaldehyde. Dispose of this product in the designated waste bottle.

The water solution left after the ether extraction is poured into a mixture of 8 mL of concentrated HCl (**Corrosive**), 5 mL of water and 3 ice cubes. Stir the mixture vigorously during the addition. Cool the resultant mixture in an ice-bath, and collect the solid benzoic acid by vacuum filtration using the 5.5cm Buchner funnels. Dry the solid with the filter paper in the brown envelopes provided at the organic prep table. **Fold, do not lick**, the envelope closed. Determine its melting point when dry. After taking the melting point remember to dispose of the benzoic acid and its envelope in the designated waste box.

Cross Cannizzaro's Reaction

If the aldehydes are different, the reaction is called the "crossed Cannizarro reaction" e.g.

$$\underbrace{HCHO + CH_3CHO}_{\substack{\text{formaldehyde + acetaldehyde} \\ \text{one molecule each}}} + NaOH \rightarrow C_2H_5OH + CH_3OH + NaCOOH + CH_3COONa$$

Schiff's Reaction: Aldehydes restore pink colour of rosaniline dye decolourized by sulphur dioxide. The reaction is used as test of aldehyde.

Chlorination

Except formaldehyde other aldehydes can be chlorinated when substituted occurs in the ethyl group.

$$\underset{\text{Acetaldehyde}}{H_3C{-}CHO} + 3Cl_2 \longrightarrow \underset{\text{Trichloroacetaldehyde}}{O{=}CH{-}CCl_3} + 3HCl$$

Choral on further treatment produces chloroform

Bromination

Aldehydes possess the property of polymerization.

Formaldehyde polymerizes spontaneously at the ordinary temperature

$$\underset{\text{Formaldehyde}}{HCHO} \xrightarrow{\text{ordinary temperature}} \underset{\text{Metaformaldehyde}}{(HCHO)_2}$$

Acetaldehyde is polymerized upon the addition of a little hydrochloric acid, sulphuric acid, or Sulphur dioxide, zinc chloride and calcium chloride etc.

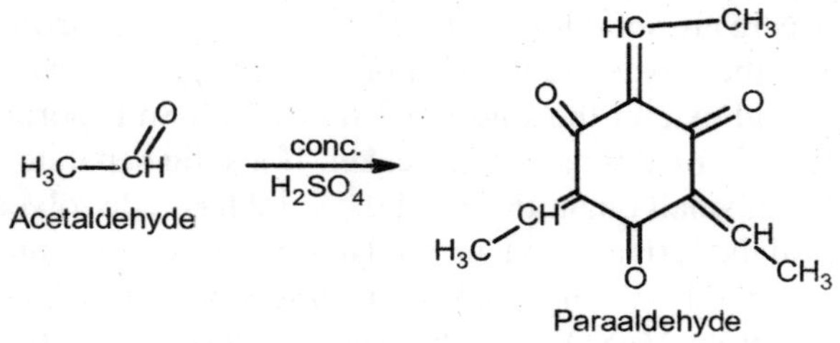

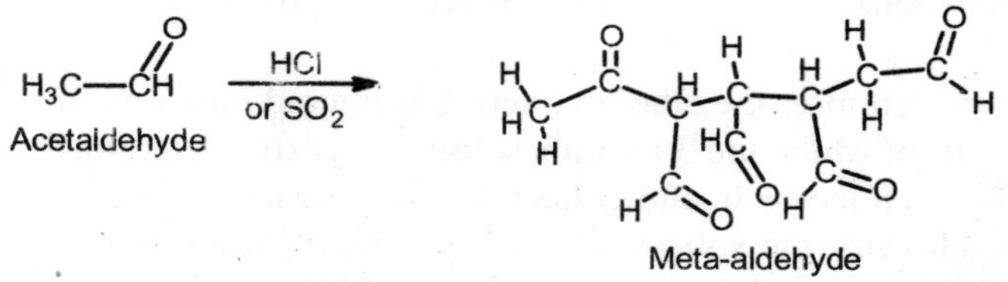

H3C—CH(=O) Acetaldehyde —conc. H2SO4→ Paraaldehyde

H3C—CH(=O) Acetaldehyde —HCl or SO2→ Meta-aldehyde

Condensation

Aldehydes also show a tendency to form condensation products with aldehydes.

(a) Condensation in the presence of dilute alkalies or KCN

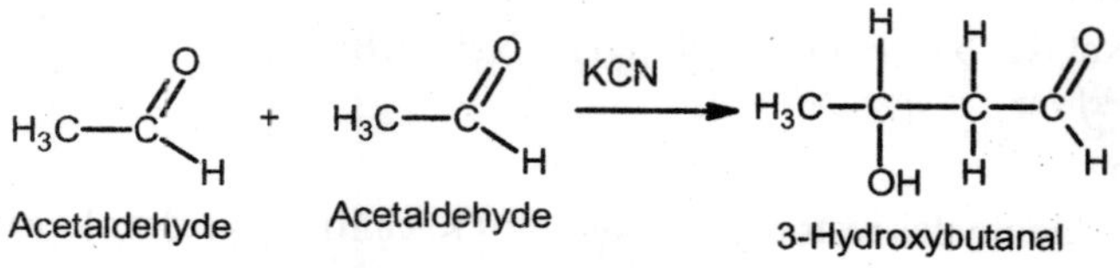

$$\underset{\text{Acetaldehyde}}{H_3C{-}CHO} + \underset{\text{Acetaldehyde}}{H_3C{-}CHO} \xrightarrow{KCN} \underset{\text{3-Hydroxybutanal}}{H_3C{-}CH(OH){-}CH_2{-}CHO}$$

The reaction is known as aldol condensation.

Aldols are β-hydroxy aldehydes and are formed by the addition of one carbonyl compound to the carbonyl group of a second compound. The reaction is catalyzed by both acids and bases, however, the latter are for more effective and are the ones generally employed in industry.

Aldehydes that undergo the aldol condensation necessarily posses at least one α-hydrogen atom. This active hydrogen atom may also be furnished by a ketone, such as acetone, which has available α-hydrogen atoms. The condensation of two or more ketone molecules is called **ketols**.

When more than one α-hydrogen atom present in the, acid catalyzed dehydration of the aldol may be readily induced to produce α-β-unsaturated aldehydes and thence by catalytic hydrogenation; primary alcohols for example, crotonaldehyde and 1-butanol form aldol.

$$\underset{}{CH_3CH(OH)CH_2CHO} \rightarrow \underset{\text{Crotonaldehyde}}{CH_3CH{=}CHCHO} \rightarrow \underset{\text{1-butanol}}{CH_3CH_2CH_2CH_2OH}$$

1-3 glycols are smoothly obtained by hydrogenation of the Aldols. These methods are used indirectly for the production of 1, 3 Butylene glycol, 2-ethyl-2-hexenal, 1-butanol etc.

A number of other reactions believed to proceed by the addition of one compound to the carbonyl group of another with the formation of aldol like intermediates are often classified as Aldol Condensations. Several such reactions:

(i) the conversion of aromatic aldehyde to acetic anhydride.
(ii) the condensation aromatic aldehyde with ketone.
(iii) The condensation of an aldehyde with malonic ester, acetic ester, or ethyl cyano acetate and
(iv) the condensation of an aldehyde or ketone with a nitro-paraffin.

(i) The Condensation of Aromatic Aldehyde to Acetic Anhydride

$$\underset{\text{benzaldehyde}}{C_6H_5CHO} + \underset{\text{acetic anhydride}}{(CH_3CO)_2O} \xrightarrow{CH_3COOK(\text{potassium acetate})/175^\circ-180^\circ C} \underset{\text{cinnamic acid}}{C_6H_5CH:CHCOOH}$$

(ii) The Condensation of Aromatic Aldehyde with Ketone

$$\underset{\text{benzaldehyde}}{C_6H_5CHO} + \underset{\text{acetic anhydride}}{CH_3CHO} \xrightarrow{NaOH} \underset{\text{cinnamic aldehyde}}{C_6H_5CH:CHCHO} + \underset{\text{aldehyde}}{CH_3COOH}$$

(iii) The Condensation of Acetoacetic acid ester with Aldehyde

$$\underset{\text{Acetoacetic ester}}{CH_3COCH_2COOC_2H_5} + \underset{\text{acetaldehyde}}{CH_3CHO} \rightarrow \underset{\text{diethyl dimethylcyclohexanone diacarbooxylate}}{CO{:}CH{:}C(CH_3).CH(COOC_2H_5)CHCH_3CHCOOC_2H_5}$$

(iv) Condensation of Nitroethane with benzaldehyde and acetaldehyde

$$\underset{\text{benzaldehyde}}{C_6H_5CHO} + \underset{\text{nitroethane}}{CH_3CH_2NO_2} \xrightarrow{n-C_4H_9NH_2} \underset{\text{[(1Z)-2-nitroprop-1-enyl]benzene}}{C_6H_5CH = C(NO_2)CH_3} + H_2O$$

$$\underset{\text{acetaldehyde}}{CH_3CHO} + \underset{\text{nitroethane}}{CH_3CH_2NO_2} \xrightarrow{\text{Zinc chloride}} \underset{\text{3-nitro-2-butanol}}{CH_3CH(NO_2)CHOHCH_3}$$

(v) Condensation in the presence of Aluminium ethoxide

$$\underset{\text{acetaldehyde}}{H_3C{-}C({=}O)H} + \underset{\text{acetaldehyde}}{H_3C{-}C({=}O)H} \xrightarrow{Al(OC_6H_5)_3} \underset{\text{ethyl acetate}}{H_3C{-}C({=}O){-}O{-}CH_2{-}CH_3}$$

The reaction is used for the preparation of ethyl acetate in large scale.

Test for Aldehydes

1. Silver mirror test: aldehydes produce silver mirror in the presence of Tollen's reagent.
2. Aldehydes give red precipitate to the Fehlling's reagent.
3. Schiff's reagent gives pick colour.
4. resinification of aldehydes produce a brown resins with hot conc. Sodium hydroxide.
5. add a drop or two of aldehyde in an alkaline (NaOH) solution of sodium nitroprusside, it turns cherry red.

FORMALDEHYDE

Formula: HCHO

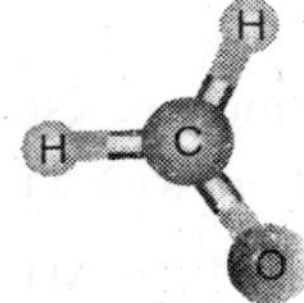

Introduction and Occurrence

The chemical "formaldehyde" is a colourless, pungent gas at room temperature with an approximate odor threshold of about 1 ppm While the term "formaldehyde" is also used to describe various mixtures of formaldehyde, water, and alcohol, the term "formalin" more precisely describes aqueous solutions, particularly those containing 37 to 50 per cent formaldehyde and 6 to 15 per cent alcohol stabilizer. Most formaldehyde enters commerce as formalin. Alcoholic solutions of formaldehyde are available for processes that require low water content. Paraformaldehyde, a solid, also serves as a source of formaldehyde gas. Formaldehyde gas per se is not available commercially. The Chemical Abstracts Service (CAS) has assigned the number "50-00-0" to formaldehyde. This number applies to both formaldehyde gas and its aqueous or alcohol stabilized solutions.

Preparation

Formaldehyde may be prepared by any one of the general methods given above. In the lab it is prepared by the following process:

Laboratory Preparation

Theory: Formaldehyde is prepared by the catalytic oxidation of methanol (methyl alcohol)

Apparatus	Reagents	Catalyst	Temperature
Two necked flask, pyrex tube, delivery tubes, with std. joints, stands, heating mantle Vacuum filter pump	Methanol, platinum, water	Platinum wire	40°C for methanol and above 250°C for platinum wire

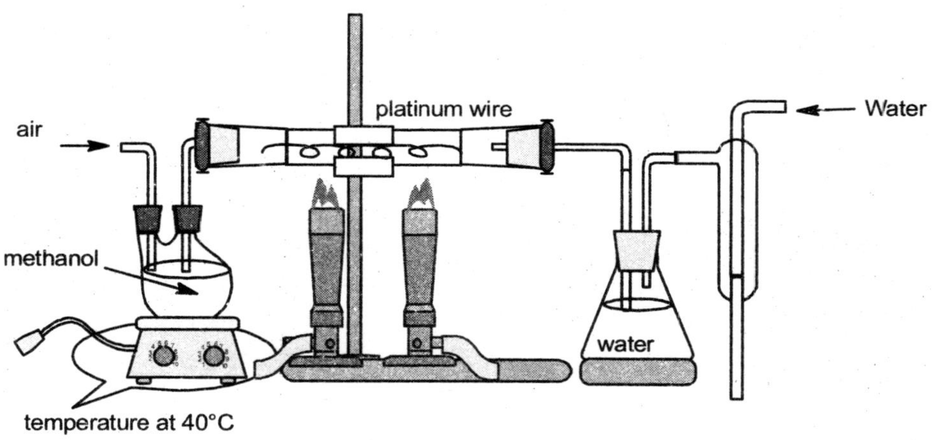

Fig. 12.1. The Lab Preparation of Formaldehyde

Reaction equation:

$$\underset{\text{methanol}}{CH_3OH} + O \xrightarrow{\text{platinum / catalyst}} \underset{\text{formaldehyde}}{HCHO} + H_2O + \text{heat}$$

Description: The mixture of methyl alcohol and air vapour is passed over hot platinum and the formaldehyde vapour is drawn in side the reservoir flask with cold water by the vacuum filter pump, here the 40% water solution is produced called formalin.

Precaution: *Here the maintenance of the temperature is absolutely necessary, and all stoppers must be maid B-24 std. Joint.*

Industrial Preparation

Industrially, formaldehyde is produced by the catalytic oxidation of methanol. The most commonly used catalysts are silver metal or a mixture of an iron oxide with molybdenum and vanadium. In the more commonly used iron oxide system (Formox process), methanol and oxygen react at 400°C to produce formaldehyde according to the chemical equation

$$CH_3OH + \tfrac{1}{2}\,O_2 \rightarrow H_2CO + H_2O$$

The silver-based catalyst is usually operated at a higher temperature, about 650°C. On it, two chemical reactions simultaneously produce formaldehyde: the one shown above, and the dehydrogenation reaction

$$CH_3OH \rightarrow H_2CO + H_2$$

Further oxidation of the formaldehyde product during its production usually gives formic acid that is found in formaldehyde solution, found in ppm values.

On a smaller scale, formalin can be produced using a whole range of other methods including conversion from ethanol instead of the normally-fed methanol feedstock. Such methods however are of less commercial importance.

Properties

Physical Properties

Properties	
Density	815 g/cm^3,
Solubility in water	Very soluble
Melting point	– 116.9°C (156.2 K)
Boiling point	– 19.0°C (254.1 K)
Molar mass	30.03 gm/mol
Appearance	colorless gas
Dipole moment	
Safety data	
EU classification	ToxicCarc. Cat. 3
R-phrases	R23/24/25, R34R40, R43
S-phrases	S1/2, S26, S36/37/39S45, S51
PEL-TWA (OSHA)	0.75 ppm
Flash point	– 53°C
Auto ignition temperature	430°C
Explosive limits	7–73%

Chemical Properties

Formaldehyde is very reactive gives all the general reactions of aldehydes. It is however differs from acetaldehyde in behaviour towards:

(*i*) ammonia

(*ii*) caustic alkalies and on

(*iii*) polymerization.

A synopsis of reaction of formaldehyde is give below:

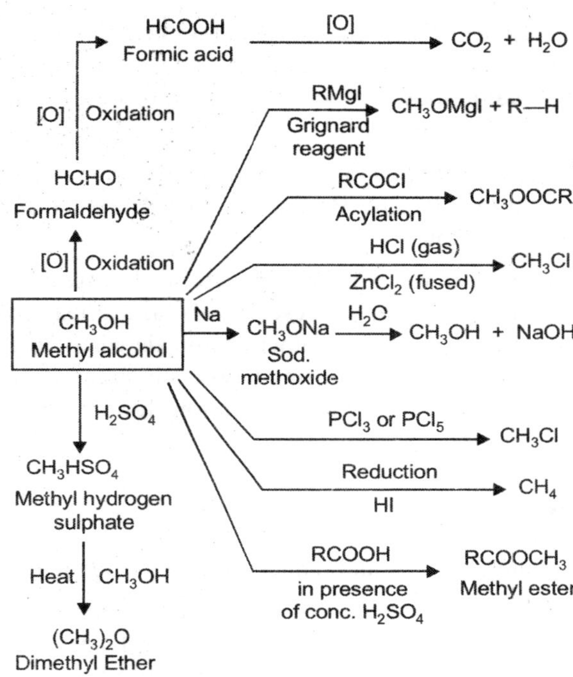

[A synopsis of the properties of Methyl Alcohol]

Uses

Formaldehyde is a major industrial chemical, ranked 24th in production volume in the United States. In 1985, 5.7 billion pounds of 37 per cent formaldehyde (by weight) was produced. Formaldehyde has four basic uses: as an intermediate in the production of resins; as an intermediate in the production of industrial chemicals; as a bactericide or fungicide; and as a component in the formulation of end-use consumer items. The manufacture of three types of resins: urea-formaldehyde, phenol-formaldehyde, and melamine formaldehyde, accounts for about 59 per cent of total consumption. An additional seven per cent is consumed in the production of thermoplastic acetal resins. About one-third is used in the synthesis of high volume chemical derivatives, including pentaerythritol, hexamethylenetetramine, and butanediol. Two per cent is used in textile treating and small amounts of formaldehyde are present as preservatives or bactericides in consumer and industrial products, such as cosmetics, shampoos and glues.

Some products prepared from formaldehyde contain unreacted formaldehyde residues which may be released from the product over its useful life. One example is urea-formaldehyde resin. Urea-formaldehyde resin is a generic name that actually represents an entire class of related formulations. Over 60 per cent of urea-formaldehyde resin production in 1977 was consumed by particleboard and plywood manufacturing, where the resin is used as glue. Urea-formaldehyde resins are also used in decorative laminates, textiles, paper, and foundry sand molds. Formaldehyde resins are used to treat textiles to impart wrinkle-resistance to clothing. About 60-85 per cent of all apparel fabric is finished with formaldehyde-containing resins. As apparel manufacture is the sixth largest industry sector in the United States, this use is the major source of widespread exposure to formaldehyde because of the large number of workers potentially exposed.

Formaldehyde destroys bacteria, fungi, molds, and yeast. Its commercial importance as a fungicide is probably its greatest use as a disinfectant. Because of its bactericidal properties, formaldehyde is used in numerous cosmetic preparations.

Formaldehyde's uses can lead to widespread exposure in downstream industries. For example, when formaldehyde is present in disinfectants, preservatives, and embalming fluid, worker exposure can occur. Although formaldehyde changes into other chemicals when urea-formaldehyde resins and concentrates are produced, decay may occur, causing workers in numerous industries including wood products and apparel manufacture to be exposed to airborne formaldehyde when it offgasses from products manufactured with these resins.

Test of Formaldehyde

(I) formaldehyde has a pungent smell and reduces Tollen's reagent and Fehlling's reagent.

(II) Schryver's Test: To a dilute (1%) solution of formaldehyde a few drops of 10% phenyl hydrazine is added, followed by the addition of .1ml of potassium ferricyanide solution. On acidifying with conc. hydrochloric acid a rose red colour is produced.

Bakelite: A phenol formaldehyde resin, the condensation reaction between phenol and formaldehyde in the reaction between phenol and formaldehyde, in the presence of certain basic catalyst (e.g. ammonia) yield a resinous fraction matter of high molecular weight. This is good thermo-setting plastic and is named as Bakelite (after the name of its discoverer Bakeland) it serves as an insulator and is extensively used for making various electrical and domestic articles.

ACETALDEHYDE

Formula: CH_3CHO

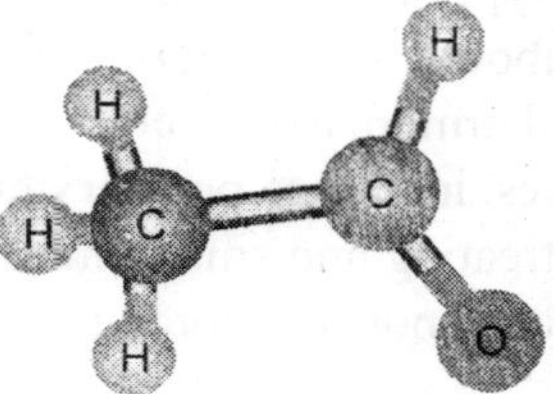

Introduction and Occurrence

Acetaldehyde, also known as ethanal, is a chemical compound, an aldehydeIt is a highly reactive flammable liquid with a strong fruity smell. It melts at -123.5°C and boils at 21°C.

It occurs naturally in ripe fruit and coffee, and is produced by plants as part of their normal metabolism. In the liver, the enzyme alcohol dehydrogenase converts ethanol into acetaldehyde, which is then further converted into the harmless acetic acid by acetaldehyde dehydrogenase. Acetaldehyde is more toxic than alcohol and is responsible for many hangover symptoms. N-acetylcysteine (NAC) is known to assist in processing acetaldehyde in the body and therefore can help to relieve hangover symptoms. Some persons from far-Eastern descent have a mutation in their acetaldehyde dehydrogenase gene, making this enzyme less effective. In these people, acetaldehyde accumulates after drinking, leading to severe and immediate hangover symptoms. These people are therefore less likely to become alcoholics. The drug antabuse (disulfiram) also prevents the oxidation of acetaldehyde to acetic acid, with the same unpleasant effects for drinkers. It is used in the treatment of certain alcoholics.The last steps of alcoholic fermentation in bacteria, plants and yeast involves the conversion of pyruvate into acetaldehyde by the enzyme pyruvate decarboxylase, followed by the conversion of **acetaldehyde** into ethanol. The latter reaction is again catalyzed by alcohol dehydrogenase, now operating in the opposite direction.

Acetaldehyde is the most typical member of the homologous series of aldehydes. It can be prepared by any of the given general process.

Laboratory Preparation

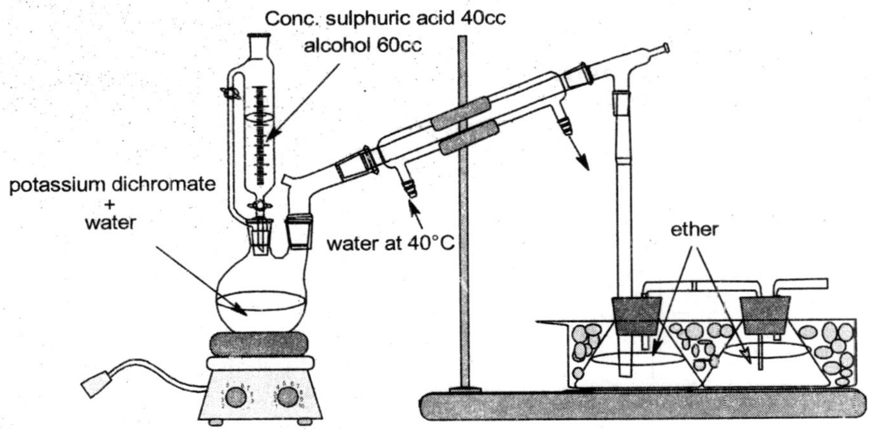

Fig. 12.2. The Lab Preparation of ACetaldehyde

***Theory*:** Acetaldehyde in the laboratory is prepared by the oxidation of ethyl alcohol with potassium dichromate and sulphuric acid.

Apparatus	Chemicals	Catalyst	temperature
Two necked flask, Liebig's condenser, stand delivery tubes conical flask (2) corks, pneumatic trough, heating mantle, dropping funnel	Ethyl alcohol, ether, freezing mixture, potassium dichromate, conc. Sulphuric acid		45°C, distillation of acetaldehyde takes place at 21°C

Reaction Equation:

$$K_2Cr_2O_7 + 4H_2SO_4 \rightarrow K_2SO_5 + Cr_2(SO_4)_3 + 4H_2O + 3O$$

Potassium Persulphate, Chromium sulphate

$$3C_2H_5OH + 3O \rightarrow 3CH_3CHO + 3H_2O$$

Ethanol, Acetaldehyde

Procedure: A mixture of potassium dichromate (50 g) and water (150 cc) us placed in a two necked fitted with Liebig's condenser slightly up wards and a hot water must circulate at 30-40°C. An alcoholic sulphuric acid mixture (60cc alcoholic + 40cc sulphuric acid) is dropped in the flask and the flask is heated over a heating mantle. Vapours of the alcohol gets condensed on the way in the condenser and comes back in the flask where the acetaldehyde (bp21°C) is absorbed in the ether kept in the freezing mixture.

Purification: Ethereal acetaldehyde solution is saturated by dry ammonia gas in the cold state when crystals of aldehyde ammonia separates out. The crystals are washed with dry ether and then dried in desiccator. The dry crystals are mixed with dilute sulphuric acid and distilled. Pure acetaldehyde separates out at 21°C and collected over ice cold water.

Precaution: *The setting of the Liebig's condenser must be as shown in the diagram; the ether must be kept under freezing mixture, as because it is flammable. The acetaldehyde must be collected over the ice cold water because its boiling point is 21°C*

Manufacture of Acetaldehyde

The hydration of acetylene became a commercial process for the manufacturing of acetaldehyde. The reaction is generally carried out at 70 - 80°C in 18 to 25% sulphuric acid solution (see Fig. 12.3)

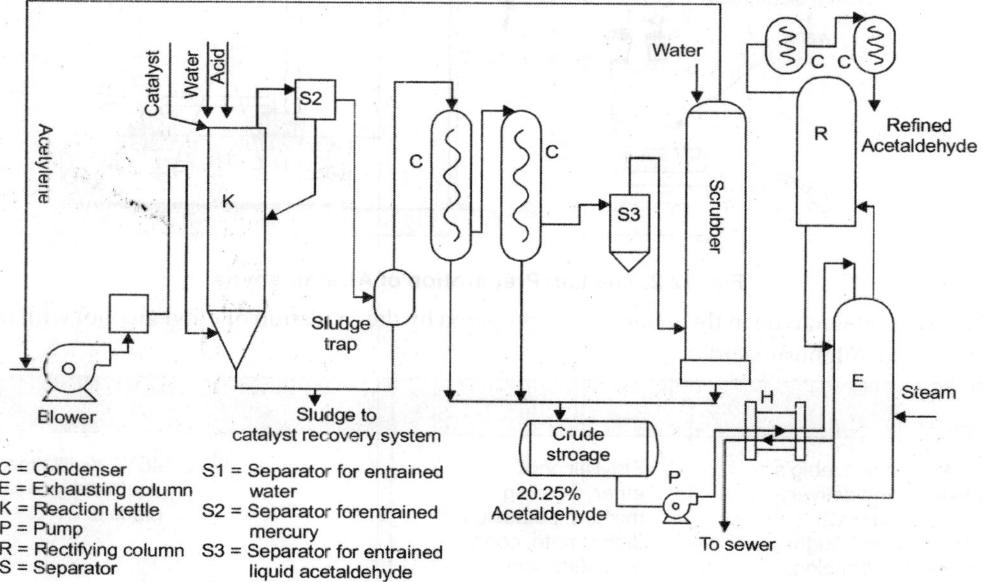

Fig. 12.3. The Manufacturing of Acetaldehyde

With mercuric sulphate as catalyst ferric or other salts are sometimes added to prolong the life of the catalyst. The reaction mechanism and intermediate compounds have been studied. Acetylene is excess sweeps out the dissolved acetaldehyde, which is condensed and purified by distillation while the excess acetylene is recycled. The sludge, consisting mainly of reduced catalyst with some tarry condensation, products of acetaldehyde is drained from the reaction kettle at intervals and resulphated. Crotoaldehyde is the chief by-product in the distillates of the acetaldehydes.

The vapour-phase reaction between acetylene and steam in the presence of various catalysts may eventually complete with or be coupled with the reaction in such ass these obtained by the decomposition of hydrocarbon

The addition of methanol to acetylene at about 15 atmos pressure, with potassium hydroxide as catalyst forms methyl vinyl ether, $CH_3OCH:CH_2$, which may be hydrolyzed with dilute sulphuric acid to give acetaldehyde of high purity, the methyl alcohol formed in the hydrolysis is separated by distillation and returned to the system.

Properties

Physical Properties

Specific Heat at 0°C	0.522
Heat capacity	15.8 g - cal/gm mole
Ration of specific heat and heat capacity $\gamma(C_p:C_v)$	1.145 at 30°C and 760 mm
Latent heat of fusion	17.6 g cal/g
Latent heat of vaporization	6.01 kg - cal/g mole
Heat of solution in water	4.98 kg - cal/g mole
Critical Temperature	181.5°C
Critical Pressure	63.2 atm
Heat of combustion	279.2 kg - cal g mole
Free energy of formation (DF°) at 25°C Liquid Gas	-31.880 g - cal/g - mole -32,000 g - cal/g - mole
Flash point	-50°C
Open cup	-35°C
Ignition temperature (in air)	165°C
Explosive limits of mixture with air	4 - 57 %
Dipole moment	2.72×10^{-18}
Dissociation constant at 0°C	$.7 \times 10^{-14}$

Chemical Properties

Acetaldehyde reacts exactly as the reaction of the general aldehydes.

Uses

1. For the manufacture of ethyl alcohol.
2. In the synthesis of dyes and medicinal chemicals.
3. Methaldehyde is used as solid fuel for portable cookers and the paraldehyde is used as hypnotic.
4. For preparing silver mirrors.
5. As an antiseptic.
6. Aldehyde ammonia is used as a rubber accelerator.

KETONES AND ALKANONES

Ketones forms a homologous series of the general formula

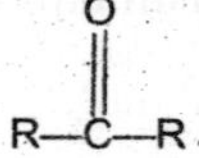

This characteristic group

present in between two similar or different alkyl groups. The ketones are isomeric with aldehydes consisting the same number of carbon atoms. The typical member of this family is Dimethyl ketone. Acetone or propanone.

The ketones closely resemble aldehydes as regard methods of preparation and may of their chemical reactions.

General Methods of Preparation

Ketones may be prepared by the following general methods:

1. From Secondary Alcohols:

(a) by oxidation with potassium dichromate and sulphuric acid air in the presence of platinum as catalyst

$$\underset{\text{Propan-2-ol}}{(H_3C)_2CH\text{-}OH} \xrightarrow{O} \underset{\text{Acetone}}{(H_3C)_2C{=}O} + H_2O$$

(b) By catalytic decomposition or dehydrogenation in the presence of copper at 300°C

$$\underset{\text{Propan-2-ol}}{(H_3C)_2CH\text{—}OH} \xrightarrow{\text{Cu } 300°C} \underset{\text{Acetone}}{(H_3C)_2C{=}O} + H_2O$$

2. From Monocarboxylic Acid:

(a) by the dry distillation of the salt of corresponding acid (except formic acid)

$$(CH_3COO)_2Ca \longrightarrow (H_3C)_2C{=}O + CaCO_3$$

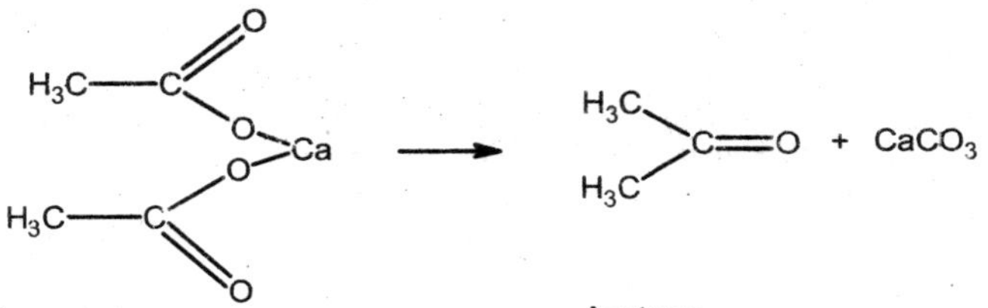

Calcium Acetate → Acetone

If a mixture of calcium salts of two different fatty acids (except formic acid) be distilled, a mixed ketone is obtained.

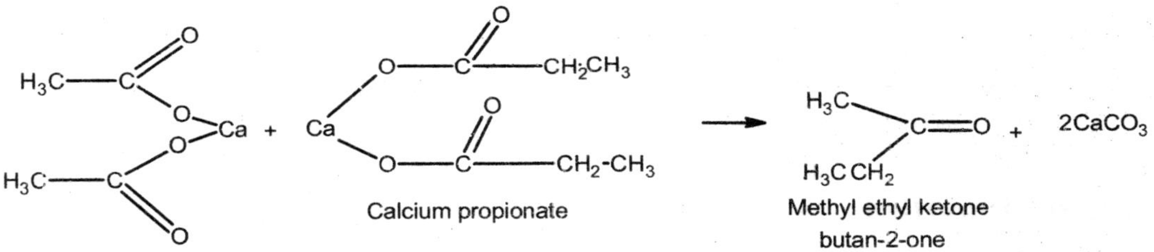

$$(CH_3COO)_2Ca + (CH_3CH_2COO)_2Ca \longrightarrow H_3C(CH_3CH_2)C{=}O + 2CaCO_3$$

Calcium Acetate + Calcium propionate → Methyl ethyl ketone (butan-2-one)

(b) By catalytic decomposition: Vapours of a fatty acid (except formic acid) are passed over heated catalyst Al_2O_3 or ZnO at 400°C.

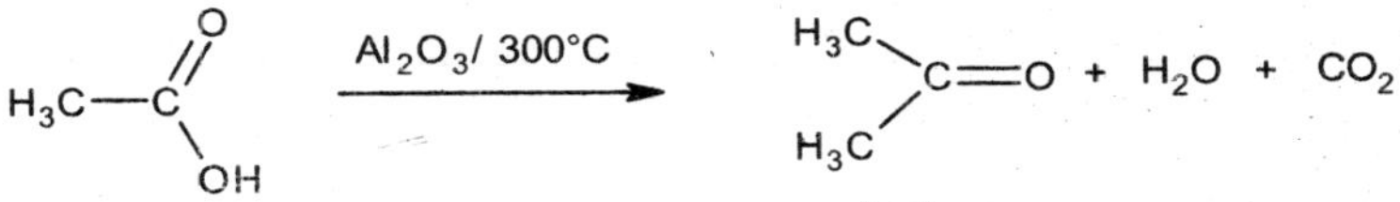

$$H_3C{-}COOH \xrightarrow{Al_2O_3/\ 300°C} (H_3C)_2C{=}O + H_2O + CO_2$$

Acetic acid → Acetone

3. From Acid Chlorides: By the action of Grignard's reagent

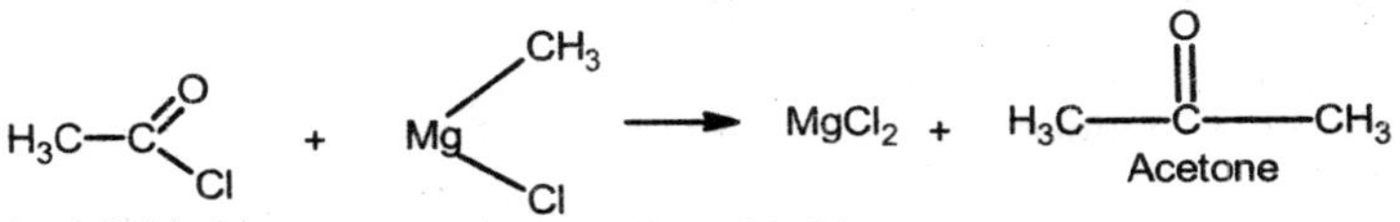

$$H_3C{-}COCl + CH_3MgCl \longrightarrow MgCl_2 + H_3C{-}CO{-}CH_3$$

Acetyl chloride + Methyl magnesium chloride → Acetone

4. From Alkaledine Halides: (with both the halogen atoms attached to the same contained carbon atom.

By hydrolysis with alkaledines or super heated steam

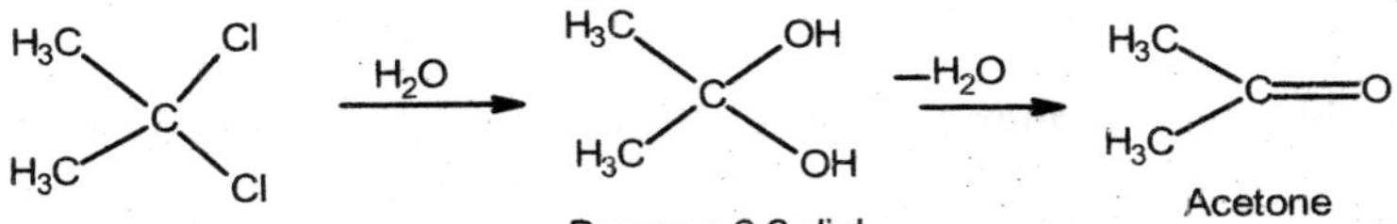

$$(H_3C)_2CCl_2 \xrightarrow{H_2O} (H_3C)_2C(OH)_2 \xrightarrow{-H_2O} (H_3C)_2C{=}O$$

2,2-Dichloropropane → Propane-2,2-diol → Acetone

5. **By the Action of Grignard reagent on Alkyl Cyanide**

$$CH_3MgBr + H_3C{-}C{\equiv}N \longrightarrow \underset{\text{N-(calcium bromide)-N-(1-methylethylidene)amine}}{(CH_3)_2C{=}N{-}MgBr} \xrightarrow{KOH} \underset{\text{Acetone}}{(H_3C)_2C{=}O} + NH_3 + \underset{\text{Magnesium hydroxy bromide}}{Mg(OH)Br}$$

6. **From Acetylene Hydrocarbon:** By passing acetylene and steam over heated catalyst ZnO at 400°C

(a)

$$\underset{\text{Acetylene}}{2H{-}C{\equiv}C{-}H} \xrightarrow[H_2SO_4\ [HgSO_4]]{+H_2O} \underset{\text{Acetone}}{(H_3C)_2C{=}O}$$

(b)

$$\underset{\text{Acetylene}}{2H{-}C{\equiv}C{-}H} + 3H_2O \xrightarrow[400°C]{ZnO\ (-CO_2)} \underset{\text{Acetone}}{(H_3C)_2C{=}O} + 2H_2O$$

Chemical Properties

Many of the reactions of ketones are similar to these of the aldehydes because of the presence of common carbonyl group (>C=O).

The important reactions of the ketones are:

1. **Oxidation**: Ketones are not affected by mild oxidizing agents therefore they are not good oxidizing agents they form corresponding acids with a few number of carbon atoms.

$$\underset{\text{Acetone}}{(H_3C)_2C{=}O} + 4O \longrightarrow \underset{\text{Acetic acid}}{H_3C{-}COOH} + CO_2 + H_2O$$

Ketones therefore do not reduce the Tollen's reagent or Fehlling's Reagent

2. **Reduction**: With sodium amalgam and dilute Sulphuric acid or Zinc dust or acetic acid. Ketones yield secondary alcohol as according to the reaction following:

$$\underset{\text{Acetone}}{(H_3C)_2C{=}O} + 2H \longrightarrow \underset{\text{Propan-2-ol}}{(H_3C)_2CH{-}OH}$$

3. **Addition Reaction**: With sodium bisulphate, hydrogen cyanide and Grignard's Reagent

(a) With sodium bisulphate: bisulphate compounds are precipitate.

$$(CH_3)_2C{=}O + NaHSO_3 \longrightarrow (CH_3)_2C(OH)\text{–}S({=}O)\text{–}O\text{–}O\text{–}Na$$

Acetone → Acetone sodium bisulphite

The ketone is recovered by warming the bisulphites with dilute alkalies. Bisulphite compounds are crystalline and hence they may be employed for the purification of ketones.

4. **With Hydrocyanic acid**: cyanohydrins are produced

$$(CH_3)_2C{=}O + HCN \longrightarrow (CH_3)_2C(OH)CN$$

Acetone → 2-hydroxy-2-methylpropanenitrile
acetone cyanohydrin

5. **With Ammonia**: At very low temperature acetone ammonia is produced. At ordinary temperature a complex condensation product is produced.
6. **With Grignard Reagent**: Tertiary alcohols are produced
7. **Condensation Reaction**: with NH_2OH, $NH_2 - NH_2$ $C_6H_5NHNH_2$ PCl_5 etc

(a) With hydroxylamine NH_4OH: - Oximes are produced

$$(CH_3)_2C{=}O + H_2N\text{–}OH \longrightarrow (CH_3)_2C{=}N\text{–}OH + H_2O$$

Acetone, Hydroxylamine, Acetone oxime

The oximes are crystalline compounds and hence their formation may be used for the isolation and identification of ketones.

(b) With Hydrazine $H_2N - NH_2$: - hydrazines are produced.

$$(CH_3)_2C{=}O + H_2N\text{–}NH_2 \longrightarrow (CH_3)_2C{=}N\text{–}NH_2 + H_2O$$

Acetone, Hydrazine, Acetone hydrazone

(c) With Phenyl hydrazine $C_6H_5NHNH_2$: Phenyl hydrazines are produced

$$(CH_3)_2C{=}O + H_2N\text{–}NH\text{–}CH_2CH_3 \longrightarrow (CH_3)_2C{=}N\text{–}NH\text{–}CH_2CH_3 + H_2O$$

Acetone, Ethylhydrazine, Acetone ethylhydrazone

(d) With PCl_5: - Alkaledine chlorides are formed

$$(CH_3)_2C{=}O + PCl_5 \longrightarrow (CH_3)_2CCl_2 + POCl_5$$

acetone → 2,2-dichloropropane acetone dichloride

(e) With Thioalcohols: - Mercaptols re produced

$$(CH_3)_2C{=}O + H{-}S{-}CH_2CH_3 \xrightarrow{-H_2O} (CH_3)_2C(S{-}CH_2CH_3)_2$$

acetone + ethanethiol → 2,2-bis(ethylthio)propane

(f) Chlorination: - Acetone can be chlorinated when substitution occurs in the alkyl group

$$(CH_3)_2C{=}O \xrightarrow{3Cl_2} Cl_3C{-}C(=O){-}CH_3 + 3HCl$$

acetone → 1,1,1-trichloroacetone

Trichloroacetone on further chlorination yields chloroform [acetone when distilled with bleaching powder gives chloroform]

Special Condensation of Acetone

Acetone undergoes special condensation on treatment with suitable dehydrating agents and gives different products.

The condensation is accompanied by the laws of water.

$$\left[(CH_3)_2C{=}O \quad H_3C{-}C(CH_3){=}O\right] \xrightarrow{\text{dry HCl}} (CH_3)_2C{=}CH{-}C(=O){-}CH_3 + H_2O$$

acetone, acetone → 4-methylpent-3-en-2-one mesityl oxide

$$\left[(CH_3)_2C{=}O \quad H_3C{-}C(CH_3){=}O \quad (CH_3)_2C{=}O\right] \xrightarrow{\text{dry HCl}} (CH_3)_2C{=}CH{-}C(=O){-}CH(CH_3)_2 + H_2O$$

acetone, acetone, acetone → 2,5-dimethylhex-4-en-3-one phorone

(c) In the presence of con. sulphuric acid mesitylene is produced:

$$\underset{\text{Acetone}}{\begin{matrix} H_3C \\ H_3C \end{matrix}\!\!>\!C{=}O} \xrightarrow[H_2SO_4]{\text{Conc.}} \underset{\text{Mesitylene}}{C_6H_3(CH_3)_3} + 3H_2O$$

ACETONE

DIMETHYL KETONE

Formula: C_3H_6O

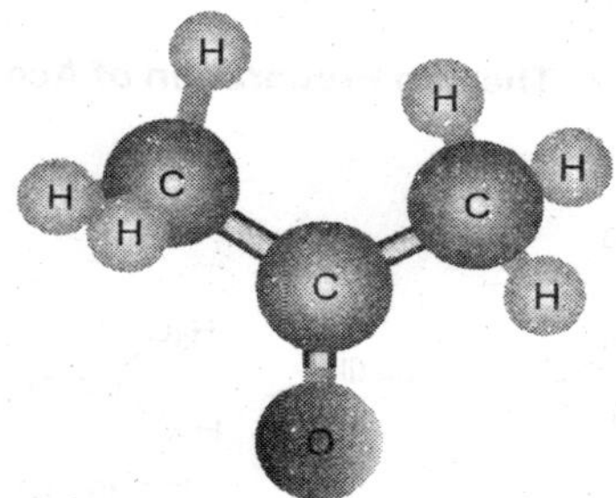

Introduction and Occurrence

Acetone is the most important member of the homologous series of ketones. Acetone occurs naturally in plants, trees, volcanic gases, forest fires, and as a product of the breakdown of body fat. It is present in vehicle exhaust, tobacco smoke, and landfil sites. Industrial processes contribute more acetone to the environment than natural processes. It is found among the products formed in destructive distillation of wood, sugar, cellulose, etc., and for this reason it is always present in crude wood spirit, from which the greater portion of it may be recovered by fractional distillation.

As a member of the ketone bodies it is present in very small quantity in normal urine and in the blood. Larger quantities can be found after starvation and in diabetic patients with severe insulin deficiency (that is untreated or inadequately treated persons); a fruity smell of the breath caused by acetone is one symptom of diabetic ketoacidosis

Laboratory Preparation

Theory: From the dry distillation of calcium acetate

Apparatus	Reagents	Catalyst	temperature
Distillation flask, adopter, stand, liebig's condenser, reservoir flask, heating mantle, pipes etc.	Calcium acetate	No	120°C and 56.5°C

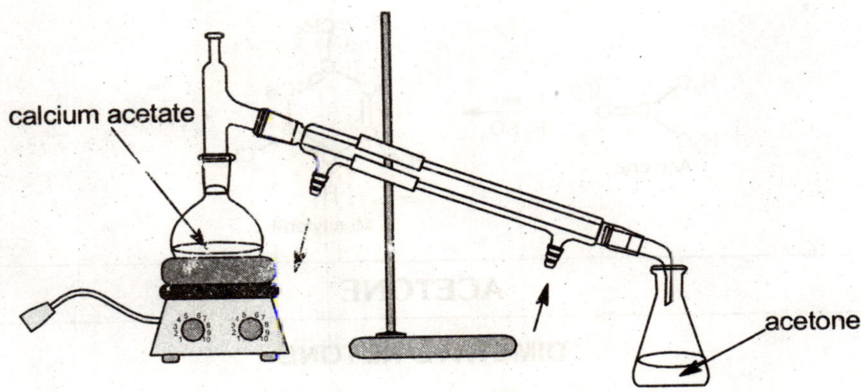

Fig. 12.4. The Lab Preparation of Acetone

Reaction Equation:

$$\underset{\text{Calcium Acetate}}{(H_3C-COO)_2Ca} \xrightarrow{\text{distil}} \underset{\text{Impure acetone}}{(H_3C)_2C{=}O} + CaCO_3$$

$$\underset{\text{Impure acetone}}{(H_3C)_2C{=}O} + \underset{\text{Sulfosodium}}{Na-SO_2-OH} \longrightarrow \underset{\text{Acetone sodium bisulite}}{(H_3C)_2C(OH)-SO_2-ONa} \xrightarrow{Na_2CO_3} 2\underset{\text{Pure acetone}}{(H_3C)_2C{=}O} + 2Na_2CO_3 + H_2O + CO_2$$

Purification: Acetone obtained is heated with a solution of sodium bisulphite, when crystals of acetone bisulphite compound separate out. These crystals are dried between folds of filter paper and distilled with sodium carbonate solution. Acetone is liberated and distills over. The distills over fused calcium chloride and redistilled. Pure acetone is collected at 56.5°C

Very Pure Acetone: It is obtained by treating the commercial sample with sodium iodide and cooling the product to -8°C in a freezing mixture crystals of the additive compound $3(CH)_3ONaI$ separate out. The crystals are removed from the mother liquor and heated to dry distillation flask filled with Liebig's condenser and a receiver very pure acetone distills over.

Precaution: *The temperature must be monitored very carefully specially in the case of the calcium acetate distillation.*

Manutacture

Acetone is prepared scale by the following four methods:

1. **From Wood Distillation**. Pyroligneous acid the aqueous distillate obtained by the dry distillation of wood Refer manufacture of methyl alcohol, contains methyl alcohol, acetone, acetic acid and water. Pyrolig-neous acid is distilled and its vapours are passed through hot milk of lime. Acetic acid is retained as calcium acetate and the other three pass over and are titer dried over CaO to remove water and further separated by fractional distillation. Calcium acetate obtained on the way may also be converted to acetone by distillation. Crude acetone so obtained is purified as described above under laboratory Preparation
2. **From Starch by Fermentation.** Large amount of Acetone is also prepared form maize and rice starch by a process of fermentation induced by an *organism-Fernbach's bacillus.* Normal butyl alcohol is obtained as a by-product and is used in making varnishes.
3. **From Acetic Acid.** By passing its vapours over $BaCO_3$. Al_2O_3, CaO. MgO or pumice stone, heated to 300-400°C.

$$2H_3C{-}C(=O){-}OH \longrightarrow H_3C{-}C(=O){-}CH_3 + CO_2 + H_2O$$

acetic acid → acetone

4. **From Alcohol**: By passing a mixture of ethyl alcohol vapour and steam over iron and calcium acetate as catalyst.

$$2\,H_3CCH_2{-}OH \xrightarrow{+H_2O} H_3C{-}C(=O){-}CH_3 + CO_2 + 4H_2$$

5. **From Petroleum cracking**: When petroleum is cracked under suitable conditions, propylene is obtained. Propylene is then converted into isopropyl alcohol which on dehydration or on oxidation yields acetone.

$$\text{petroleum} \longrightarrow H_3C\text{-}CH{=}CH_2 \longrightarrow H_3C{-}CH_2{-}CH_2{-}I \xrightarrow{H_2O} H_3C{-}CH(OH){-}CH_3 \longrightarrow H_3C{-}C(=O){-}CH_3 + H_2O$$

prop-1-ene; 1-iodopropane; propan-2-ol; acetone

Properties

Physical Properties

Molar mass	58.08 g/mol
Appearance	clear colourless liquid
Properties	
Density and phase	0.79 g/cm³, liquid
Solubility in water	miscible
Melting point	-94.9 °C (178.2 K)
Boiling point	56.3 °C (329.4 K)
Viscosity	0.32 centipoise at 20 °C
Hazards	
Main hazards	flammable
Flash point	< -20 °C

Chemical Properties

Acetone behaves generally like ketones and the properties has been discussed in the general properties discussion.

Test of Acetone: Acetone can be identified by following tests:

(*i*) *Legal's Test*: To a solution of acetone add sodium hydroxide solution and add sodium nitroprusside solution, which turns deep red, this fades on standing.

(*ii*) *Iodoform Test*: Acetone when warmed with potassium hydroxide solution iodine, gives a yellow precipitate iodoform which passes a characteristics odour. The reaction is also given by ethyl alcohol.

(*iii*) *Indigo Test:* To a warm mixture of acetone and orthonitrobenzaldehyde add sodium hydroxide solution and dilute with water. Blue colour is produced.

Uses and Heat Effects

An important industrial use for acetone its reaction with phenol, used for the manufacture of bisphenol A. Bisphenol A is an important component of many polymers such as polycarbonates, polyurethanes and epoxy resins. Acetone is often the primary (or only) component in nail polish remover. Swallowing very high levels of acetone can result in unconsciousness and damage to the skin in the mouth. Skin contact can result in irritation and damage to the skin.

The smell and respiratory data are known mostly from animal studies. Kidney, liver, and nerve damage, increased birth defects, and lowered reproduction ability of males (only) occurred in animals exposed long-term. It is not known if these same effects would be exhibited in humans.

Other Ketones

Ketones are organic compounds containing a carbonyl (keto) group attached to two hydrocarbon radicals and having the general formula RCOR'. If Rand R' are the same radical, the ketone is a symmetrical or *simple ketone;* if they are different, it is an unsymmetrical or *mixed ketone.* The carbon atom of the carbonyl group may also be a member of a ring, as in cyclohexanone $CH_2(CH2)_4.CO$. Owing to the presence of the carbonyl group, the ketones are similar to the aldehydes, RCHO, in their reactions and methods of preparation. Ketones are decidedly less reactive than aldehydes in additions to the carbonyl group because of the absence of hydrogen attached directly to this group, and the reactivity of individual ketones depends on the size and character of the hydrocarbon substituents. However, some diketones (con-taining two carbonyl groups) are very reactive. Cyclic compounds containing two carbonyl groups and having the so-called quinonoid structure are called *quinones* (q.v.); anthraquinone (q.v.), $C_6H_4.CO.C_2H_4.CO$, sometimes called diphenylene diketone, is an example. α-Hydroxy ketones (keto alcohols), RCH(OH)COR', are called *acyloins* or *a-ketols,* for example, benzoin, $C_6H_5CH(OH)COC_6H_5$.

The largest use for the lower ketones is as solvents (q.v.) for vinyl resins, cellulose esters and ethers, etc. Ketones are also important in the synthesis of many industrially important compounds and resins.

Nomenclature

The generic name "ketone" is derived from the first discovered and simplest member of the series, acetone (q.v.), CH_3COCH_3. There are three principal types of names for monoketones: (1) radical names containing the word "ketone", (2) names in which the suffix "-one" designates the keto function, as the I.U.P.A.C. (Geneva) names, and (3) names in which the ketone function is expressed by the prefix "keto-" or "oxo-" (I.U.P.A.C. and *C.A.)* or other ketonic prefix (as acetyl- and acetonyl-).

Radical names are particularly convenient for naming mixed ketones and ketones with rings attached to both sides of the carbonyl group, for example, methyl ethyl ketone, cyclohexyl phenyl ketone, di-I-naphthyl ketone or simply 1-naphthyl ketone.

In the I.U.P.A.C. names the longest straight chain including the carbonyl group is taken as the parent hydrocarbon and the position of the group is indicated by an Arabic numeral, as 3-pentanone or pentanone-3. Cyclic ketones in which the carbonyl group is a ring member are similarly named, as 3-cyclohexen-l-one.

Some analogous heterocyclic compounds, though they are not ketones (because the carbon of the carbonyl group is a ring member attached on one or both sides *to* a hetero atom), are also conveniently named with the suffix "-one," as 5-pyrazolone (I). If a cyclic ketone or analogous heterocyclic com-pound must be regarded as formed by substitution of a hydro derivative rather than the parent com-pound itself, the position of the "extra" hydrogen is indicated by "*xH*" in parentheses following the number for the position of the carbonyl group; thus 1 (4H)-naphthalenone is used for 1,4-dihydro-l-oxonaphthalene (II) and 4(l H)-quinolone for 1,4-dihydro-4-oxoquinoline (III).

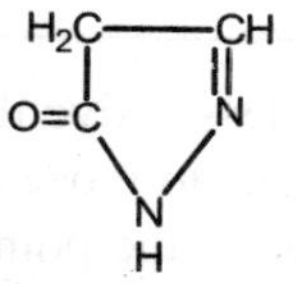

2,4-dihydro-3*H*-pyrazol-3-one

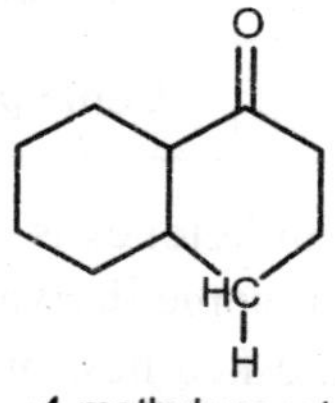

4-methyleneoctahy-dronaphthalen-1(2 *H*)-ONE

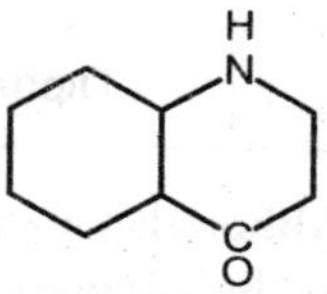

octahydroquinolin-4(1*H*)-one

The suffix "-one" is also used in naming some aliphatic ketones from the acids from which they can be prepared, as butyrone, $(CH_3CH_2CH_2)_2CO$. Many mixed aliphatic-aromatic ketones, RCOAr, in which R is an alkyl group and Ar is a benzene or naphthalene ring, are usually named by adding "-phenone" or "-naphthone," respectively, to the stem of the appropriate aliphatic acid (RCOOH), for example, acetophenone, $CH_3COC_6H_5$. Benzophenone, $C_6H_5COC_6H_5$, is a similar name.

The names in which the ketone function is expressed by a prefix are used primarily for compounds with mixed functions, as a-keto(or oxo-)butyraldehyde, CH_3CH_2CO-CRO; o-benzoylbenzoic acid, $C_6H_5COC_6H_4COOH$. Although the prefix "keto-" is more common, the prefix "oxo-" is recommended by the IU.P.A.C. and has been used by *C.A.* beginning with because it applies equally well to ketones and to hetero-cyclic compounds that are not strictly ketones.

Physical and Chemical Properties

The lower aliphatic ketones are mobile liquids, soluble in water, and have a pleasant, slightly aromatic odour which progressively decreases with increasing molec-ular weight. Symmetrical dialkyl ketones beginning with dihexyl ketone are crystal line solids, whereas the still higher-molecular-weight ketones are waxlike and closely resemble the paraffins. The unsymmetrical ketones have somewhat lower melting points and higher boiling points than the corresponding symmetrical ketones. The symmetrical and unsymmetrical ketones are soluble in all organic solvents, the solubility decreasing with increasing molecular weight of the ketone and increasing polarity of the solvent. Alkyl aryl ketones exhibit physical and chemical properties corresponding to the equivalent dialkyl ketone and are distinguished from the latter in that they possess some properties characteristic of the aromatic group. The physical constants for some of the better known ketones are listed in the table below:

Formula	Ketone	"-one" (I.U.P.A.C.)	"Keto-" ("oxo-")
CH_3COCH_3	Dimethyl ketone	2-Propanone; acetone	Ketropropane
CH_3COC2H_5	Methyl ethyl ketone	2-Butanone	2-Ketobutane
$C_2H_5COC_2H_5$	Diethyl ketone	3-Pentanone	3-Ketopentane
$[CH_2(CH_2)_{10}]_2CO$	Di-n-undecyl ketone	12-Tricosanone; laurone	12-Ketotricosane
$CH_3COCH{:}CH_2$	Methyl vinyl ketone	3-Buten-2-one	3-Keto-1-butane
$CH_3COC_6H_5$	Methyl phenyl ketone	Acetophenone	Acetylbenzene
$(C_6H_5)_2CO$	Diphenyl ketone	Benzophenone	Benzoylbenzene

Reactions. Although the carbonyl group in ketones is considerably less active than the terminal carbonyl group of the aldehydes, ketones form some derivatives analogous to those of aldehydes. In some reactions common to both aldehydes and ketones, on the other hand, the point of attack with ketones is at one of the carbon atoms (a-carbon) adjacent to the carbonyl group rather than at the car-bonyl group. Ketones having at least one a-hydrogen are in equilibrium with the enol forms.

Oxidation. Ketones are not oxidized by the usual oxidizing agents, although they are capable of oxidative fission when treated with strong oxidizing agents such as hot nitric acid. If both carbon atoms adjacent to the carbonyl group contain an available hydrogen atom, four oxidation products are formed.

$$RCH_2COCHR' \xrightarrow{HNO_3} RCOOH + RCH_2COOH + R'COOH + R'CH_2COOH$$

It is probably the enol form of the ketone that is oxidized. The enol form is removed from the system by irreversible oxidation and in this manner all of the ketone is oxidized.

$$RCH_2COCH_2R' \rightarrow RCH{:}C(OH)CH_2R' \xrightarrow{(O)} RCOOH + R'CH_2COOH$$

Table 12.1. Physical Constants of Representative Monoketones

Common name	Synonyms	Formula	M.p., °C	B.p., °C	d_4^{34}	n_D^{20}
		Aliphatic				
	Saturated					
Acetone (*q.v.*)	2-Propanone; dimethyl ketone	$(CH_3)_2CO$	–94.6	56.5	0.7898	1.3591
Methyl ethyl ketone	2-Butanone	$CH_2COC_2H_5$	–86.4	79.6	0.8061	1.3789
Methyl *n*-propyl ketone	2-Pentanone	$CH_3COCH_2CH_2CH_3$	–77.8	102.1	0.8120	1.3895
Methyl isopropyl ketone	3-Methyl-2-butanone	$CH_3COCH(CH_3)_2$	–92	95	0.8030	1.3862
Diethyl ketone	3-Pentanone	$(C_2H_5)_2CO$	–42	101.7	0.8136	1.3927
Methyl *n*-butyl ketone	2-Hexanone	$CH_3CO(CH_2)_3CH_3$	–56.9	127.2	0.8300	1.4007
Methyl isobuty ketone	4-Methyl-2-pentanone; iso-propylacetone; Hexone	$CH_3COCH_2CH(CH_3)_2$	–84.7	115.8	0.8010	1.3959
Pinacolone	Methyl *text*-buty! ketone; 3, 3-dimethyl-2-butanone; 1,1,1-trimethylacetone	$CH_3COC(CH_3)_3$	–52.5	106.2	0.8110	—
Methyl *n*-amyl ketone	2-Heptanone; methyl pentyl ketone	$CH_3CO(CH_2)_4CH_3$	–35.5	151.5	0.8170	1.4083
Ethyl *n*-butyl ketone	3-Heptanone	$C_2H_5CO(CH_2)_3CH_2$	–39	149	—	—
Di-*n*-propyl ketone	H-Heptanone; butyrone	$(CH_3CH_2CH_2)_2CO$	–32.6	144.2	0.821	1.4081
Disopropyl ketone	2-4-Dimethyl-3-pentanone	$(CH_3)_2CH)_2CO$	—	125	0.806	1.4170
Methyl *n*-hexyl ketone	2-Octanone	$CH_3CO(CH_2)_5CH_3$	–16	172.9	0.819	1.4161
6-Methyl-2-heptanone (see *Perfumes*)	Methyl isohexyl ketone; iso-amylacetone	$CH_3CO(CH_2)_3CH(CH_3)_2$	—	170-171	0.8151	
Diisobutyl ketone	2-6-Dimethyl-4-heptanone; isovalerone	$(CH_3)_2CHCH_2)_2CO$	—	166	0.805	1.4123
Methyl *n*-nonyl ketone	2-Undecanone; 2-hendecanone	$CH_3CO(CH_2)_8CH_3$	13.5	223	0.833	1.4291
Stearone	Di-*n*-heptadecyl ketone; 18-pentatriacontanone	$(CH_3(CH_2)_{16})_2CO$	88	–	—	—

(Contd...)

Table 12.1 (Contd...)

Common name	Synonyms	Formula	M.p., °C	B.p., °C	d_4^{34}	n_D^{20}
		Alicyclic				
	Saturated					
Cyclopentanone	—	$CH_2.(CH_2)_3.CO$	—	130	0.9480	1.4366
Cyclohexanone (*q.v.*)	Pimelic ketone	$CH_2.(CH_2)_4.CO$	–16.4	155	0.9478	1.4507
Cyclopentadecanone	—	$CH_2.(CH_2)_{13}.CO$	63.0	120_3	—	—
Camphor (*q.v.*)	2-Camphanone; 1,7,7,-tri-methyl-2-oxobicyclo-[2.2.1]heptane	$CH_2.CO.C(CH_3).C(CH_3)_2.$ $CH.CH_2.CH_2$	174	204-205.3	$0.990\,^{25}_{25}$	—
Fenchone	1,3,3-Trimethylnorcamphor	$C(CH_3)_2.CO.C(CH_3).$ $CH_2.CH.CH_2.CH_2$	5-6	192-193	—	—
	Unsaturated					
Isophorone	3,5,5,-Trimethyl-2-cyclo-hexen-l-one	$CH_2.C(CH_3)_2.CH_2.$ $C(CH_3):CHCO$	—	215.2	0.9255	1.4766^{18}
α-Ionone	1,1,3-Trimethyl-2-(γ-keto-butenyl)-3-cyclohexane; a-cyclocitrylideneacetone	$C(CH_3):CH.(CH_2)_2.C(CH_3)_2$ $CHCH:CHCOCH_3$	—	147.5_{28}	$0.9298^{21.2}$	1.5041
β-Ionone	1,1,3-Trimethyl-2-(γ-keto-butenyl)-2-cyclohexane; b-cyclocitrylideneacetone	$C(CH_3)_2.(CH_2)_3.C(CH_2):$ $CCH:CHCOCH_3$	—	150_{24}	$0.9445^{9.6}$	1.5210
Carvone	6,8(9)-*p*-Menthadien-2-one	$CO.CH_2.CH(C(CH_3):$ $CH_2).CH_2.CH:C(CH_3)$	—	230-231	—	—
		Aromatic				
Acetophenone (*q.v.*)	Phenyl methyl ketone; acetylbenzene	$CH_3COC_6H_5$	20	202	1.026	—
Benzophenone	Diphenyl ketone; benzoyl-benzene	$(C_6H_5)_2CO$	48	307	$1.0976\,^{50}_{50}$	—
Acrylophenone	Phenyl vinyl ketone	$C_6H_5COCH:CH_2$	—	—	—	—

Crotonophenone	Phenyl propenyl ketone	$C_6H_5COCH{:}CHCH_3$	20-21	135^{20}	1.0223^{18}	—
Benzylidenaecetone	4-Phenyl-3-buten-2-one; benzalacetone; *b*-acetyl-styrrene	$C_6H_5CH{:}CHCOCH$	41	262	—	—
Chalcone	α-Benzylideneacetophenone; β-phenylacrylophenone; benzalacetophenone	$C_6H_5CH{:}CHCOC_2H_3$	58	346	—	—
Benzoin	α-Hydroxy-α-phenylaceto-phenone; benzohylphenyl-carbinol	$C_6H_3CH(OH)COC_6H_5$	134	344	1.310	—
Michler's ketone	4,4'Bis(dimethylamino)-benzophenone; *N, N'*-tetramethyl-4, 4'-diamino-benzophenone	$((CH_3)_2NC_6H_4)_2CO$	173	360	—	—
Methyl 2-naphthyl ketone	2-Acetonaphthone; 2-cetyl-naphthalene	$CH_4COC_{10}H_7$	51	172	—	—
9-Fluorenone	Biphenylene ketone	$C_6H_4.C_6H_4.CO$	84	341	—	—
		Heterocyclic				
Kojic acid	5-Hydroxy-2-(hydroxy-methyl)-4-pyrone	O.CH:C(OH).CO.CH:C CH2OH	153	—	—	—
Flavone	2-Phenylchromone; 2-phenyl-1, 4-benzopyrone	$C_6H_4.O.C.\ (C_6H_5){:}CH.CO$	97	—	—	—
Xanthone	9-Xanthenone; dibenzo-pyrone	$O.C_6H_4.CO.C_6H_4$	174	250	—	—

Enolization also accounts for the formation of dinitroparaffins, which are by-products in the action of concentrated nitric acid on ketones.

$$CH_3CH_2C(OH):CHCH_3 + 2HNO_3 \rightarrow CH_3CH(NO_2)_2 + CH_3CH_2COOH + H_2O$$

With hydrogen peroxide, ketones form polymeric ketone peroxides.

Cyclic ketones undergo a simple carbon-to-carbon fission at the carbonyl group and form the corresponding dibasic acids. Methyl ketones are oxidized at the methyl group with sodium hypochlorite and form a carboxylic acid and carbon dioxide. This reaction is specific for the methyl group and unsaturated ketones are oxidized to form unsaturated acids; for example, methyl isopropenyl ketone is oxidized to meth-acrylic acid.

$$CH_2:C(CH_3)COCH_3 \xrightarrow{NaOCl} 4CH_2:C(CH_3)COOH + CO_2$$

Reduction. Secondary alcohols are formed by the *catalytic hydrogenation* of ketones using Raney nickel catalyst.

$$(CH_3)_2CO + H_2 \xrightarrow{Ni\ (Raney)} (CH_3)_2CHOH$$

Ketones are also reduced to secondary alcohols by heating the carbonyl compound with an equivalent amount of aluminum isopropoxide in isopropyl alcohol and distilling off acetone from the equilibrium mix-ture. Lithium aluminum hydride in ether likewise reduces ketones to secondary alcohols:

$$4R_2CO + LiAlH_4 \rightarrow (R_2CHO)_4LiAl \xrightarrow{H_2O} 4R_2CHOH + LiOH + Al(OH)_3$$

The reduction of aliphatic and aromatic ketones with hydrogen produced *in situ* by the reaction of a metal with an acid or a base is a *bimolecular reduction* because the product is composed of two molecules of the starting material. For example, acetone is reduced only in part to isopropyl alcohol by metal combinations and forms, in addition, a considerable amount of the bimolecular reduction product, pinacol.

$$2(CH_3)_2CO + 2H^+ \xrightarrow{MgHg} (CH_3)_2C(OH)C(OH)(CH_3)_2$$

Similarly, benzophenone forms good yields of benzopinacol in sunlight in the presence of isopropyl alcohol, which functions as a hydrogen donor. Bimolecular reduction occurs with metal combinations, such as zinc and a mineral acid, and with sodium, magnesium, or aluminum amalgams.

The carbonyl group in mixed aliphatic aromatic ketones and wholly aromatic ketones is reduced to $—CH_2—$ by the *Clemmsen reduction,* which comprises refluxing the ketone with amalgamated zinc and concentrated hydrochloric acid. Acetophen-one, for example, is reduced to ethylbenzene.

$$C_6H_5COCH_3 \xrightarrow{ZnHg+HCl} C_6H_5CH_2CH_3$$

The Clemmsen reduction is not applicable to acid-sensitive compounds and can be replaced by the modified *Wolf-Kishner reduction,* which is not applicable to alkali sensitive compounds. In this latter method, the ketone is refluxed in a high-boiling, water-miscible solvent with aqueous hydrazine and sodium hydroxide to form the hydrazone. Water is then distilled out until the base temperature reaches 200°C, and the mixture is refluxed several hours to complete decomposition of the hydrazone.

$$>C:O \xrightarrow{H_2NNH_2} >C:NNH_2 \xrightarrow[200°C]{NaOH} >CH_2 + N_2$$

Enol Acetylation

All ketones containing at least one available hydrogen atom on a carbon atom adjacent to the carbonyl group form enol acetate. Isopropenyl acetate, CH_2: $C(CH_3)OOCCH_3$, the enol acetate of acetone, is formed by reaction of ketene with acetone in the presence of an acid catalyst. Enol acetylation is an equilibrium reaction and the enol acetates of ketones above acetone are generally formed by acidolysis with isopropenyl acetate. In this reaction the enol form of the ketone is continuously removed from the equilibrium by acetylation and the acetone by distillation

$$\underset{\text{ethyl acetate}}{RCH_2COOR'} \, f \, \underset{\text{ethyl acohol}}{RCH:C(OH)R'} + \underset{\text{isoprcpenyl acetate}}{CH_2:C(CH_3)OOCCH_3}$$

$$\underset{\text{isopropenyl acetate}}{CH_2:C(CH_3)OOCCH_3} \xrightarrow{H_2SO_4/300°C} \underset{\text{isopropenyl alcohol}}{RCH:C(OOCCH_3)R'} + \underset{\text{acetone}}{CH_3COCH_3}$$

These enol acetates are unique acetylating agents. Some physical constants of enol acetate are listed in Table 12.2.

Table 12.2 : Physical Constent of Erol Acetate

Ketone	Enol acetate	B.p., °C	n_D^{20}
Acetone	Isopropenyl acetate	96.6_{748}	1.4001
Methyl ethyl ketone	1-Buten-2-ol acetate	—	—
	2-Buten-2-ol acetate	$118\text{-}119_{740}$	1.4120
Methyl isobutyl ketone	4-Methyl-1-penten-2-ol acetate	$143\text{-}145_{740}$	1.4164
	4-Methyl-2-penten-2-ol acetate	—	—
Cyclobexanone	Cyclohexenyl acetate	180_{732}	1.4573
Acetophenone	α-Acetoxystyrene	85_2	1.5330
Biacetyl	Methyl α-acetoxyvinyl ketone	32_5	1.3911
	2, 3-Diacetoxy-1, 3-butadiene	53_5	1.4372
Acetylacetone	1-Methyl-2-acetylvinyl acetate	84_{10}	1.4521
	2, 4-Diacetoxypiperylene	114_{10}	1.4694
Methyl vinyl ketone	2-Acetoxy-1, 3-butadiene	54_{40}	1.4420
Methyl propenyl ketone	2-Acetoxypiperylene	67_{40}	—
Methyl isopropenyl ketone	2-Acetoxyisoprene	62_{40}	1.4450

Condensation Reactions

(1) *Basic catalysis.* Ketone α-hydrosen atoms have comparable activation to aldehyde hydrogens. Lower aliphatic ketones will form aldol condensation products *(ketols)* in the presence of barium hydroxide as, for example, in the formation of diacetone alcohol from acetone. The

condensation is carried out by forming an equilibrium mixture in the presence of the catalyst, separating the components in the absence of the catalyst, and recycling the acetone.

$$2C_6H_5COCH_3 \xrightarrow{Ba(OH)_2} CH_6COCH_2C(OH)(CH_3)_2$$

Acetophenone undergoes homocondensation in the presence of various catalysts, particularly aluminum *tert-butoxide,* to form the dehydrated aldol product, dypnone (β-methylchalcone)

$$2C_6H_5COCH_3 \rightarrow C_6H_5C(CH_3):CHCOC_6H_5 + H_2O$$

Ketones and aldehydes condense to form mixed aldol products with the ketone adding to the more reactive carbonyl group of the aldehyde to the exclusion of the reverse process. Benzalacetone is produced by the condensation of benzaldehyde with acetone using alcoholic potassium hydroxide.

$$C_6H_5CHO + CH_3COCH_3 \xrightarrow{alc.\ KOH} C_6H_5CH:CHCOCH_3 + H_2O$$

(2) ***Acid catalysis.*** Although for basic catalysis the carbonyl group in ketones has less additive power than the carbonyl group in aldehydes, this situation is somewhat reversed for acid catalysis. Two and three molecules of a methyl ketone condense in the presence of acid catalysts such as hydrochloric acid, sulphuric acid, and zinc chloride to form the dehydrated aldol products. For example, two molecules of acetone condense in the presence of hydrochloric acid with the elimination of water to form mesityl oxide, which condenses with a third molecule of acetone to form phorone.

$$2CH_3COCH_3 \xrightarrow{HCl} (CH_3)_2C:CHCOCH_3$$

$$(CH_3)_2C:CHCOCH_3 \xrightarrow{HCl/CH_3COCH_3} (CH_3)_2C:CHCOCH:C(CH_3)_2$$

The aldol intermediates first formed are dehydrated under the influence of the mineral acid. The strong dehydrating action of sulphuric acid leads to a third type of condensation product, mesitylene.

$$3CH_3COCH_3 \overset{H_2SO_4}{f} 3\ CH_3C(OH):CH_2.$$

$$3CH_3C(OH):CH_2 \rightarrow \text{(1,3,5-trimethylbenzene: } CH_3,\ H_3C,\ CH_3) + 3H_2O$$

This is undoubtedly due to the effectiveness of sulphuric acid as an enolization catalyst. Acetophenone condenses in the same manner to form 1,3,5-triphenylbenzene.

Condensation with Reactive Aromatic Compounds

Ketones condense with reactive aromatic compounds to form derivatives of the type Ar2CRR'. For example, acetone and phenol condense in the presence of sulphuric acid to form 2,2-bis (p-hydroxyphenyl)-propane (Diane), a useful intermediate in the synthesis of Bakelite-type resins.

$$(CH_3)_2CO + C_6H_5OH \xrightarrow{H_2SO_4} (HOC_6H_4)_2C(CH_3)_2$$

Reaction with α-Halo Esters. The reaction which takes place between a ketone and an α-halo ester in the presence of zinc is commonly known as the *Reformatsky reaction.* Isolation of the organozinc compound is unnecessary

$$R_2CO + BrCH_2COOC_2H_5 \xrightarrow{Zn} R_2C(OZnBr)CH_2COOC_2H_5$$

$$R_2C(OZnBr)CH_2COOC_2H_5 \xrightarrow{H_2O} R_2C(OH)CH_2COOC_2H_5$$

Condensation with Ketene. In the presence of condensation catalysts such as $ZnCl_2$ and H_3BO_3, ketene adds across the double bond of the ketone carbonyl group to form β-lactones. At 50-70°C. The ketone and ketene react via the lactone to form low-molecular-weight, linear polyesters which are depolymerized by distillation to form α, β-unsaturated acids.

$$(CH_3)_2CO + CH_3{:}C{:}O \xrightarrow{ZnCl_2} -OC(CH_3)_2CH_2CO(OC(CH_3)_2CH_2CO)_nOC(CH_3)_2CH_2CO-$$

ketene αβ unsaturatedacid

$$-OC(CH_3)_2CH_2CO(OC(CH_3)_2CH_2CO)_nOC(CH_3)_2CH_2CO- \xrightarrow{heat} (CH_3)_2C{:}CHCOOH$$

αβ unsaturated acid ethyl acetate

Diketones and keto esters also condense with ketene to form β-lactones. α, β-Un-saturated ketones add ketene to form both β and δ-lactones

$$RCH{:}CRCOR + CH_2{:}C{:}O \xrightarrow{ZnCl2} RCH{:}CRCR.CH_2.CO.O + RCH.CR{:}CR.CH_2.CO.O$$

Condensation with Malonic Acid Derivatives. Like aldehydes, ketones condense with malonic acid esters in the presence of glacial acetic acid, giving products which, on hydrolysis followed by heating, give unsaturated acids.

Reactions with Ammonia and Amines. Aliphatic, aromatic, and aliphatic heterocyclic ketones can be used in the reductive alkylation of ammonia or primary or secondary amines in a process known as the *Leuckart reaction.* The reduction is accomplished by heating a mixture of the ketone with a formic acid salt or formyl derivative of ammonia or the amine. Where primary and secondary amines are pr.o-duced in the reaction, they are usually obtained as the formyl compounds and are isolated by hydrolysis. For example, the reaction of ammonia with acetone can be illustrated as follows:

$$(CH_3)_2CO + 2HCOONH_4 \rightarrow (CH_3)_2CHNHCHO + 2\,H_2O + NH_3 + CO_2$$

acetone formyl amide ethyl amide

$$(CH_3)_2CHNHCHO + H_2O \rightarrow (CH_3)_2CHNH_2 + HCOOH$$

ethyl amide methyl amide formic acid

Primary amines are formed by the reaction of ketones with ammonia and hydrogen in the presence of barium or copper chromite. Methyl ethyl ketone reacts with ammonia.

$$C_2H_5COCH_3 + NH_3 + H_2 \rightarrow C_2H_5CH(NH_2)CH_3 + H_2O$$

methyl ethyl ketone → sec-butylamine

Acetone and aniline react in the presence of iodine to form trimethyldihydro-quinoline in a manner analogous to the Skraup synthesis.

aniline + acetone $\xrightarrow{I_2}$ 2,2,4-trimethylcyclohexanimine compound with benzene (1:1) + H_2O

Ketones also react with ammonium carbonate via the cyanohydrin to form 5,5-dialkyl-hydantoins. Thus acetone reacts with hydrogen cyanide and ammonium carbonate to form 5,5-dimethylhydantoin.

$H_3C{-}CO{-}CH_3$ (acetone) + HCN $\xrightarrow{KCN}$ $H_3C{-}CH_2{-}CH(OH)C{\equiv}N$ (2-hydroxybutanenitrile) → 5-ethylimidazolidine-2,4-dione

Ketal Formation

Ketones react so much less readily than aldehydes with alcohols in the presence of anhydrous hydrogen chloride that the preparation of ketals by the direct route is seldom used. With orthoformic esters in the presence of alcohols and a small amount of an acid catalyst, the ketones react irreversibly to form ketone acetals which are analogous to the aldehyde acetals.

$$(CH_3)_2CO + HC(OC_2H_5)_3 \xrightarrow{HCl} (CH_3)_2C(OC_2H_5)_2 + HCOOC_2H_5$$

Thioalcohols (mercaptans) react with ketones in the presence of an acid catalyst to form mercaptals.

$$(CH_3)_2CO + 2\,HSC_2H_5 \xrightarrow{HCl} (CH_3)_2C(SC_2H_5)_2 + H_2O$$

Ketones do condense quite readily in the presence of strong mineral acid catalysts with glycols and polyhydroxy compounds to form cyclic ketals.

$$R_2CO + HOCH_2CH_2OH \rightarrow R_2C.O.CH_2.CH_2.O + H_2O$$

Grignard Reaction

Alkyl magnesium halides add to ketones to form prod-ucts which are decomposed to tertiary alcohols with dilute hydrochloric acid.

$$R_2CO + R'MgI \rightarrow R_2R'COMgI \xrightarrow{H_2O} R_2R'COH$$

This useful synthesis involves the addition of an alkyl- or aryl magnesium halide to a carbonyl group and is applicable with few exceptions to all ketones. The limiting factor in this reaction is the presence of highly branched or large hydrocarbon radicals in either the ketone or the Grignard's Reagent, which restrict the free space available for the formation of the addition complex and thus impose hindrance. Substituted ketones containing hydroxyl or carboxyl groups react if sufficient Grignard's Reagent is used to allow for that portion destroyed by the other functional group.

Zinc alkyls do not react with ketones, but tertiary alcohols are formed by the simultaneous action of zinc and alkyl iodides on ketones.

Pyrolysis

The lower aliphatic ketones when pyrolyzed at 500-1000°C form ketene *(q.v.)* as the pyrolysis product.

$$CH_3OCH_3 \rightarrow CH_3{:}C{:}O$$

Methyl ethyl ketone and diethyl ketone also give largely ketene together with a smaller amount of methylketene.

Addition Reactions

Hydrogen cyanide adds to the carbonyl group in ketones in the presence of sodium cyanide to form cyanohydrins.

$$R_2CO + HCN \xrightarrow{NaCN} R_2C(OH)CN$$

Acetone cyanohydrin is a commercial intermediate in the manufacture of methyl methacrylate.

Sodium bisulfite adds to ketones containing a methyl group to form ketone bisulfites, which can be regarded as the sodium salts of hydroxy sulfonic acids.

$$RCOCH_3 + NaHSO_3 \rightarrow RC(CH_3)(OH)SO_3Na$$

The percentage of bisulphite compound present at equilibrium decreases rapidly as the second alkyl group increases in size or branching. Acetylene adds to ketones in the presence of basic catalysts to form acetylenic alcohols. Thus, acetone adds acetylene in the presence of sodium hydroxide at 20 atm. pressure to produce dimethylethynylcarbinol (2-methyl-3-butyn-2-01).

$$CH_3COCH, + HC{:}CH \xrightarrow{pressure} (CH_3)_2C(OH)C:CH$$

Chloroform adds to ketones in the presence of sodium hydroxide to yield a trichloro tertiary alcohol which hydrolyzes to form an α-hydroxy acid.

$$R_2CO + CHCl_3 \xrightarrow{NaOH} R_2C(OH)CCl_3, \rightarrow R_2C(OH)COOH$$

Replacement Reactions

Phosphorus pentachloride and phosphorus pentabromide react with ketones to replace the carbonyl oxygen with two chlorine or two bromine atoms.

$$R_2CO + PBr_5 \rightarrow R_2CBr_2 + POBr_2$$

The hydrogen atoms of the alkyl groups are easily replaced by chlorine and bromine.

Characterization

Ketones are usually identified by reaction with hydroxylamine, hydrazine, phenyl hydrazine, and semicarbazide to form oximes, hydrazones and azines, phenylhydrazines, and semicarbazines. The condensation with amine derivatives proceeds through the initial addition to the double bond of the carbonyl group followed by the elimination of water. The semicarbazines which form readily at ordinary temperatures are particularly suited to the identification of ketones because of their excellent crystallization properties.

$$R_2CO + H_2NNHCONH_2 \rightarrow R_2C{:}NNHCONH_2 + H_2O$$

All of the reactions mentioned are reversible, and the nitrogen derivative can be hydrolyzed by boiling with dilute mineral acids. It is possible to prepare very pure samples of ketones by isolating the ketone as the oxime, phenylhydrazine, or semi-carbazine and then recovering the ketone by acid hydrolysis.

Occurrence and Preparation

Dialkyl ketones occur in certain oils and waxes, and in the products formed by the action of bacteria or molds on fats and starch-containing grains. Lower dialkyl ketones (acetone, methyl ethyl ketone, and diethyl ketone) are formed in the dry distillation of wood and in the thermal decomposition of coal and coal tar. They are also formed in the saccharification of cellulose, and by the dry distillation with calcium oxide of sugars, citric acid, and pyroligneous acid. Dialkyl ketones are formed as by-products in the hydroformylation of lower alkenes, in the Fische-Tropsch synthesis of paraffins from water gas, in the pressure oxidation of hydrocarbons, and in the oxidation of dialkyl acetaldehydes.

Ketones are prepared by a wide variety of reactions and some of the more common methods are described here. The preferred synthesis usually varies according to the particular type of ketone to be prepared and the methods of manufacture are reported by classes: aliphatic ketones, alicyclic ketones, unsaturated ketones, and aromatic ketones.

Aliphatic Ketones

Pyrolysis of Metal Salts

The most common method for the formation of ketones is the pyrogenic decomposition of alkaline earth metal salts of the fatty acids. Most of the known symmetrical dialkyl ketones have been produced by this procedure. Substantial yields of the ketones result when the fatty acids or their esters are passed in the vapour phase over the alkaline earth oxides, and thoria and ceria gels are very effective catalysts for this transformation.

$$2RCOOH \xrightarrow{ThO_2\ gel/400\text{ - }500°C} RCOOR + CO_2$$

Unsymmetrical dialkyl ketones are produced by the dry distillation of a mixture of the barium salts of two fatty acids, or by passing the vapours of the acids over an alkaline earth oxide at 400-500°C. The product contains the unsymmetrical ketone and two symmetrical ketones. Formation of either symmetrical ketone can be sup-pressed by employing an excess of the other reactant.

Oxidation of Secondary Alcohols

Ketones are resistant to oxidation and can be formed in high yields from secondary alcohols by a variety of oxidation methods. Lower aliphatic ketones are produced in essentially quantitative yields by oxidation of Secondary alcohols with air over silver and copper metals or combinations thereof. This reaction actually combines oxidation and dehydrogenation, and the ratio of air to alcohol used is regulated by the quantity of oxidation required to maintain the reaction temperature.

$$RR'CHOH \xrightarrow[Ag]{O_2\ (air)} RR'CO + H_2O + H_2$$

In the oxidation of isopropyl alcohol and see-butyl alcohol to acetone and methyl ethyl ketone with air at dull-red heat, the quantities of ketone formed by oxidation and by dehydrogenation are approximately equal. Secondary alcohols above see-butyl alcohol tend to dehydrate at the temperatures required for air oxidation; so other oxidation methods are preferred. Procedures that are often used for the oxidation of secondary alcohols include the gradual addition of potassium dichromate to a dilute sulphuric acid solution of the alcohol, and for water-insoluble alcohols, the addition of chromic anhydride or an acetic acid solution thereof to an acetic acid solution of the alcohol.

The Oppenauer reaction for the oxidation of secondary alcohols with aluminum tert-butoxide is specific for the hydroxyl group and finds particular application in the oxidation of unsaturated alcohols. It is an equilibrium reaction in which a lower aliphatic ketone such as acetone functions as a hydrogen acceptor and is used in large excess to force the reaction to completion.

$$RR'CHOH + (CH_3)_2CO \xrightarrow{Al(OC(CH_2)_3)_3} RCOR' + (CH_3)_2CHOH$$

Methyl alkyl ketones are formed by the oxidation of the ammonium or alkali metal salts of fatty acids with hydrogen peroxide to form β-keto acids which decarboxylate to yield the ketone.

Dehydrogenation of Secondary Alcohols

An alternative procedure for the formation of ketones from secondary alcohols is to pass the alcohol vapours over a suitable catalyst bed heated to the required reaction temperature. Catalysts useful in the dehydrogenation of secondary alcohols include copper, silver, platinum, palladium metals, and the oxides or sulfides of the transition elements of groups IV, V, and VI of the periodic table on suitable supports. Recommended temperatures for maximum conversion with minimum decomposition losses are in the range 400-600°C.

$$RR'CHOH \xrightarrow{CrO / 500°C} RCOR' + H$$

Hydration of Acetylenes

Ketones can also be manufactured by the direct hydration of acetylenes. Acetone is manufactured by passing a mixture of acetylene and steam over a catalyst bed comprised of Fe20a-ZnO at temperatures above 233°C

$$2\,C_2H_2 + 3H_2O \rightarrow (CH_3)_2CO + CO_2 + 2H_2$$

This process is not competitive with the dehydrogenation and/or selective oxidation of secondary alcohols for the lower saturated aliphatic ketones such as acetone and methyl ethyl ketone.

Decomposition of Hydroperoxides

The newest industrial process for the production of the lower aliphatic ketones such as acetone and methyl ethyl ketone is the acid catalyzed decomposition of the alkylbenzene hydroperoxides. In this process alkyl benzene is oxidized with air to form the hydroperoxide, which is decomposed to form mole-equivalent amounts of phenol and a ketone Propylbenzene produces phenol and acetone.

$$\text{isopropylcyclohexane} \xrightarrow[H_2O]{\text{air}} \text{2-cyclohexyl-2-methylpropanoic acid} \xrightarrow[\Delta]{H_2SO_4} \text{cyclohexanol} + H_3C{-}C(=O){-}CH_3 \ \text{(acetone)}$$

Methyl ethyl ketone is made in a similar fashion from sec-butylbenzene.

Grignard Reaction

Aliphatic nitriles, amides, and anhydrides form addition products with Grignard reagents which are hydrolyzed with dilute hydrochloric acid to form ketones.

$$RC : N + RMgX \rightarrow RR'C : NMgX$$

$$RR'C: NMgX + 2HCl + H_2O \rightarrow RCOR' + MgXCl + NH_4Cl$$

This is a very satisfactory procedure for the preparation of high-molecular-weight ketones.

Reaction of Zinc Alkyls

Acid chlorides react with zinc alkyls to form a similar intermediate which reacts with a second molecule of the acid chloride to form the ketone and zinc chloride.

$$2\,CH_3COCl + ZnR \rightarrow 2\,CH_3COR + ZnCl_2$$

Condensation Reactions

The ketonic cleavage of the β-keto esters produced by the condensation of esters of aliphatic acids using sodium alcoholate is an excellent method of preparing higher symmetrical dialkyl ketones and can also be used for unsymmetrical ketones.

$$2\,RCH_2COOC_2H_5 \xrightarrow{NaOC_2H_5} RCH_2COCHRCOOC_2H_5 + C_2H_5OH$$

$$RCH_2COCHRCOOC_2H_5 + H_2O \xrightarrow{\text{aq. NaOH}} RCH_2COCH_2R + CO_2 + C_2H_5OH$$

Similarly, acid chlorides react with ferric chloride and split out hydrochloric acid to form the chloride of a 8-keto carboxylic acid which readily decomposes with water and decarboxylates to form the ketone.

$$2\,RCH_2COCl \xrightarrow{FeCl_3\ H_2O} RCH_2COCHRCOCl \xrightarrow{H_2O/\ -CO_2} RCH_2COCH_2R$$

Unsymmetrical ketones are formed by reaction of the sodium derivative of acetoacetic ester with an alkyl halide, hydrolyzing the alkyl-substituted acetoacetic ester and decarboxylating to form the ketone. Thus methyl *n*-amyl ketone is produced by using *n*-butyl bromide.

$$Na+[CH_3C(O-):CHCOOC_2H_5] + C_4H_9Br\ \ CH_3COCH(C_4H_9)COOC_2H_5 + NaBr$$

$$CH_3COCH(C_4H_9)COOC_2H_5 + H_2O \xrightarrow{\text{aq. NaOH}} CH_3CO(CH_2)_4CH_3 + CO_2 + C_2H_5OH$$

Methyl isobutyl ketone and diisobutyl ketone are formed in the ratio of 5:3 by the condensation of isopropyl alcohol with acetone at 250°C using a freshly reduced copper-zinc oxide-aluminum oxide catalyst.

$$CH_3COCH_3 + HOCH(CH_3)_2 \rightarrow CH_3COCH_2CH(CH_3)_2 + H_2O$$

$$CH_3COCH_3 + 2HOCH(CH_3)_2 \rightarrow (CH_3)_2CHCH_2)_2CO + 2H_2O$$

Aldehydes plus Diazomethane

Methyl alkyl ketones are formed by the reaction of diazomethane with aldehydes.

$$RCHO + CH_2N_2 \rightarrow RCOCH_2 + N_2$$

Unsaturated Ketones

Aldol Condensation

The aldol condensation products of aldehydes with ketones easily give up water to form α, β-unsaturated ketones. Methyl isopropenyl ketone is produced in high yields by the condensation of formaldehyde with methyl ethyl ketone in the vapor phase.

$$CH_3COCH_2CH_3 + CH_2O \xrightarrow[300°C]{\text{Pb zeolite}} CH_3COC(CH_3){:}\ CH_2 + H_2O$$

Olefin and Acetylene Condensations

Several unsaturated ketones, including ethyl vinyl ketone, are prepared in 20-25% yields by reaction of olefins with an acyl bromide in the presence of aluminum bromide.

$$CH_2{:}\ CH_2 + CH_3CH_2COBr \xrightarrow{AlBr_3/\text{-}HBr} CH_2{:}CHCOCH_2CH_3$$

Acyl chlorides and olefinic hydrocarbons react in the vapour phase at 200°C. to form unsaturated ketones. As an example of this reaction, acetyl chloride and diiso-butylene react in the absence of a catalyst to form "methyl octenyl ketone" (4,6,6-tri-methyl-3-hepten-2-one). Although the product contains both the α, β, and β, γ-isomers, the α, β-isomer predominates.

$$CH_3COCl + C_8H_{16} \xrightarrow{200°C} CH_3COCH{:}C(CH_3)CH_2C(CH_3)_3$$

Methyl vinyl ketone and methyl isopropenyl ketone are formed by the catalytic hydration of vinylacetylene and isopropenylacetylene, respectively.

$$CH_2{:}C(R)C{:}CH + H_2O \xrightarrow{CH_3COOH/Hg(OOCCH_3)_2} CH_2{:}C(R)COCH$$

Methyl isopropenyl ketone is also formed by an indirect route which comprises condensing acetylene with acetone to form dimethylethynylcarbinol. The carbinol is treated with acetic acid to form dimethyl(acetyl)carbinol acetate and pyrolyzed to form methyl isopropenyl ketone.

$$HC{:}CH + CH_3COCH_3 \xrightarrow{KOH/160°C.} CH_3)_2C(OH)C{:}CH \xrightarrow{Hg(OOCCH_3)_2\ /\ CH_3COOH}$$

$$(CH_3)_2C(OOCCH_3)COCH_3 \xrightarrow{500°C} CH_2{:}C(CH_3)COCH_3 + CH_3COOH$$

Reaction of Diketene with Aldehydes

The reaction of diketene with aldehydes probably takes place through the formation of α β-lactone addition product unstable to heat, which undergoes decarboxylation to form the unsaturated ketone. Di-ketene condenses with acetaldehyde to form methyl propenyl ketone.

$$CH_3COCH{:}C{:}O + CH_3CHO \rightarrow CH_3COCH_2CH(CH_3).O.CO \rightarrow CH_3COCH{:}CHCH_3 + CO_2$$

In a similar fashion doubly unsaturated ketones are formed by the condensation of unsaturated aldehydes with diketene. Crotonaldehyde and diketene react to form 3,5-heptadien-2-one.

$$CH_3COCH{:}C{:}O + CH_3CH{:}\ CHCHO \rightarrow CH_3COCH{:}CHCH{:}CHCH_3 + CO_2$$

Alicyclic Ketones

Pyrolysis of Metal Salts

Carbocyclic ketones are formed by the decarboxylation of dibasic acids, their anhydrides, or their alkaline earth salts. The highest yields of cyclic ketones are obtained by the dry distillation of the thorium or cerium salts of di-basic acids, or by passing the acid in the vapour phase over either thoria or ceria gels. Cyclopentanone, for example, can be produced from adipic acid.

$$HOOC(CH_2)_4COOH \xrightarrow{ThO_2,\ gel/450°C} CH_2.(CH_2)_3.CO + CO_2 + H_2O$$

Cyclic ketones containing 9 to 30 carbon atoms are obtained by this method.

Dieckmann Reaction

Carbocyclic ketones are formed by an adaptation of the classic method for the formation of acetoacetic ester by ester condensation applied to esters of dibasic acids. For example, diethyl adipate cyclicizes under the influence of metallic sodium to form 2-carbethoxycyclopentanone, which is hydrolyzed and decarboxylated by refluxing with dilute sulphuric acid to form cyclopentanone.

$$C_2H_5OOC(CH_2)_4COOC_2H_5 \xrightarrow{Na} C_2H_5OOCCH.(CH_2)_3.CO \xrightarrow{dil\ H_2SO_4} CH.(CH2)_3CO$$

For large-ring ketones a modification of this synthesis employing dinitriles and an ether-soluble condensing agent, lithium ethylanilide, is used. The activating influence of the nitrile group leads to the formation of a lithium ketimino nitrile, which is hydrolyzed to a β-keto acid and decarboxylated.

$$NCCH_2(CH_2)_nCH_2CN \xrightarrow{Li(C_2H_5)NC_6H_5} NCCH.(CH_2)_n.CH_2.C{:}NLi \rightarrow HOOCCH.(CH_2)_n.CH_2.C{:}O$$

$$HOOCCH.(CH_2)_n.CH_2.C{:}O \rightarrow H_2C.(CH_2)_n.CH_2.C{:}O$$

Aromatic Ketones

Friedel-Crafts Synthesis

This important method for the formation of aromatic ketones usually involves the condensation of an aromatic or aliphatic acid halide with an aromatic hydrocarbon in the presence of a molecular equivalent of anhydrous aluminum chloride.

$$RCOCl + C_6H_6 \xrightarrow{AlCl_3} RCOC_6H_6 + HCl$$

The reaction proceeds through the formation of an acid chloride-aluminum chloride addition complex which acts as the acylating agent. Acid anhydrides generally react in the same manner as the acid chlorides, as illustrated in the production of acetophenone.

$$(CH_3CO)_2O + C_6H_6 \xrightarrow{2AlCl_3} CH_3COC_6H_6 + CH_3COOH$$

Cyclic ketones are produced by intramolecular condensation between an aromatic ring and an acid chloride, as exemplified by the preparation of α-tetralone.

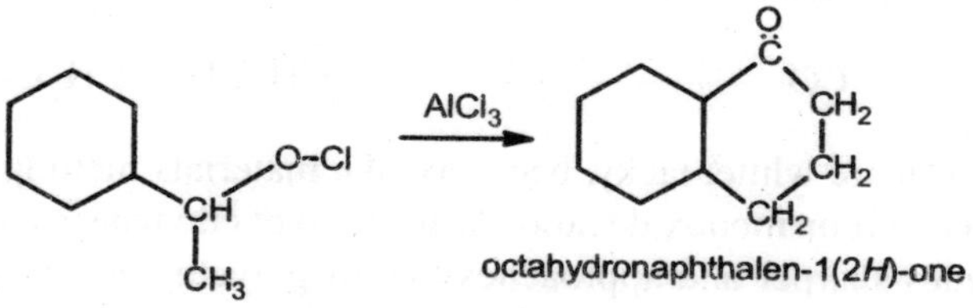

$\xrightarrow{AlCl_3}$

1-cyclohexylethyl hypochlorite

octahydronaphthalen-1(2*H*)-one

Fries Reaction

Acyl derivatives (esters) of phenols when heated with anhydrous aluminum chloride are converted into a mixture of the 0- and p-hydroxy ketones. The aluminum chloride salts are formed as intermediates and are decomposed with dilute hydrochloric acid.

$$C_6H_5OOCCH_3 \xrightarrow{AlCl_3/HCl} CH_3COC_6H_4OAlCl_2 \xrightarrow{H_2O/HCl} CH_3COC_6H_4OH \text{ (o- and p-)}$$

Hoesch Synthesis

Aliphatic imido chlorides condense with polyhydric phenols in the presence of zinc chloride or aluminum chloride to form ketimine hydrochlorides, which are decomposed by hot water to form ketones. The reaction is not applicable to phenol itself.

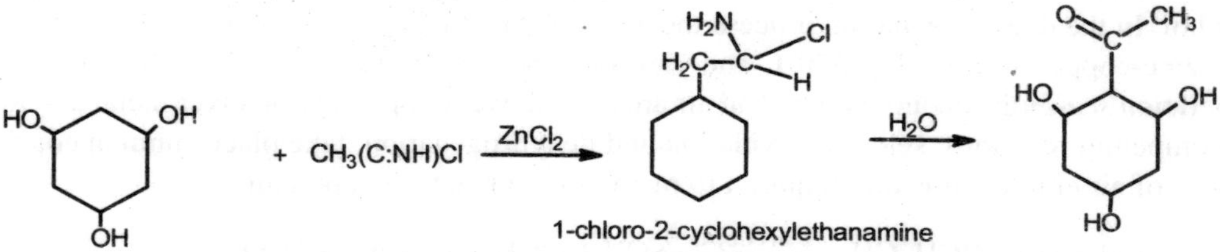

$+ CH_3(C{:}NH)Cl \xrightarrow{ZnCl_2}$ $\xrightarrow{H_2O}$

cyclohexane-1,3,5-triol

1-chloro-2-cyclohexylethanamine

1-(2,4,6-trihydroxycyclohexyl)ethanone

Miscellaneous Syntheses

Michler's ketone, p, p'-bis(dimethylamino)benzophenone, is produced by the condensation of dimethylaniline with phosgene.

$$4(CH_3)_2NC_6H_5 + COCl_2 \xrightarrow{ZnCl_2} (CH_3)_2NC_6H_4COC_6H_4N(CH_3)_2 + 2C_6H_4N(CH_3)_2HCl$$

The condensation of citral, $C_9H_{15}CHO$, with acetal by means of sodium ethoxide yields pseudoionone, which can be cyclized with sulphuric acid to α- and β-ionone 9-Fluorenone is formed by the dry distillation of diphenic acid with calcium oxide.

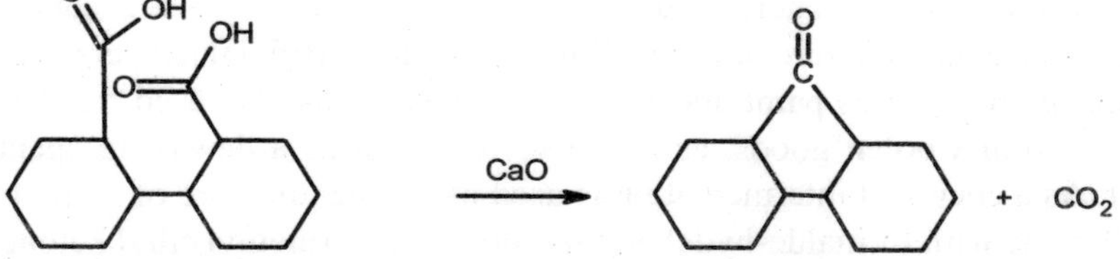

$\xrightarrow{CaO}$ $+ CO_2$

1,1'-bi(cyclohexyl)-2,2'-dicarboxylic acid

dodecahydro-9*H*-fluoren-9-one

Polyketones are produced by the copolymerization of ethylene and carbon mon-oxide at 100-1000 atm. and 130-140°C with a peroxide catalyst in a suitable solvent

$$CH_2{:}CH_2 + CO \xrightarrow{140°C/\text{tert-butyl peroxide}} (CH_2CH_2)_xCH_2CH_2CO)_y$$

The polymeric ketones are white, tacky, beeswax-like materials melting over a fairly wide range depending on the ethylene-carbon monoxide ratio. As the ratio of ethylene to carbon monoxide increases, the melting point becomes sharper and approaches that of polyethylene. Carbon monoxide has been copolymerized with a variety of ethanoid monomers to produce polyketones. The polyketones are particularly useful both alone and as additives to waxes and surface-coating compositions.

Aliphatic Monoketones

Methyl ethyl ketone (2-butanone, MER), $CH_3COCH_2CH_3$, formula weight 72.10, is an acetone-odoured, colorless liqulid; f.p., -86.4°C.; b.p., 79.57°C.; density, 0.8047; *refractive index,* 1.3789; vapour pressure at 20°C, 71.2 mm; latent heat of vapourization, 106.0 cal./gram; coeff. of expansion at 20°C., 0.00142;°C.; flash point (Tag closed cup), 28°F.; explosive limits of vapour with air, 1.97-10.2%; solubility at 20°C, 26.8 wt. % of ketone in water, 12.5 wt. % of water in ketone. The azeotrope with water contains 88.7% by weight of methyl ethyl ketone and boils at 73.41°C.

Methyl ethyl ketone is produced commercially either by dehydrogenation or selective oxidation of *see-butyl* alcohol. In the dehydrogenation process the alcohol is passed over a catalyst containing zinc oxide or over zinc-copper alloy at 450-550°C and yields of about 88% are achieved. In the preferred method of operation *see-butyl* alcohol vapors and air are passed over copper and/or silver metal at 450-500oe. Two competing reactions, selective oxidation and dehydrogenation, take place simultaneously and the quantity of air employed is that required to maintain the reactor temperature.

$$CH_3CH(OH)CH_2CH_2 \xrightarrow{air/Cu,Ag} CH_3COCH_2CH_3 + H_2 + H_2O$$

The yields of methyl ethyl ketone are essentially quantitative.

Methyl ethyl ketone was produced at the rate of 140 million pounds annually. Older methods of preparation, including the pyrogenic decomposition of a mixture of acetic and propionic acids over thorium oxide gel at 450°C, and the distillation of methyl acetoacetate in the presence of sulphuric acid, are no longer used.

Methyl ethyl ketone is rated as having about the same toxicity as acetone. The chief commercial use of methyl ethyl ketone is as a solvent and it is used to a lesser degree as a chemical intermediate. As a solvent it is used by the surface coatings industry with nitrocellulose and vinyl resins and it is finding increasing use in film-forming materials such as cellulose acetate, cellulose acetate butyrate, and acrylic resins. It is used in the manufacture of nitrocellulose cements, artificial leathers, photographic film, and cleaning compositions such as paint and tarnish removers. It is also used for the removal of oil and fatty substances from woolen goods, as a dye solvent, and as a dewaxing agent in the refining of lubricating oil. As a chemical intermediate it is used in the manufacture of methyl isopropenyl ketone (methyl ethyl ketone with formalde-hyde) and *see*-butylamine (methyl ethyl ketone with ammonia and hydrogen).

Methyl isobutyl ketone (4-methyl-2-pentanone), $CH_3COCH_2CH(CH_3)_2$ formula weight 100.16, is a medium-boiling ketone; m.p., -80.26°C.; b.p.115.9°C.; density, 0,8004; *refractive index*, 1.3958; vapour pressure at 20°C., 15.2 mm.; flash point (Cleveland open cup), 75°F., (Tag closed cup), 60°F.; A.P.H.A. colour 15 (max.); solubility at 20°C., 1.7-2.0 wt. % of ketone in water, 1.9-2.2 wt. % of water in ketone.

Methyl isobutyl ketone is made commercially by the selective hydrogenation of the double bond in mesityl oxide by means of various catalysts; for !iquid-phase hydrogenation, Raney nickel may be used as catalyst under neutral conditions, or, in the vapor phase, nickel at 165°C. or reduced copper on asbestos at 120°C. A technical grade is available in tank cars, drums, and coils.

Methyl isobutyl ketone is a good solvent for nitrocellulose, ethyl cellulose, and a variety of natural and synthetic resins. It is particularly satisfactory in protective coatings based on nitrocellulose and vinyl-type resins, because of its ability to produce low-viscosity solutions of high solids concentration. It evaporates 1.6 times as fast as *n*-butyl acetate and may be used to replace this ester. It is also one of the principal active solvents in spraying thinners. Its use in certain denatured alcohol formulas has been authorized. Vapours of methyl isobutyl ketone are irritating to the eyes and nose in concentrations below 0.1 vol % and are lethal to gunea pigs at 1.0 vol in about 4 hr Methyl *n*-amyl ketone (2-heptanone), $CH_3CO(CH_2)_4CH_3$, formula weight 114.18; high-boiling ketone, with a fruity odour; b.p., 150.6°C.; density 0.8166; vapour pressure 20°C., 2.6 mm.; flash point (Cleveland open cup), 120°F.; A.P.H.A. color, 18 max.); solubility at 20°C., 0.43 wt. % of ketone in water, 1.5 wt. % of water in ketone. It is miscible with the usual organic solvents.

Methyl amyl ketone occurs in oil of cloves and in Ceylon cinnamon oil and is produced commercially by dehydrogenation of 2-heptanol. It is available in tank cars, drums, and cans.

It is a useful solvent in synthetic resin finishes, especially for metal roll coating; imparts good blush resistance in lacquers. It is also a solvent for rubber and litrocellulose. Prolonged breathing of the vapour or contact of the liquid with the skin should be avoided.

Ethyl *n*-butyl ketone (3-heptanone), $C_2H.COCH_2CH_2CH_2CH_3$, formula weight 114.18, is a stable, colourless liquid; f.p., -36.7°C.; b.p., 147.8°C., m.p, 69.0°C.; density 0.8197; *refractive index* 1.4085; vapour pressure at. 20°C., 3.9 mm.; absolute viscosity at 20°C., 0.76 centipoise; coeff. of expansion at 20°C., 0.00107 ;CC.; flash point (Cleveland open cup), 115°F.; A.P.H.A. colour,15 (max.); solubility. at 20°C., 1.43 wt. % of ketone in water, 0.78 wt. % of water in ketone.

This ketone was first made by Desgrez in 1894 by hydrating 2-heptyne with 5 parts of water in a sealed tube at 325°C. Ethyl butyl ketone is now made commercially by the catalytic dehydrogenation of 3-heptanol.

Ethyl butyl ketone is a high-boiling solvent of particular interest for lacquers and synthetic resin coatings. It evaporates more slowly than butyl acetate and methyl isobutyl ketone and faster than "Cellosolve" acetate, diisobutyl ketone, and methyl amyl ketone. Ethyl butyl ketone is particularly useful in solvent mixtures for "forced dry" nitrocellulose finishes and vinyl baking enamels.

Ethyl butyl ketone is useful in formulating organosols, producing fluid dispersions that have good viscosity stability on aging. The medium evaporation rate of ethyl butyl ketone retards film cracking and facilitates fusion.

Ethyl butyl ketone has been condensed with ethyl formate and cyanoacetamide to form a substituted pyridone.

6-Dimethylamino-4,4-diphenyl-3-heptanone hydrochloride (methadone hydro-chloride, N.N.R.), $C_2H_6COC(C_6H_6)_2CH_2CH(CH_3)N(CH_3)_2$.HCl, is an analgesic similar in action to morphine. The base was introduced in Germany under the name Amidone.

Diisobutyl ketone (2,6-dimethyl-4-heptanone, isovalerone), $(CH_3)_2CHCH_2CO\text{-}CH_2CH(CH_3)_2$, formula weight 142.23, is an oily liquid with a peppermint odour; density, 0.8089; vapor pressure at 20°C., 1.7 mm.; flash point (Cleveland open cup), 140°F., A.P.H.A. color, 40 (max.); solubility at 20°C., less than 0.06 wt. % of ketone in water, 0.45 wt. % of water in ketone.

It can be obtained by the reduction of phorone, which is formed by the condensation of acetone in the presence of acid, or as a byproduct in the manufac-ture of methyl isobutyl ketone from acetone via diacetone alcohol and mesityl oxide. It has also been prepared from isovaleric acid and its alkaline earth salts. A technical grade is available in tank cars, drums, and cans and requires an M.C.A. caution label.

As a relatively slow-evaporating solvent, diisobutyl ketone is useful in nitrocellulose, vinyl-type, and other synthetic coatings. It is also a good solvent for milled crepe rubber.

2,6,8-**Trimethyl-4-nonanone,** $(CH_3)_2CHCH_2COCH_2CH(CH_3)CH_2CH(CH_3)_2$, formula weight 184.19, is a high-molecular-weight ketone with a pleasant, fruity odor; b.p., 207-228°C.; density, 0.8165, *refractive index,* 1.4273; absolute viscosity at 20°C., 1.91 centipoises; flash point (Cleveland open cup); 195°F.

Trimethylnonanone was first prepared by Bodroux and Taboury in 1909 by the catalytic hydrogenation at 280°C. of the corresponding twelve-carbon unsaturated ketone obtained by condensing two moles of methyl isobutyl ketone in the presence of calcium carbide.

Trimethylnonanone is an exceptionally effective vinyl chloride resin dispersant. Small quantities added to plastisols and organosols produce drastic re :\uctions in vis-cosity, and the high boiling point of trimethylnonanone prevents the formation of bubbles during fusion. It is a good solvent for organic stabilizers, helping to insure their even distribution in resins.

Mesityl oxide (4-methyl-3-penten-2-one), $CH_3COCH: C(CH_3)_2$, formula weight 98.14, is an oily, colorless liquid of strong odor; m.p., -52. 9°C.; b.p., 129.5°C.; density, 0.8548; *refractive index* 1.4475; vapor pressure at 20°C., 7.9 mm.; flash point (Tag open cup), 98°F., (Tag closed cup), 79°F.; A.P.H.A. color, 40 (max.); solubility at 20°C., 2.8-3.1 wt. % of ketone in water, 3.1:-3.4 wt. % of water-in ketone. Prolonged contact with air promotes polymerization as well as oxidation, accompanied by discoloration; stabilizers' such as diisopropylamine may be used. If peroxides have formed on long standing in air, there is danger of explosion during distillation.

Mesityl oxide is produced commercially by the dehydration of diacetone alcohol (see below); among the agents that haye been used are iodine, mineral acids such as sulphuric acid, sodium hydrogen sulfate, and chlorides or bromides of iron, cobalt, and nickel. Other products such as mesitylene (1,3,5-trimethylbenzene) are also formed in some cases. Mesityl oxide is also formed in the auto condensation of acetone: along with **isophorone** (3,5,5-trimethyl-2-cyclohexen-1-one) and mesitylene

(1,3,5,trimethylbenzene) by passing acetone over a mixed oxide catalyst at 200-400°C or along with diacetone alcohol and isophorone by the action of aqueous alkali on acetone. Technical-grade mesityl oxide and isophorone are both available in technical grades in tank cars, drums, and cans.

Mesityl oxide and isophorone are particularly good solvents for nitrocellulose and vinyl-type resins. The evaporation rate of mesityl oxide is between those of acetone and diacetone alcohol, whereas isophoron has a very slow evaporation rate. The acute toxicity of isophorone is greater than that of mesityl oxide vapour, which is lethal to rats in a concentration of 1000 p.p.m. in -t hr. Mesityl oxide is also used commercially in the manufacture of methyl isobutyl ketone.

Diacetone alcohol (4-hydroxy-4-methyl-2-pentanone, dimethylacetonylcarbinol, "diacetone"), $CH_3COCH_2C(CH_3)_2OH$, formula weight 116.16, is a pleasant-odoured, colorless liquid completely miscible with water and most organic liquids; f.p., - 42.8 °C. ; b.p., 169.1°C.; vapour pressure at 20°C., 1.1 mm. (about 1/40, 1/50 the evaporation rate of acetone); absolute viscosity at 20°C., 3.2 centipoises; coeff. of expansion at 20°C., 0.00097; flash point (Cleveland open cup), 155°F. The azeotrope with water contains 13% by weight of diacetone alcohol and boils at 99.6°C. The commercial product is supplied as an "acetone-free" grade; 0.9370-0.9420; flash point (open cup), 170°F.; and as a "technical" grade con-taining about 14% aeetone; density, 0.9150-0.9200; flash point (Cleveland open cup), below 60°F.The highest concentration considered safe for 8-hr. exposure is 50 p.p.m.

As both a ketone and an alcohol, diacetone alcohol takes part in reaetions associated with both the earbonyl and hydroxyl groups. Although stable when pure, diacetone alcohol immediately decomposes when treated with alkalies, forming an equilibrium mixture which is mostly acetone. It tends to revert to acetone when distilled at atmospheric pressure, but at reduced pressures it can be distilled without decomposition. When the alcohol is heated with iodine or sulphuric acid, dehydration takes place and mesityl oxide, $(CH_3)_2C{:}CHCOCH_3$, is formed.

Another commercially available derivative, 2-methyl-2,4-pentanediol is made by catalytic hydrogenation of the carbonyl group.

Diacetone alcohol is prepared by aldol condensation of two molecules of acetone flowing over a solid alkali or alkaline earth catalyst such as barium hydroxide

$$2(CH_3).CO \xrightarrow{Ba(OH)_2} (CH_3).C(OH)CH_2COCH_3$$

The concentration of diacetone alcohol in contact with the catalyst is greatly influenced by the temperature; thus at 15-20°C. the concentration at equilibrium is of the order of 12%, and at 0°C. it is about 16% (by weight). To secure good yields, it is necessary to remove the diacetone alcohol as fast as it is formed. Once the resulting solution is out of contact with the catalyst, the dimeric condensation product has little tendency to revert to acetone, but accumulates. The more volatile acetone then is continually removed by distillation and recycled. Another method of preparing diacetone alcohol consists of flowing acetone by gravity through iron drums containing lime. The ace-tone is distilled from the equilibrium mixture thus formed, and the diacetone alcohol is tapped off.

The acetone-free grade is obtained by fractional distillation under vacuum, or by stripping out the acetone by means of a stream of inert gas at an elevated temperature.

Diacetone alcohol provides an excellent means of reducing the viscosities of heavy organic liquids and at the same time minimizes the effect of low temperatures upon their viscosities. On this property is based its patented use as a component of castor oil-base hydraulic brake fluids, where large quantities have been required. Other desirable properties of diacetone alcohol in this connection are its minimum effect on rubber parts in the brake system, its high boiling point, low vapor pressure, and low freezing point.

Closely resembling acetone in its solvent properties, diacetone alcohol is an active but slow-drying solvent for both cellulose nitrate and acetate, as well as alkyd and vinyl types of resins and a large number of oils, fats, and waxes. It has one of the highest aromatic hydrocarbon dilution ratios of nitrocellulose solvents, with comparable evaporation rates, tolerating 2.8 volumes of toluene and 2.3 of xylene. As a high boiler, it retards blushing and produces smooth, brilliant lacquer films. Because of the high boiling point and solvent power for oils and grease, diacetone alcohol is a vehicle for the fungicides and insecticides used in wood preservation to pro-tect fence posts, telegraph poles, and other wood products, from rot and termites. Diacetone alcohol is also used as an ingredient in non-grain-raising wood stains, in metal-cleaning compounds, in paint and rust removers, and in printing-roll ink cleaner.

Diketones

1,2- Diketones (*a*-Diketones)

1,2- or *a*-Diketones contain two adjacent carbonyl groups neither of which terminates a carbon chain. They may be represented by the general formula RCOCOR′ in which Rand R′ represent alkyl, aryl, or heterocyclic radicals. 1,2-Diketones may also be cyclic, having the general formula:

H_2C—C=O

H_2C—C=O

Nomenclature. There are three principal methods of naming *a*-diketones: (1) by the I.U.P.A.C. or Geneva system, characterized by the addition of "-dione" to the name of the parent hydrocarbon (the position of the carbonyl groups being designated by the smallest possible numbers); (2) by the "glyoxal system", in which the compounds are named as derivatives of glyoxal; (3) by naming the compound from the two acyl radicals formed by splitting the molecule between the carbonyl groups (if the two radicals are the same, the prefix "bi-" is considered preferable to "di-," as $CH_3COCOCH_3$, biacetyl). By these methods, $CH_3COCOCH_2CH_3$ can be named: (1) 2,3-pentanedione, (2) methylethylglyoxal, and (3) acetyl propionyl.

A fourth method is sometimes used when R and R′ are the same aromatic or heterocyclic radical. In these cases, the ending "-il" is added to the stem of the radical. Examples are: $C6H_5COCOC_6H_5$, benzil, $CH_3OC_6H_4COCOC_6H_4OCH_3$, anisil, $C_4H_3OCOCOC_4H_3O$, furil.

Physical and Chemical Properties

The aliphatic 1,2-diketones are yellow liquids having boiling points that increase regularly with increasing molecular weight. The presence of branching in the carbon skeleton tends to lower the boiling point. Most of the cyclic and aromatic diketones are yellow solids. All of the unsubstituted aliphatic α-diketones have densities less than unity, while most of the aromatic dike tones are slightly heavier than water. The water solubility decreases rapidly with increasing molecular weight, only the first two members of the series (biacetyl and acetyl propionyl) being water-soluble. Odours of liquid α-diketones are sharp and penetrating (somewhat resembling quinones), while the solids, in general, are odourless. Since at high dilution the liquid diketones have a pleasant odour, traces of biacetyl and certain other aliphatic diketones have been used as flavouring agents.

Examples of 1,2-diketones are given in Table 12.3.

The carbonyl groups of α-diketones undergo many of the reactions observed with simple ketones. For example, a-diketones undergo reduction, form mono- and di-oximes, phenylhydrazines and osazines, and react once or twice with Grignard re-agents.

In addition, α-diketones undergo many reactions which are specific for this group of compounds.

Reactions with Aromatic o-Diamines. A method used for distinguishing α-diketones from other diketones is the reaction with aromatic o-diamines. The *a*-diketones react readily to give quinoxalines.

2-aminophenylamine + biacetyl → 3-methyl-2-methylene-1,2-dihydroquinoxaline + $2H_2O$

This reaction is used qualitatively to identify both 1,2-diketones and 1,2-diamines.

Oxidation. α-Diketones are cleaved by sodium peroxide or hydriodic acid to give two carboxylic acids. This reaction is specific for a-diketones and therefore can be used for purposes of identification.

$$\text{RCOCOR}' \rightarrow \text{RCOOH} + \text{R}'\text{COOH}$$

Reactions with Ketones. In the presence of alkali, benzil reacts with ketones having active methyl or methylene groups to form diacylstyrenes and hydroxycyclo-pentenones.

bicyclo[4.2.0]octane-7,8-dione + o-diketone → quinooxalines

bicyclo[4.2.0]octane-7,8-dione + pentan-3-one → hydrocyclopentanone

Table 12.3. 1, 2-Diketones

Common name	Synonyms	Formula	M.p., °C	B.p., °C	d_4
		Aliphatic			
Biacetyl	2, 3-Butanedione; dimethylglyoxal; diacetyl	$CH_2COCOCH_2$	—	87.5-88	$0.9808^{18.5}$
Acetyl propionyl	2, 3-Pentanedione; methylethylglyoxal	$CH2COCOCH_2CH_3$	—	108	0.09565^{10}
Biptopionyl	3, 4-Hexanedione; diethylglyoxal	$CH_3CH_2COCOCH_2CH_3$	—	129	—
Acetyl isobutyryl	4-Methyl-2, 3-pentanedione; methyliso-propylglyoxal	$CH_3COCOCH(CH_3)_2$	—	115-116	0.9215^{11}
Pripionyl butyryl	3, 4-Heptanedione; ethyl-*n*-propylglyoxal	$CH_2CH_1COCOCH_2CH_2CH_3$	—	147	0.885°
Acetyl isovaleryl	5-Methyl-2, 3-hexanedione; methylisobutyl-glyoxal	$CH_3COCOCH_2CH\ (CH_3)_2$	—	138	0.9082^{11}
Bibutyryl	4, 5-Octanedione; di-*n*-propylglyoxal	$CH_2(CH_2)_2COCO(CH_2)_2CH_3$	—	168	0.934°
Busobutyryl	2, 5-Dimethyl-3, 4-hexanedione; disopropyl-glyoxal	$(CH_3)_2)CHCOCOCH(CH_2)_2$	—	144-145	0.912°
Hexamethylbiacetyl	2, 2, 5, 5-Tetramethyl-3, 4-hexanedione; bipivaloyl	$(CH_3)_3CCOCOC(CH_3)_3$	—	169-170	0.895°
		Alicyclic			
1, 2-Cyclopentanedione	—	$CH_2.CO.CO.CH_2.CH_2$	—	9710	—
3-Methyl-1, 2-cyclopentanedione	—	$CH_2.CO.CO.CH(CH2).CH_2$	106	—	—
1, 2-Cyclobexanedione	—	$CH_2.CH_2.CO.CO.CH_2.CH_2$	104	—	—
		Aromatic			
Benzil	Bibenzoyl; diphenylglyoxal	$C_4H_5COCOC_6H_4$	95	—	1.25^{15}
Furil	Bi-2-furoyl; di-2-furfuroyl; di-2-furylglyxal	O.CH:CH.CH:CCOCO C:CH.CH:CH.O	165-166	—	—
Acetyl benzoyl	Methylphenylglyoxal	$C_6H_5COCOCH_2$	—	222	1.1006^{30}
Bensoyl phenylacetyl	Phenylbenzylglyoxal; 1, 3-diphenyl-1, propanedione	$C_6H_6CH_2COCOC_6H_5$	35-36	—	—
Anisil	4, 4-Dimethoxybenzil	$CH_2OC_5H_4COCOC_6H_4OCH_2$	133	—	—

With acidic dehydrating agents, either a cyclopentadienone or a bimolecular product is obtained.

If α-diketones possessing α-hydrogen atoms are used in place of benzil, the reaction is complicated by the self-condensation of the α-diketones to form quinones.

Reactions with Ammonia and Aldehydes. Imidazoles are formed in low yields by the interaction of a-diketones with ammonia and aldehydes.

$$\underset{\text{aldehyde}}{RCOCOCR'} + 2NH_3 + \underset{\text{diketones}}{R''CHO} \longrightarrow \text{(imidazole: } R_1\text{, R, } R_2\text{)} + 3H_2O$$

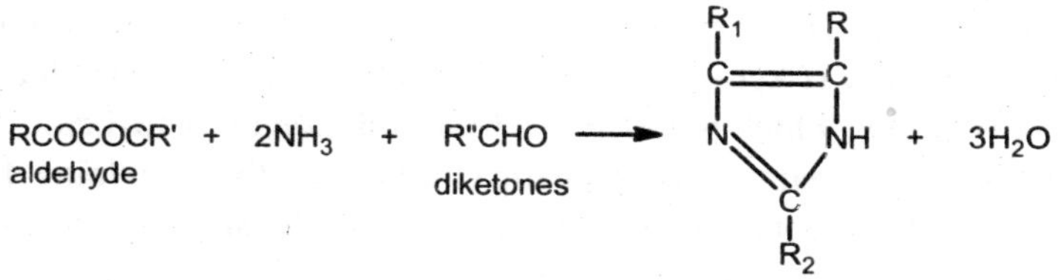

2,4,5-trimethyl-1 *H*-imidazole

Reactions with Alkali. There are many reactions which occur when an α-diketone is treated with alkali. Probably the most familiar is the *benzilic acid rearrangement.*

$$C_6H_5COCOC_6H_5 \rightarrow (C_6H_5)_2C(OH)COOH$$

Although the reaction seems to be general for α-diketones, low yields are frequently obtained with compounds containing a-hydrogen atoms, probably because of accompanying self-condensation.

1,2-Diketones containing active methylene groups undergo a remarkable condensation in the presence of alkali. Aldols are first produced, and these react further to give quinones.

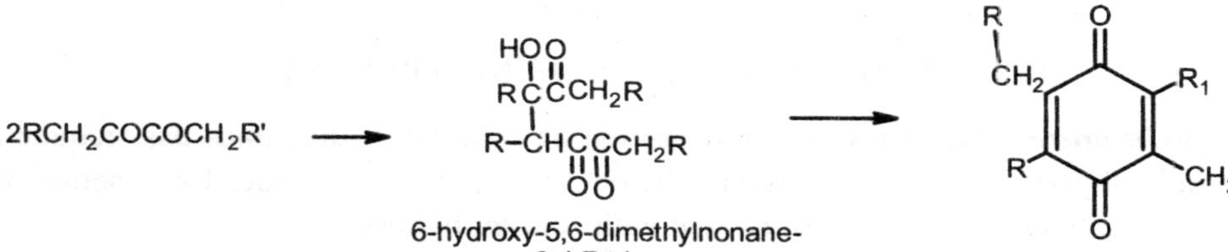

$$2RCH_2COCOCH_2R' \longrightarrow \text{RC(OH)(COCH}_2\text{R)–CH(R)COCOCH}_2\text{R} \longrightarrow \text{quinone}$$

6-hydroxy-5,6-dimethylnonane-3,4,7-trione

2,5-diethyl-3,6-dimethylbenzo-1,4-quinone

Traces of basic reagents have been found to catalyze the tautomrism between the enol and keto forms of α-diketones. The enol concentration of equilibrium mixtures has been determined at various temperatures for phenylbenzylglyoxal.

$$RCH_2COCOCH_2R' \rightarrow RCH{:}C(OH)COCH_2R'$$

Alkylation and Acylation. α-Diketones which are capable of enolization frequently demonstrate complete oxygen acylation or alkylation. This tendency is much more pronounced with α-diketones than with 3-diketones.

Preparation

N one of the α-diketones occurs in significant amounts among natural products, but they may be prepared by a wide variety of general methods.

Oxidation of Monoketones. Monoketones containing at least one methylene group are oxidized to α-diketones by means of selenium dioxide, nitric acid, or nitrogen tetroxide. A commercial method involves oxidation at 150-450°C., using as a catalyst the oxide of a heavy metal possessing more than one oxide (for example, copper, lead, iron, manganese, or vanadium.

$$RCOCH_3R' \rightarrow RCOCOR'$$

Oxidation of α-Ketol. This is the most important method for preparing aromatic or heterocyclic α-diketones, since the necessary α-ketols are readily available from the benzoin condensation. A very useful oxidizing agent is iodine, which is added to a basic solution of the α-ketol. This method has not proved as satisfactory for the preparation of aliphatic α-diketones from acyloins. However, up to 15% yields of α-diketones have been obtained as by-products from the acyloin reaction.

$$RCOCH(OH)R \xrightarrow{O} RCOCOR'$$

Oxidation of α-Diols. Very similar to the oxidation of α-ketols is the oxidation of a-diols. The oxidation has been carried out on a commercial scale using copper chromite catalyst in mineral oil at 250°C.

$$RCH(OH)CH(OH)R' \xrightarrow{O} RCOCOR'$$

From α-Bromo Ketones. Treatment of an *α-bromo* ketone with sodium azide and decomposition of the resulting azide with hydrochloric acid furnishes a general laboratory method for the preparation of α-diketones. The method has also been used commercially.

$$RCOCHR'Br + NaN_2 \rightarrow RCOCHR'N_2 \rightarrow RCOCOR'$$

From α, β-Unsaturated Ketones. The preparation of complex α-diketones has frequently been achieved in excellent yield from *a,* β-unsaturated ketones and sodium peroxide. The reaction proceeds by formation of an epoxide, which is then treated with sodium hydroxide.

$$H_3C{-}CH{=}CH{-}\overset{O}{\overset{\|}{C}}{-}CH_3 \longrightarrow H_3CHC\overset{O}{\frown}CH{-}\overset{O}{\overset{\|}{C}}{-}CH_3 \xrightarrow{NaOH} H_3C{-}\overset{O}{\overset{\|}{C}}{-}\overset{O}{\overset{\|}{C}}{-}CH_3$$

pent-3-en-2-one 1-(3-methyloxiran-2-yl)ethanone biacetyl

From Diazonium Compounds and Isonitrosoacetone. A good method for the preparation of methyl aryl diketones is the addition of a solution of a diazonium chloride to isonitrosoacetone in 10% sodium hydroxide solution, followed by hydrolysis with dilute sulphuric acid of the monoxime formed.

$$CH_3COCH{:}NOH + RN_2Cl \rightarrow CH_3COC({:}NOH)R \xrightarrow[H^+]{H_2O} CH_3COCOR$$

Dieckmann Condensation. The preparation of cyclic α-diketones from ethyl oxalate and glutaric or adipic esters has been realized in good yield.

methyl 4-(acetyloxy)butanoate + 2,2'-dimethyl-2,2'-bioxirane-3,3'-dione $\xrightarrow{C_2H_5ONa}$ methyl 4-(acetyloxy)-2,3-dioxocyclopentanecarboxylate

Friedel-Crafts Reaction (q.v.). Aromatic compounds containing electron-releasing groups (such as OR or OR) undergo a Friedel-Crafts reaction with oxalyl chloride to give derivatives of benzil.

$$2\ ROC_6H_5 + ClCOCOCl \xrightarrow{AlCl_3\ /\ CS_2} ROC_6H_5COCOC_6H_4OR$$

Uses

Biacetyl and other aliphatic α-diketones remove undesirable odours from dairy products. Certain α-diketones are useful as specific analytical reagents and as catalysts for polymerization.

Biacetyl is a better preservative against various yeasts and torulae than benzoic acid or sodium benzoate. It also has a bactericidal action on various pathogenic bacteria. Benzil can be used as a chigger repellent.

Biacetyl or diacetyl (2,3-butanedione, diketobutane, dimethylglyoxal), $CH_3CO\text{-}COCH_3$, formula weight 86.09, occurs naturally in bay and other essential oils, from which it can be separated by steam distillation. It also occurs in butter and cheese. Biacetyl is a greenish-yellow, mobile liquid with a strong odor similar to quinone. In extremely dilute solutions, however, it has a distinct butter-like odour. It is soluble in water, ether, and alcohol, and is volatile in steam.

Biacetyl exhibits the following physical properties: m.p. -3 to -4°C.; b.p. 88-°C-91°C.; density 2.2 0.98311, *refractive index 8.5* 1.39105.

Biacetyl undergoes reactions typical of α-diketones. In addition it undergoes reduction to 2,3-butanediol by means of yeast and forms a crystalline compound, $C_4H_6O_2.2HaPO_4$, on treatment with sirupy phosphoric acid. Biacetyl can be detected, even in small amounts, by its conversion to dimethylglyoxime; the addition of a soluble nickel salt produces a typical red colour.

Biacetyl is made commercially by passing vinyl acetylene into a solution of mercuric sulphate in sulphuric acid and decomposing the insoluble product with hydrochloric acid at 95-100°C. It is also made commercially by treating methyl ethyl ketone . With nitrous acid to form an isonitro so derivative (biacetyl monoxime), $CH_3COC\text{-}(:NOH)CH_3$, with subsequent decomposition with either sodium hydrogen sulfate or dilute sulphuric or hydrochloric acid. Fractionation of the crude biacetyl yields a pure food grade. Other methods include the oxidation of 2,3-butanediol with oxygen at 270°C. in the presence of a copper catalyst or the dehydrogenation of 2,3-butanediol with a copper-copper oxide catalyst at 325°C. in the absence of air or oxygen (30% yield) or a brass catalyst at 450°C. (60% yield).

Biacetyl is available commercially as the technical grade. It is sold in 10-gram, 100-gram, and 1-lb. glass bottles and in 5-lb. and 30-lb. aluminum containers. Since it is a flammable material, shipping regulations require that red warning labels be attached.

Biacetyl has been used as an odorant for butter, cream, milk, margarine (hydrogenated fats and oils), cheese, coffee, confectionery (rum and butter taffies), fats, honey, vinegar, essences, flavouring agents, blended perfumes, perfume materials, ice cream mixes, and tobacco. Biacetyl has also been used in the photographic and ad-hesive industries as a hardening agent for gelatin.

BENZIL

Benzil (bibenzoyl, dibenzoyl, diphenyl diketone, diphenylglyoxal), $C_6H_5COCO\text{-}C_6H_6$, formula weight 210.22, is a yellow solid which crystallizes from alcohol in hexagonal prisms. It is soluble in alcohol and ether and insoluble in water.

Benzil has the following physical constants: m.p. 95°C.; b.p. 346-348°C (very little decomposition). Perhaps the most familiar reaction of benzil is the benzilic acid rearrangement. Benzil is nitrated by a mixture of nitric and sulphuric acids to form either 3,3'-dinitrobenzil or 3,3',5,5'-tetranitrobenzil. It is reduced to benzoin, $C_6H_5CHOHCOC_6H_5$ by acid reducing agents and to hydrobenzoin by sodium amalgam. There are four isomeric dioximes of benzil known as the α, β, γ, and δ modifications. The δ form can be used for the estimation of nickel. Benzil forms eutectic with phenol, resorcinol, pyrocatechol, naphthalene, stilbene, and hydrobenzoin. Pure benzil in alcohol gives no color on addition of potassium hydroxide in the cold, but on heating, a purple-red colour is produced which is stable in air. If any benzoin is present, however, the colour appears in the cold and disappears on shaking in air. Benzil is prepared commercially by the oxidation of benzoin. Either nitric acid or air is employed as oxidizing agent, with copper sulfate as catalyst and pyridine as solvent. Benzil is available in tin containers as the technical grade. Benzil is used chiefly in organic syntheses. Recently it has been found to be of value as an insecticide and chigger repellent.

1,3-Diketones (β-Diketones)

1,3- or β-Diketones contain two carbonyl groups separated by one methylene group and may be represented by the general formula $RCOCH_2COR'$.

Nomenclature. β-Diketones are variously named (1) by the I.U.P.A.C. or Geneva system, using the suffix "-dione" with the name of the parent hydrocarbon; (2) as acyl derivatives of a methyl ketone; and (3) as diacyl derivatives of methane. Thus, $CH_3COCH_2COCH_3$ can be named: (1) 2,4-pentanedione, (2) acetyl acetone , and (3) diacetylmethane.

Physical and Chemical Properties

The aliphatic β-diketones are colourless liquids. The boiling points tend to rise as the length of the alkyl chain increases and tend to decrease in isomers with branching in the chain. Their odours resemble those of esters, but are not sweet or sharp. A few of the lower-molecular-weight 1,3-diketones have appreciable water solubility (as acetylacetone, 1: 10), but the solubility decreases rapidly with ascending molecular weight. The alicyclic and aromatic β-diketones are mostly solids; however, many of them are best purified by vacuum distillation. Examples of β-diketones are given in Table 12.4. Most of the reactions of β-diketones are analogous to those of β-keto esters.

Enolization. β-Diketones normally exist in the enol form to an unusually great extent (as I); their vapour phase is sometimes 100% enolic. Ferric chloride imparts an intense red colour to their alcoholic solutions. The enolic form as such, or in many of its metallic derivatives, has a chelate structure.

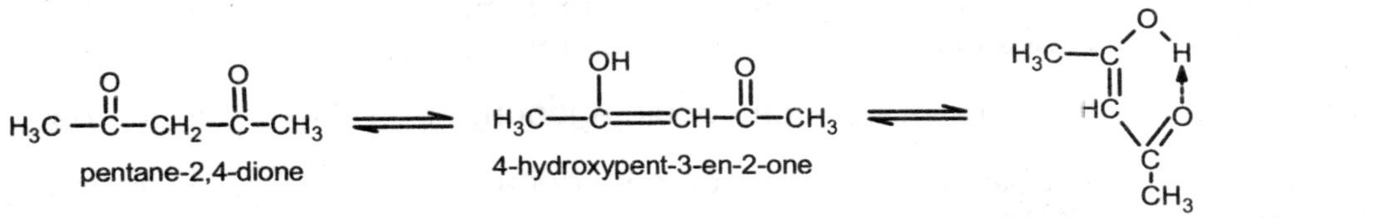

$$H_3C-\overset{O}{\overset{\|}{C}}-CH_2-\overset{O}{\overset{\|}{C}}-CH_3 \rightleftharpoons H_3C-\overset{OH}{\overset{|}{C}}=CH-\overset{O}{\overset{\|}{C}}-CH_3 \rightleftharpoons \text{(chelate)}$$

pentane-2,4-dione — 4-hydroxypent-3-en-2-one — (3Z)-4-hydroxypent-3-en-2-one

The copper derivative (as II) is a convenient means of isolating the b-diketone during synthesis it is also useful as a means of identification.

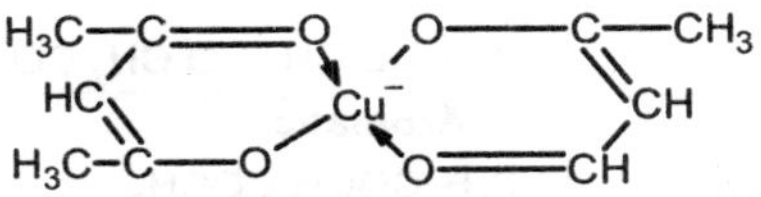

Alkylation. Like β-keto esters, β-diketones can be alkylated by means of their alkali metal derivatives.

$$Na^+[CH_3COCH{:}C(O^-)CH_3] + CH_3I \rightarrow CH_3COCH(CH_3)COCH_3$$

In certain difficult cases the use of silver oxide with the diketone may be advantageous however, this occasionally leads to a mixture of the O-alkylated and C-alkylated products.

Acylation. β-Diketones can be acylated with acid anhydrides, ketene, or iso-propenyl acetate, in a manner similar to that in which they are alkylated; *O*-acylation predominates, however, with the formation of esters of the enol form.

$$Na^+[CH_3COCH{:}C(O^-)CH_3] \xrightarrow{(CH_3CO)_2O.orCH_3{:}CO{:}+H_2SO_4} (CH_3CO)_3CH$$

C-Acylated products can be formed by rearrangement by heat and/or mineral acid catalysts, similar to the pyrolysis of isopropenyl acetate.

$$CH_3COCH{:}C(CH_3)OOCCH_3 \rightarrow (CH_3CO)_3CH$$

With esters of chloroformic acid, β-diketones form unsaturated carbonic esters

$$Na^+\left[CH_3COCH:C(O^-)CH_3\right] + ClCOOC_2H_5 \rightarrow CH_3COCH:C(CH_3)OCOOC_2H$$

Reaction with Phenylhydrazine, Semicarbazide and Hydroxylamine. β-Diketones react with the usual carbonyl group reagents, but the reaction goes further to produce heterocyclic derivatives, which

Table 12.4. 1, 3-Diketones

I.U.P.A.C. Name	Synonym	Formula	M.p., °C	B.Pp. °C
		Aliphatic		
2, 4-Pentanedione	Acetylacetone	$CH_2COCH_2COCH_2$	—	$134\text{-}136_{740}$
2, 4-Hexanedione	Pripionylacetone	$CH_2COCH_2COCH_2CH_2$	—	$154\text{-}157_{740}$
2, 4-Heptanedione	Butyrylacetone	$CH_2COCH_2CO(CH_2)_2CH_2$	—	$87\text{-}90_{88}$
5-Methyl-2, 4-hexanedione	Isobutyrylacetone	$CH_3COCH_2COCH(CH_2)_2$	—	$161\text{-}164_{740}$
2, 4-Octanedione	Valerylacetone	$CH_3COCH_2CO(CH_2)_3CH_3$	—	$79\text{-}81_{17}$
5, 5-Dimethyl-2, 4-hexanedione	Pivalylacetone	$CH_3COCH_2COC(CH_3)_2$	—	$164\text{-}167_{740}$
3-Thyl-2, 4-pentanedione	Ethylacetylacetone	$CH_3COCH(C_2H_5)COCH_3$	—	$175\text{-}178_{740}$
3, 3-Diethyl-2, 4-pentanedione	Diethylacetyl-acetone	$CH_2COC(C_2H_5)_2COCH_2$	—	$95\text{-}100_{10}$
2, 2-Dimethyl-3, 5-nonanedione	Pivalylvaleryl-methane	$(CH_3)_3CCOCH_2CO(CH_2)_3CH_2$	—	$113\text{-}117_{11}$
		Alicyclic		
1-Cyclohexyl-1, 3-butanedione	—	$CH_3COCH_2COCH.(CH_2)_4.CH_2$	—	$105\text{-}107_7$
5, 5-Dimethyl-1, 3-	—	$(CH_2)_2C.CH_2.CO.CH_2.CO.CH_2$	147-148	—
		Aromatic		
1-Phenyl 1, 3-butanedione	Acetylbenzoyl-methane	$C_6H_5COCH_2COCH_2$	59-60	$134\text{-}136_{16}$
1-(4-Biphenylyl)-1, 3-butanedione	—	$C_6H_5C_6H_4COCH_2COCH_3$	156-157	—
1-Phenyl-1, 3-pentanedione	Benzoylpropionyl-methane	$C_4H_5COCH_2COCH_2CH_3$	—	$128\text{-}132_7$
1-Phenyl-5, 5-dimethyl-2, 4-hexanedione	Phenylacetyl-pivalylmethane	$C_6H_5CH_2COCH_2COC(CH_3)_3$	—	$155\text{-}163_{10}$
1, 3-Diphenyl-1, 3-propanedione	Dibenzoylmethane	$C_6H_5COCH_2COC_6H_5$	77-78	—
1, 4-Diphenyl-1, 3-butanedione	Benzoylphenyl-acetylmethane	$C_6H_5COCH_2COCH_2C_6H_6$	—	$210\text{-}217_{10}$
1-Phenyl-2-benzyl-1, 3-butanedione	—	$C_6H_5COCH(CH_2C_6H_5)COCH_3$	55-56	$135\text{-}138_{10}$
1-Phenyl-3-(2-methoxyphenyl)-1, 3-propanedione	2-Methoxydibenz-oylmethane	$C_6H_5COCH_2COC_6H_4OCH_3$	119	—
1-(4-Nitrophenyl-1, 3-butanedione	p-Nitrobenzoyl-acetone	$O_2NC_6H_4COCH_2COCH_2$	234-244	—
		Heterocyclic		
1-(2-Furyl)-1, 3-butanedione	—	$CH_3COCH_2COC:CH.CH:CH.O$	—	$107\text{-}110_{10}$
1-(Tetrahydro-2-furyl)-1, 3-butane-dione	—	$CH_3COCH_2COCH.(CH_2)_3.O$	—	$95\text{-}97_8$

can be used as a means of identification (15). Acetyl-acetone and phenylhydrazine give 3,5-di,nethyl-1-phenylpyrazole.

$$CH_3COCH_2COCH_3 + C_6H_5NHNH_2 \xrightarrow{-H_2O} H_3C-C(=NNH-C_6H_5)-CH_2-C(CH_3)=O \rightarrow \text{3,5-dimethyl-1-phenyl-4,5-dihydro-1}H\text{-pyrazol-5-ol} \rightarrow \text{5-methyl-1-phenyl-1}H\text{-pyrazole}$$

(2Z)-pentane-2,4-dione phenylhydrazone

3,5-dimethyl-1-phenyl-4,5-dihydro-1*H*-pyrazol-5-ol

5-methyl-1-phenyl-1*H*-pyrazole

Scmicarbazide reacts in a similar manner to yield pyrazolecarboxamides (20). Certain nitroso-substituted pyrazoles show considerable promise as fungicides. Hydroxylamine also yields cyclic products known as isoxazoles (11).

$$CH_3COCH_2COCH_3 \quad H_2NOH \xrightarrow{-H_2O} H_3C-C(=N-NH-C_6H_5)-CH_2-C(CH_3)O$$

(2Z)-pentane-2,4-dione phenylhydrazone

Combes Reaction. B-Diketones react with aniline to give anils which can bp cyclized to quinoline derivatives with sulphuric acid (14).

$$CH_3COCH_2COCH_3 + C_6H_5NH_2 \longrightarrow C_6H_5-N=C(CH_3)-CH_2-C(=O)CH_3 \xrightarrow{H_2SO_4} \text{2,4-dimethyl-1,2-dihydroquinoline}$$

(4Z)-4-(phenylimino)pentan-2-one

2,4-dimethyl-1,2-dihydroquinoline

Cleavage. When, B-diketones are treated with 1% aqueous sodium hydroxide they are cleaved to form an acid and a ketone.

$$RCOCH_2COR' \xrightarrow{1\%\ NaOH} RCOOH + R'COCH_3$$

In this reaction the stronger of the two possible acids is formed, that is, cleavage occurs at the more reactive of the two carbonyl groups. Diketones are also cleaved when treated with ethanol containing hydrogen chloride.

$$CH_3COCH_3COCH_3 + C_2H_5OH \xrightarrow{HCl} CH_3COOC_2H_5 + CH_3COCH_3$$

As in the case of alkaline cleavage, fission occurs predominantly at the more reactive carbonyl group (7); for example, with benzoylacetone, the sole products are acetophenone and ethyl acetate (1).

Acyl Exchange. The sodium enol ate reacts with an ester to give another 13-dike tone (18).

$$Na^+[RC(O^-):CHCOR'] + R''COOC_6H_5 \rightarrow Na^+[RC(O^-):CHCOR''] + R'COOC_2H.$$

Preparation

β-Diketones are not available as natural products. They are prepared synthetically by the following methods:

The Claisen Condensation. *13-Diketones* are most frequently prepared from a ketone and an ester in the presence of an alkaline condensing agent; the highest yields are obtained with sodium amide.

$$RCOOC_2H_5 + HCH_2COR' \rightarrow RCOCH_2COR' + C_2H_5OH$$

From a,13-Unsaturated Ketones. This method is used when the sodium amide procedure is unsatisfactory, that is, when the desired diketone contains a nitro or other reactive group. The unsaturated ketone is brominated, and the dibromide is converted with sodium methoxide to the enol ether, which is hydrolyzed in the presence of an acid.

$$NO_2C_6H_4COCH{:}\ CHC_6H_5 \xrightarrow{Br_2} NO_2C_6H_4COCHBrCHBrC_6H_5$$

$$NO_2C_6H_4COCHBrCHBrC_6H_5 \xrightarrow{NaOCH_3} NO_2C_6H_4C(OCH_3){:}CHCOC_6H_5$$

$$NO_2C_6H_4C(OCH_3){:}CHCOC_6H_5 \xrightarrow{H^+/H_2O} NO_2C_6H_4COCH_2COC_6H_5$$

Boron Trifluoride Method. The synthesis of 13-diketones by the condensation of ketones with acid anhydrides in the presence of boron trifluoride has found much favour in recent years (12a).

$$CH_3COCH_3 + (CH_3CO)_2O \xrightarrow{BF_3} CH_3COCH_2COCH_3 + CH_3COOH$$

The yields are high, and the method is suitable for use in the presence of nitro groups

Sodium Ethoxide Method. Acetylacetone may also be made by condensation of acetone with ethyl acetate by means of sodium ethoxide, but the yield is only about half that in the boron trifluoride method.

From β-Keto Esters by Acylation. The ester is acylated by an acid chloride in an inert solvent, using sodium methoxide, to give an a, a-diacyl ester, which is then selectively cleaved to the β-diketone.

Carboxylic anhydrides may likewise be used as acylating agents in the presence of magnesium or a nonacidic magnesium compound to give 80% or better yields of β-diketones pyrolytic condensation of an a,a-diacyl ester with a monocarboxylic acid in the presence of a catalyst such as boric or sulphuric acid also gives good yields of a β-diketone.

$$\text{phenyl 3-oxobutanoate} \xrightarrow{CH_3COCl} \text{phenyl 2-formyl-3-oxobutanoate} \xrightarrow[H^+]{H_2O} \text{pentane-2,4-dione}$$

phenyl 3-oxobutanoate

phenyl 2-formyl-3-oxobutanoate

pentane-2,4-dione

Condensation of Acetyl Chloride with Aluminum Chloride. This reaction may be formulated in the following manner.

$$CH_3COCl \xrightarrow{AlCl_3} (CH_3CO)_2CHCCl_2O - AlCl_2 \xrightarrow{H_2O} (CH_3CO)_2CHCOOH_3$$

$$(CH_3CO)_2CHCOOH_3 \xrightarrow{-CO_2} CH_3COCH_3COCH_3$$

Pyrolysis of I sopropenyl Acetate. When isopropenyl acetate is heated with or without catalysts at 450-500°, acetylacetone is formed (32).

$$CH_2{:}C(CH_3)OOCCH_3 \xrightarrow{450\text{ - }500^\circ C} CH_3COCH_2COCH_3$$

Other β-Diketones may also be prepared by this method.

Uses

β-Diketones may be used as couplers in photography, an azomethine dye being formed during development.

Acetylacetone (2,4-pentanedione), $CH_3COCH_2COCH_3$, formula weight 100.11, is a colourless liquid having an ester-like odor; b740, 134-136°,density 0.9771; refractive index 1.4518; vapor pressure at 20°C., 7.0 mm.; flash point, 105°F.; solubility at 20°C., 16.G wt.% of ketone in water, 4.5 „'t. % of water in ketone; completely miscible with most organic solvents. The vapor in 1000 p.p.m. is lethal to rats in four hours.

Acetylacetone was first made by Claisen and Ehrhardt in 1889. It is prepared in highest yield (85%) from acetone and acetic anhydride „'ith boron trifluoride ac catalyst, whereas yields by the Claisen method are only -15% (2,5). It may also be made by the sodium ethoxide method, by acetylation of acetoacetic esters, by pyrolysis of isopropenyl acetate, and by condensation of acetyl chloride with aluminum chloride (see above). A novel method involving the action of anhydrous aluminum chloride on the mixed vapors of ethanol, chlorine, and benzene has also been patented. It is useful in organic synthesis, and its complexes with some bivalent and trivalent metals have been suggested as fungicides, insecticides, driers for paints and tarnishes, and as colours for inks.

1,4-Diketones ('Y-Diketones)

In 1,4- or γ-diketones the two carbonyl groups are separated by two methylene groups: $RCOCH_2CH_2COR$. Relatively few γ-ketones were reported in the literature before 1942, since no convenient procedure existed for their synthesis. The only diketone now available in commercial quantities is acetonylacetone.

Nomenclature. Like the β-diketones, the γ-diketones can be named (1) by the I.U.P.A.C. system; (2) as acyl derivatives of a methyl ketone; and (3) as diacyl derivatives of the appropriate hydrocarbon (ethane). Thus, $CH_3COCH_2CH_2COCH_3$ can be named: (1) 2,5-hexanedione, (2) acetonylacetone, and (3) 1,2(or sym)-diacetyl ethane.

Physical and Chemical Properties

The 1,4-diketones possess very faint odors, much weaker than the corresponding monoketones of the same chain length. The physical properties of a few 1,4-diketones are given in Table V.

Table 12.5. Properties of Aliphatic 1, 4-Diketones

Name	Formula	M.p., °C	B.p., °C
2, 5-Hexanedione	$CH_2COCH_2CH_2COCH_2$	−9	194
2, 5-Octanedione	$CH_3COCH_2CH_2CO(CH_2)_2CH_3$	34-35	98_{14}
6-Methyl-2, 5-heptanedione	$CH_3COCH_2CH_2COCH(CH_3)_2$	—	91_{12}
2, 5-Decanedione	$CH_3COCH_2CH_2CO(CH_2)_4CH_3$	—	132_{17}
2, 5-Undecanedione	$CH_3COCH_2CH_2CO(CH_2)_4CH_3$	33	141_{14}
2, 5-Dodecanedione	$CH_3COCH_2CH_2CO(CH_2)_6CH_3$	40.5	148_{12}
3, 6-Dodecanedione	$CH_3CH_2COCH_2CH_2CO(CH_2)_6CH_3$	41	150_{16}
2, 5-Octadecanedione	$CH_2COCH_2CH_2CO(CH_2)_{12}CH_5$	71	170_1
11-Methoxy-2, 5-undecanedione	$CH_2COCH_2CH_2CO(CH_2)_5CH_2OCH_2$	23	167_{11}

reactions. The 1,4-diketones react with the usual carbonyl reagents; the diphenylhydrazones, in contrast to the osazones of 1,2-diketones, are colorless. They do not form salts, and consequently are insoluble in alkali. The most characteristic property of the 1,4-compounds is the readiness with which they form ring compounds derived from furan, thiophene, and pyrrole. In all these reactions it may be assumed that the diketone form is enolized, and the enolic gro,ups I are then replaced by oxygen, sulfur, or imino (NH) to give the corresponding furan, I thiophene, or pyrrole compound. 1,4-Diketones also undergo other types of cyclization

Preparation

Of the methods now known, the first is the most flexible and gives good yields. The electrosynthesis and perhaps the acid hydrolysis of furans are also convenient and useful general methods. Acetonylacetone is obtained as a by-product of the manufacture of acetic acid from acetylene (4), as well as by hydrolysis of a furan.

Acid Hydrolysis of a-Acyllevulinic Esters. This type of ester is readily secured by acylation of sodium acetoacetic ester, cleavage by sodium methoxide, and fractionation to isolate the acylacetic ester. The latter, in the form of its sodium derivative, is then alkylated with hromoacetone to yield an intermediate ester which upon cleavage gives the desired product.

$$Na^{+}[CH_3C(O-):CHCOOC_2H_5] \xrightarrow{RCOCl} CH_3COCH(COR)COOC_2H_5$$

$$CH_3COCH(COR)COOC_2H_5 \xrightarrow{NaOCH_3} RCOCH_2COOCH_3$$

Hydrolysis of Furans. Acetonylacetone may be prepared conveniently in 88% yield from 2,5-dimethylfuran, $C(CH_3)$:CH.CH:$C(CH_3)$.O, but, since other 2,5-disubstituted furans are not readily available, the method is not of general applicability. An attempt has been made to extend this synthesis by use of furfural, which is easily converted to other substances containing the furan nucleus. For example, condensation of 5-methylfurfural with methyl n-propyl ketone gives (5-methylfurfurylidene)methyl n-propyl ketone. Reduction of this ketone gives 5-n-hexyl-2-methyl-furan, which on acid hydrolysis gives 2,5-undecanedione. In the following equation, $R = n\text{-}C_3H_7$.

5-methyl-2-furaldehyde + acetone → 3-(5-methyl-2-furyl)butan-2-one

The over-all yield for this reaction, however, is not especially good. A modification of this reaction has been used by Alder and Schmidt (1).

From a-Hydroxyisobutyric Acid and Succinyl Chloride. This method is useful for obtaining symmetrical homologs of acetonylacetone. a-Hydroxyisobutyric acid is condensed with succinyl chloride to give a succinoxy fatty acid. The chloride of this acid reacts with a zinc alkyl to form a dilactone, from which the desired diketone is obtained by alcoholysis. 3,5-0ctanedione, for example, is obtained by the following reactions (3):

$$(-CH_2COCl)_2 + 2HOC(CH_3)_2COOH \rightarrow [-CH_2COOC(CH_3)_2COOH]_2$$

$$[-CH_2COOC(CH_3)_2COOH]_2 \xrightarrow{PCl_5} \rightarrow (-CH_2COC_2H_5)_2$$

$$[-CH_2COOC(CH_3)_2COCl]_2 \xrightarrow{ZnC_2H_5I} [CH_2C(C_2H_5).O.C(CH_3)_2.CO.O]_2$$

Dehydrogenation of Secondary 1,4-Diols. One method of synthesis of the diols involves the reaction of ethynylenebis(magnesium iodide), IMgCa"CMgI, with a carbonyl compound in the normal fashion to give an acetylenic diol. On catalytic reduction the triple bond is reduced, and by means of a suitable dehydrogenation catalyst, the 1,4-diketone is formed. The yields are poor.

Electrolysis of a Mixture of o-Acetonyllevulinic Acid and an Aliphatic AC1'd. Electrolysis in methanol containing a small amount of sodium methoxide gives a mixture of three products from which the desired 1,4-diketone can be separated in 30-45% yield by adequate fractionation. Thus, a mixture of δ-acetonyllevulinic anc caproic acids gives 2,5-dodecanedione, along with n-decane and 2,5, 10, 13-tetradecanetetrone. The method is limited in that only simple acids can be employed.

Ketone Hydrolysis of Ethyl α, β-Diacetylsuccinate and Analogous Compounds. The diacetylsuccinate results from the coupling of two molecules of the sodium derivative of acetoacetic ester by means of iodine (7), and cn hydrolysis is it gives acetonylacetone.

diphenyl 2,3-diacetylsuccinate → hexane-2,5-dione

This method may be applied with useful results only to compounds having equivalent acyl residues; moreover, from the higher homologs, cyclopentenone derivatives are formed for the most part.

Hydrolysis of the Addition Product of an Aliphatic Nitro Compound and an a,{3-Unsaturated Ketone. For example, phenylnitromethane may be added to phenyl vinyl ketone to give γ-nitro-' γ-phenylbutyrophenone, which on treatment with sodium . methoxide and subsequent hydrolysis with hydrochloric acid gives an 80% yielc of , 1,2-dibenzoylethane

$$CH_2{:}CHCOC_6H_5 + C_6H_5CH_2NO_2 \rightarrow CH_2CH_2COC_6H_5 \xrightarrow{NaOCH_3\,/\,HCl} [-CH_2COC_6H_5]_2$$

$$C_6H_5CHNO_2$$

The 1,4-diketones are important intermediates in the manufacture of perfumes *(q.v.)*. For example, the valuable perfume dihydrojasmone is obtained by the basecatalyzed cyclization of 2,5-undecanedione to give 3-methyl-2-amyl-2-cycloperten-lone (dihydrojasmone). Jasmone has been synthesized in a similar manner (5).

Of still greater industrial significance is the synthesis of compounds of the pyrethrin type. Methylglyoxal and sodium caproylacetate react to give a hydroxy 1,4-diketone, which is readily cyclized to dl-dihydrocinerolone. In the following equation, $R = n - C_4H_9$.

$$CH_3COCHO + RCH_2COCH_2COONa \longrightarrow$$ 3-hydroxyheptane-2,5-dione → 4-hydroxy-2,3-dimethyl-cyclopent-2-en-1-one

Acetonylacetone is a solvent for cellulose acetate with a xylene-nitrocellulose dilution ratio of 1:8. A small percentage reduces the viscosity of nitrocellulose and vinylite resin solution. It is also used as a tanning agent and in roll-coating inks, lacquers, and stains.

Comparison between formaldehyde, acetaldehyde and aceone

No.	Property	Formaldehyde	Acetoldehyde	Acetone
	Difference			
1.	formula	HCHO	Ch_3CHO	$(CH_3)CO$
2.	physical state	Colourless gas	Colour less mobile inflammable liquid	colourless mobile inflammable liquid
3.	smell	Pungent odour	Pungent, disagreeable odour	aromatic odour
4.	oxidation	Formic acid is formed. It reduces ammonical $AgNO_3$ and Fehlling's solution.	Acetic acid is formed. It also reduces both Tollen's and Fehlling's solution	Not easily oxidised
5.	Reduction	It gives methyl alcohol (primary)	It gives ethyl alcohol (primary)	Isopropyl Alcohol (secondary)
6.	With Grignard's reagent	It gives primary alcohol	It gives secondary alcohol	It gives secondary alcohol
7.	With ammonia	Urotropine is produced	Acetaldehyde ammonia is given	It does not give addition compounds in normal state however produces complex condensation
8.	With alcohol	It gives methanal	It gives acetal	No reaction
9.	With hot caustic alkalies	Cannizzaro's Reaction	Resins are produced	No reaction
10.	Schiff's reagent	Pink colour is restored	Pink colour is restored	No colour
11.	Polymerization	It gives para, meta form	It gives para, meta form	It does not polymerize
12.	Chlorination	Cannot be chlorinated	Chloral is formed	Trichloro acetone formed

No.	Property	Formaldehyde	Acetoldehyde	Acetone
	Resemblance	All these three compounds contain carbonyl group and therefore show many common properties		
1.	With Sodium bi sulphite	Crystalline bisulphite compound is produced	Same as formaldehyde	Same as Aldehydes
2.	With HCN	Cyanohydrin is formed	Same as formaldehyde	Same as Aldehydes
3.	With NH_2OH	Oxime are Produced	Same as formaldehyde	Same as Aldehydes
4.	With Hydrazone	Hydrazones are produced	Same as formaldehyde	Same as Aldehydes
5.	With $C_6H_5NHNH_2$	Phenyl hydrazones produced	Same as formaldehyde	Same as Aldehydes
6.	With Phosphorous pentachloride	Methylene chloride is produced	Ethylidine chloride is produced	Isopropyldine chloride is produced

CHAPTER

13

Derivatives of Carboxylic Acids

Introduction

The derivation of monocarboxylic acid discussed in this chapter are those which are formerly obtained by replacing the OH group of the carbonyl function by another reactive atom or group G (say)

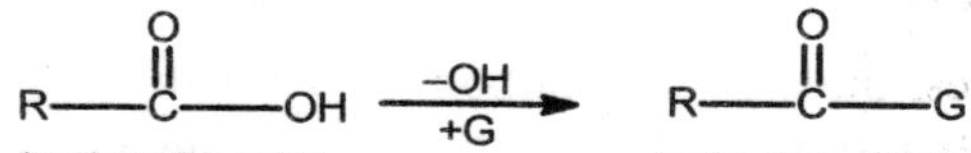

$$\underset{\text{(carboxylic acid)}}{R-\overset{O}{\overset{\|}{C}}-OH} \xrightarrow[+G]{-OH} \underset{\text{(acid derivative)}}{R-\overset{O}{\overset{\|}{C}}-G}$$

here G can be halogen, —X(Cl, Br, and I) —O—CO—R, —OR, or —NH_2 and the respective classes of carboxylic acid derivatives are designated as acyl halides, anhydrides, esters and amides. The structural relationship among the classes of compounds are summarized in the table below:

Classes		General formula—G	Specific example
Anhydride or acid halide	$H_3C-\overset{O}{\overset{\|}{C}}-X$	Where X=Cl, Br, I	$H_3C-\overset{O}{\overset{\|}{C}}-Cl$ acetyl chloride; $H_3C-\overset{O}{\overset{\|}{C}}-Br$ acetyl bromide; $H_3C-\overset{O}{\overset{\|}{C}}-I$ acetyl iodide
Acid anhydride	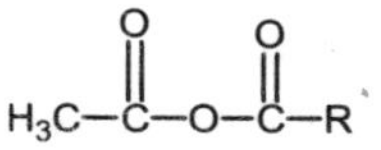 $H_3C-\overset{O}{\overset{\|}{C}}-O-\overset{O}{\overset{\|}{C}}-R$	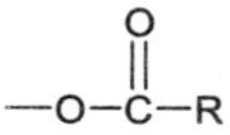 $-O-\overset{O}{\overset{\|}{C}}-R$	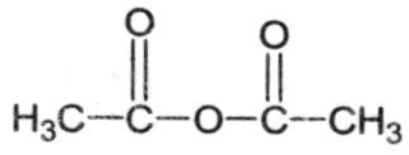 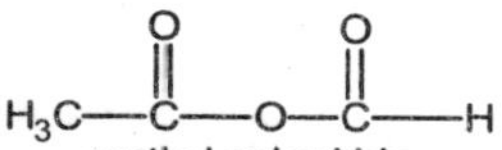 $H_3C-\overset{O}{\overset{\|}{C}}-O-\overset{O}{\overset{\|}{C}}-CH_3$ acetic anhydride; $H_3C-\overset{O}{\overset{\|}{C}}-O-\overset{O}{\overset{\|}{C}}-H$ methyl anhydride

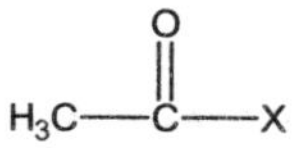

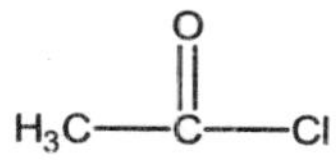

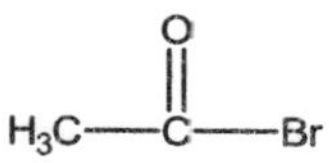

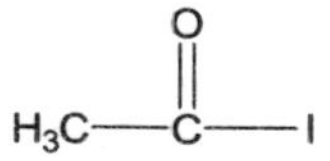

Ester	$H_3C—C(=CH_2)—O—R$	—O—R (R=CH_3, H)	$H_3C—C(=O)—O—CH_2—CH_3$ ethyl acetate; $H_3C—C(=O)—O—CH_2—H$ methyl acetate
Acid amide	$H_3C—C(=O)—N(H)—H$ $H_3C—C(=O)—NH_2$	$—NH_2$	$H_3C—C(=O)—NH_2$ acetamide

Orbital Structure

As described above, the carboxylic derivatives are characterised by the presence of a carboxyl group —CO—and a group G which may be X, —O—CO—R_1, —OR, or —NH_2

The structure of those derivatives is primarily the structure of the carbonyl group, through this structure is modified some what by the group G attached to carbonyl-carbon. Let us discuss the orbital structure of acid derivatives by taking the example of acyl halide

Like aldehydes and ketones, the carbonyl-carbon in acyl halides us sp^2 hybridized. It's one sp^2 orbital overlaps with a sp^3 orbital of the carbon of the alkyl group and the remaining two sp^2 overlap with half-filled p-orbitals on the halogen (say, Cl, $3p^2{}_z$ $3p^2{}_y 3p^2{}_x$) and oxygen ($2p^2{}_x$ $2p^1{}_y$ $2p^1{}_z$). The hybridized p orbital on the carboxyl carbon and on the oxygen atom being parallel overlap or this joined by a double bond (on s bond and p bond) This orbital structure of an acid halide is developed in fig: 13.1.

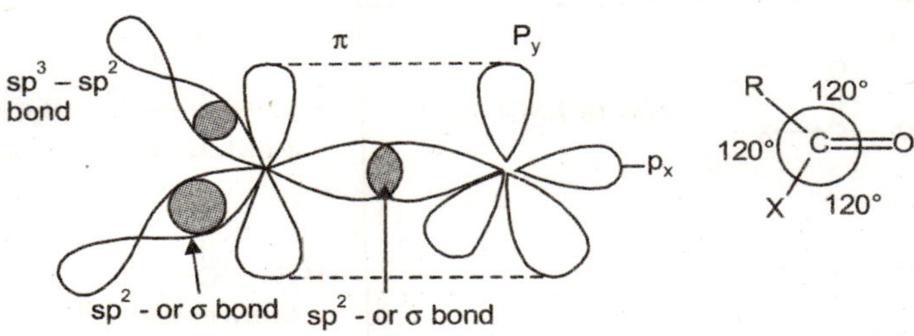

Fig. 13.1. Showing the model of acyl halide

Since the carbonyl carbon utilizes its sp^2 orbital to form these σ-bonds. The angles LRCO and LRCO and LXCO are approximately 120°. Evidently the part of the molecule (see fig. 13.2)

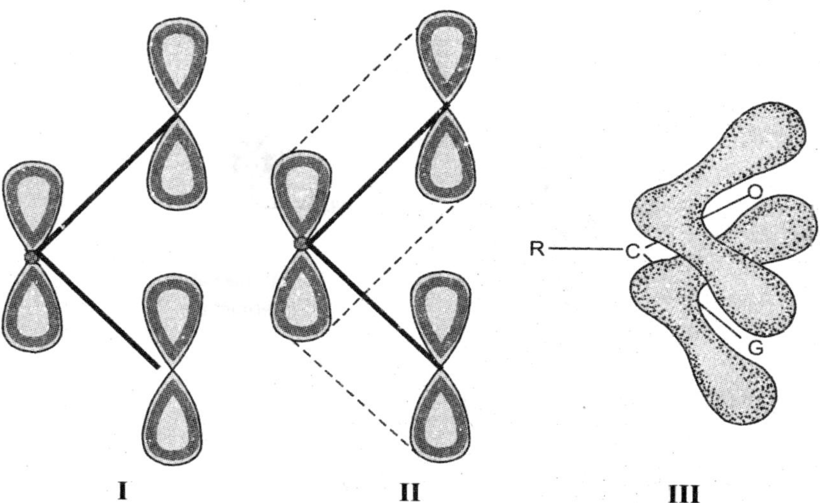

Fig. 13.2. Orbital structure R—CO—G showing the unhybridised p-orbital of carbonyl C and O a also of group G.

Immediately surrounding the carbonyl group, the group R and the group G (–X) are all in one place. The orbital model of the anhydrides, esters and amides exactly similar to that acyl halides. Thus in general, the orbital model of an acid derivative R – CO – G could be represented as in fig. 13.2.

The atom of the group G (Cl in the acyl halide, O in the anhydrides and esters N in the amides) directly bonded to the carbonyl carbon has lone pair of electrons in p-orbitals. This p-orbital may also overlap with p-orbitals on carbon and oxygen (of C = O) to provide additional bonding as shown in the II and III of fig. 13.2. This additional bonding depends on the electronegativity of the chain of G directly bonded to the carbonyl group. *Greater the electronegativity lower the additional overlapping.* The group mentioned above contains a halogen or oxygen or nitrogen which is directly bonded to the carbonyl group. Since the electronegativity of those atoms follow the pattern halogen> oxygen>nitrogen. This additional bonding is important as only with amides.

In valance bond terms the additional bonding implies that carboxylic acid derivatives may be represented by resonance structure as shown below:

$$-H_3C-C(=O)-G \longleftrightarrow H_3C-C(-O^-)=G$$

Reaction Mechanism

Acid derivatives undergo nucleophilic stabilization reactions because the carbonyl carbon is electron deficient on account of greater electronegativity of the oxygen atom bonded to it. These reactions take place through the following steps.

(1) Attack by nucleophilic (Nü) on R—C(=O)—G

R—C(=O)—G + :$\bar{Nu}$ ⇌ H_3C—C(O$^-$)(Nu)—G

Tetrahedral Intermediate

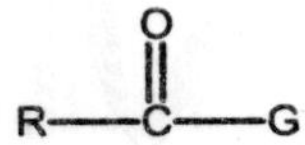

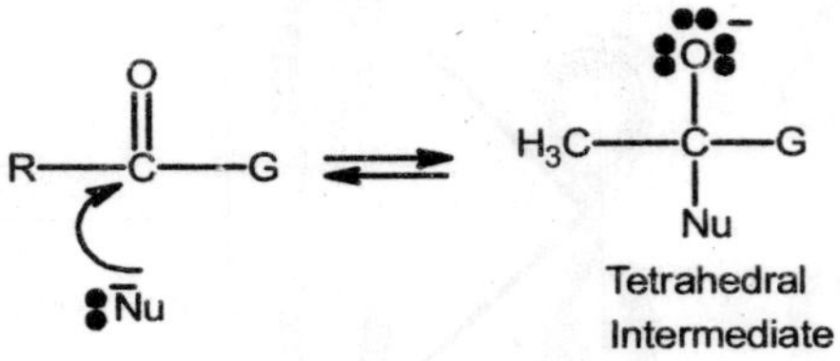

(2) Loss of $\bar{G}$ with formation of R—C(=O)—Nu

O—C—Nu ⟶ H_3C—C(=O)—CH_3 + :$\bar{G}$

Substituted product

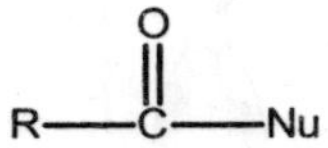

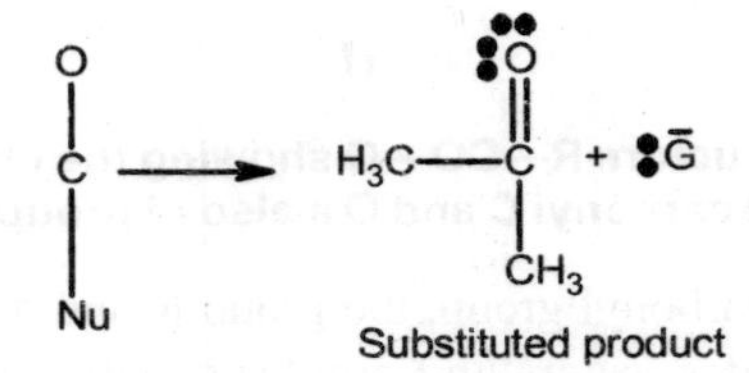

G = X,-O – COR –OR, and $-NH_2$; Nu many by R – Ö – H : N_2 or H_2O in the reaction of acyl halides or anhydrides to form esters amides and carboxylic acid.

Relative Reactivities

Experience have shown that the acyl halides are much reactive than the other derivatives in respect of their nucleophilic substitution reactions. Let is illustrate it by taking example of basic hydrolysis of acid derivatives which takes place as

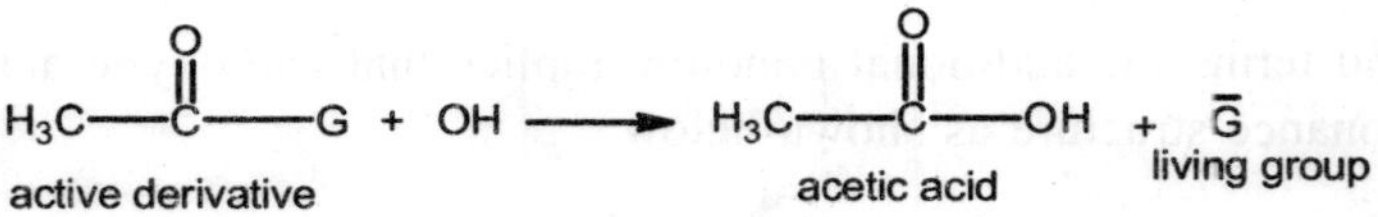

H_3C—C(=O)—G + OH ⟶ H_3C—C(=O)—OH + $\bar{G}$

active derivative — acetic acid — living group

The difference in reactivity of the various acid derivatives presumably due to the basicity of the leaving group as is clear from the following table.

The ease with which a nucleophilic substitution in a derivative takes place depends on two factors

(1) The Electronegativity of G: Greater the electronegativity of group G more polar would be the C – G bond and the carbon group becoming more electrons deficient thereby. This facilities the nucleophilic attack. We know that the electronegativities of the group G follow the sequence —X> —O—CO—R>—OR>—NH_2. Therefore, acid halides should be more reactive than anhydrides; anhydrous in turn should be more reactive then esters which gain should be more reactive than amides.

Table. 13.1. Relative reactivities of acid derivatives $H_3C-C(=O)-G$ and basicity G

Active derivative	G	Relative reactivity towards basic hydrolysis	K_p of G^{-a}	
$H_3C-C(=O)-Cl$ acetyl chloride	$\overline{Cl}$	Reacts explosively at room temperature	10^{-21}	Decreasing reactivity towards hydrolysis
$H_3C-C(=O)-O-C(=O)-CH_3$ acetic anhydride	$H_3C-C(=O)-O^-$	Reacts rapidly at room temperature	10^{-9}	
$H_3C-C(=O)-O-CH_2-CH_3$ ethyl acetate	$H_3C-C(=O)-O$	Reacts slowly at room temperature	10^{+4}	
$H_3C-C(=O)-NH_2$ acetamide	NH_2	Hydrolysis requires prolong	10^{+20}	

(2) Resonance Stabilization: Lesser the electronegativity of the group G more would be the resonance stabilization as represented below:

$$H_3C-C(=O)-G \longleftrightarrow H_3C-C(-O^-)=G^+$$

Greater the double bond character between the carbonyl-carbon owing to resonance, greater would be the electron density at the carbonyl-carbon. Thus a high degree of stabilization results in slower nucleophilic attack. Hence the reactivity of the acid derivatives with nucleophilic reagents should also follow the order, Acid halide> anhydride>ester>amides. Both the 'electronegativity and 'stabilization' factor reinforce each other and support.

$$\underset{\text{acyl halide}}{R-C(=O)-X} > \underset{\text{anhydride}}{R-C(=O)-O-C(=O)-R} > \underset{\text{ester}}{H_3C-C(=O)-CH_3} > \underset{\text{amide}}{R-C(=O)-NH_2}$$

ACYL HALIDE

Acyl halides (sometimes called acid halides) are the derivatives of carboxylic acids in which the OH of COOH group has been replaced by a halogen atom(x)

$$R-\overset{O}{\overset{||}{C}}-OH \xrightarrow[-X]{-H} R-\overset{O}{\overset{||}{C}}-X$$ where x = Cl, Br, I

Further they are named as acyl chloride, bromide and iodide according a X = Cl, Br, I respectively because of the exceptional reactivity of the acyl chloride, it is rarely necessary to prepare and use the more reactive acyl derivative have not received much attention and our discussion of acyl halides will be limited almost exclusively to the acyl chlorides.

The typical formula of acyl chloride is

$R-\overset{O}{\overset{||}{C}}-Cl$, or R – CO – Cl or ROCl and —CO – Cl is the functional group

Nomenclature

Acid chloride is named after name of the carboxylic acid. Both in the common and IUPAC systems, the oxidizing –ic is replaced by –yl chloride. The common and IUPAC names of a acyl chlorides are given below.

Formula	Common Name	IUPAC Name
$H-\overset{O}{\overset{||}{C}}-Cl$	formyl chloride	Methanoyl chloride
$H_3C-\overset{O}{\overset{||}{C}}-Cl$	acetyl chloride	Ethanoyl chloride
$H_3C-CH_2-\overset{O}{\overset{||}{C}}-Cl$	propanoyl chloride	propanoyl chloride
$H_3C-CH_2CH_2-\overset{O}{\overset{||}{C}}-Cl$	n-butyryl chloride	butanoyl chloride
$(H_3C)_2CH-\overset{O}{\overset{||}{C}}-Cl$	iso butyryl chloride	2-methylpropanoyl chloride

The common names of acyl chlorides more than five carbon atoms in a chain obtained by replacing the ending –ic acid by –oryl chloride. Thus lauric acid gives rise to lauryl chloride and not lauroyl chloride. In avoid confusion it is advisable to use IUPAC names for higher acyl chloride.

Methods of Preparation

(i) By the action of Phosphorous Pentachloride and Thionyl chloride on Carboxylic acid: Acid chloride are prepared by the action of phosphorous trichloride (PCl_5) phosphorous pentachloride (PCl_5) or thionyl chloride ($SOCl_2$) on carboxylic acid

$$3R-CO-OH + PCl_5 \rightarrow 3R-CO-Cl + H_3PO_3$$
acyl chloride — meta phosphoric acid

$$R-CO\,OH + PCl_5 \rightarrow R-CO-Cl + POCl_3 + HCl$$
Phosphoryl chloride

$$R-OC-OH + SOCl_2 \rightarrow R-O-Cl + SO_2\uparrow + HCl$$
Thionyl chloride

The choice of reagent for the conversion of a particular carboxylic acid is governed by the boiling points of the products, PCl_5 phosphorous pentachloride can be used during the fractional distillation gives first $POCl_3$(b.p 107°C) and then acyl chloride. With propionyl chloride, which has an intermediate boiling point (60°C) thionyl chloride is suitable as the gases SO_2 and HCL pass off first leaving the pure acyl chloride behind.

Examples:

$$3CH_3-CO-OH + PCl_5 \rightarrow 3CH_3-CO-Cl + H_3PO_3$$
acyl chloride

$$CH_3CH_2CH_2-CO-OH + SOCl_2 \rightarrow CH_3CH_2CH_2-CO\,Cl + SO_2\uparrow + HCl\uparrow$$

(2) By the distillation of salts of carboxylic acid with phosphorous trichloride, phosphorous oxychloride or sulphur monochloride: Acyl chloride may be prepared by distilling the salt of a corresponding carboxylic acid with either phosphorous trichloride (PCl_3); phosphoryl chloride ($POCl_3$) or sulphuryl chloride (SO_2Cl_2)

Thus

$$3CH_3-CO-ONa + PCl_3 \rightarrow 3CH_3-CO-Cl + Na_3PO_3$$
sodium acetate — acetyl achloride — sodium metaphosphotate

$$2CH_2CH_3-CO-ONa + POCl_3 \rightarrow 2CH_3CH_2-CO-Cl + 2NaCl + NaPO_3$$

$$(CH_3-CO-O)_2Ca^{2+} + SO_2Cl_2 \rightarrow 2CH_3-CO-Cl + CaSO_4$$
calcium acetate — acetyl chloride

This method is particularly useful for industrial preparation of acyl chlorides since salts of acid are cheaper than the acids.

Physical Properties (General)

(I) The lower acyl chlorides are mobile, colourless liquids while higher members are colourless solids.

> The boiling point of acetyl chloride is low than propionic acid

(II) Acyl chlorides have very pungent odour and are strong (lachrymators) tear producers.

(III) Since they contain no hydrogen attached to oxygen, hydrogen bonding between their molecular cr with water is not possible. Eventually the boiling point of acyl chlorides are much lower than those of carboxylic acids having equal number of electrons. Thus propionic acid with 40 electrons boils at 141°C while acetyl chloride with 42 electrons boils at 52°C

(IV) The boiling points of lower members record a rise in molecules weight and the higher acyl halides being solids have higher melting points. Thus,

Acyl chloride	Propyl Chloride	n-butyl chloride	Steryl chloride
b.p 52°C	b.p 80°C	b.p 102°C	m.p 32°C

Chemical Properties

Acyl halides are most reactive of all acid derivatives of carboxylic acids. The high electronegative character of chloride atom increases the partial positive change on the carboxyl carbon by withdrawing electrons from it. This makes the carbonyl carbon is thus enhanced to such an extent that even weak nucleophiles such as water and alcohols can also successively attach the carbonyl group. As a result acyl chlorides readily give a variety of nucleophilic substitution reaction in which the Cl atom is replaced by other atoms or groups.

$$R-C(=O)-Cl + :Y-H \longrightarrow \left[R-C(-O^-)(-Cl)(-\overset{+}{Y}-H)\right] \xrightarrow{-Cl^-} \left[R-C(=O)-\overset{+}{Y}-H\right] \xrightarrow{-H^+} R-C(=O)-Y$$

Hence y – H is any nucleophilic reagent such as water, alcohol, etc. *the process is where by an acyl group substitutes as a replaceable H atom in an organic molecule (y – H) is called* ***Acylation***. *Thus introduction of an acetyl group in an organic molecule is called* ***acetylation***

$$CH_3-CO-Cl+H-y \rightarrow CH_3-CO-y+H-X$$

Some important nucleophilic substitution reactions of acyl chlorides are given below:-

(I) Hydrolysis: Acyl chloride undergoes hydrolysis to give back the parent carboxylic acid when treated with water.

$$\underset{\text{acyl chloride}}{R-C(=O)-Cl} + H_2O \longrightarrow \underset{\text{carboxylic acid}}{R-C(=O)-OH} + HCl$$

$$\underset{\text{acetyl chloride}}{H_3C-C(=O)-Cl} + H_2O \longrightarrow \underset{\text{acetic acid}}{H_3C-C(=O)-OH} + HCl$$

Thus when the stopper is removed from a bottle of acetyl chloride, white fumes is seen owing to the formation of HCl with most air. This reaction is of no practical value since acyl chloride themselves are prepared from carboxylic acid.

(II) Esterification: Acyl chlorides react readily with alcohols and phenols to form esters.

$$R-\overset{O}{\overset{\|}{C}}-Cl + R_1-OH \longrightarrow R-\overset{O}{\overset{\|}{C}}-O-R_1 + HCl$$

acyl chloride; alcohol or phenol; esters

$$H_3C-\overset{O}{\overset{\|}{C}}-Cl + H_3C-CH_2\text{-}OH \longrightarrow H_3C-\overset{O}{\overset{\|}{C}}-O-CH_2-CH_3 + HCl$$

acetyl chloride; ethanol; ethyl acetate

$$H_3C-\overset{O}{\overset{\|}{C}}-Cl + C_6H_5OH \longrightarrow C_6H_5COCH_3 + HCl$$

acetyl chloride; phenol; 1-phenylethanone

Unlike the acid – catalyst esterification of carboxylic acid this reaction is irreversible and the esters is obtained in excellent yield.

(III) Amonolysis of Amide Formation: Acyl chlorides react vigorously with ammonia to form acid amides

$$R-\overset{O}{\overset{\|}{C}}-Cl + NH_3 \longrightarrow R-\overset{O}{\overset{\|}{C}}-NH_2 + HCl$$

$$H_3C-\overset{O}{\overset{\|}{C}}-Cl + NH_3 \longrightarrow H_3C-\overset{O}{\overset{\|}{C}}-NH_2 + HCl$$

acetyl chloride; acetamide

HCl produced in the reaction reacts with a second molecule of ammonia give ammonium chloride.

$$NH_3 \quad + \quad HCl \quad \rightarrow \quad NH_4^+Cl^-$$

Primary and secondary amines react similarly to produce mono and n substitution amide

$$R-\overset{O}{\overset{\|}{C}}-Cl + R-NH_2 \longrightarrow R-CH_2-R + HCl$$

$$R-NH_2 + HCl \longrightarrow R-\overset{+}{N}H_3\,Cl^-$$

$$H_3C-\overset{O}{\overset{\|}{C}}-Cl + H_3C-NH_2 \longrightarrow H_3C-\overset{O}{\overset{\|}{C}}-NH-CH_3 + HCl$$

acetyl chloride; methanamine; *N*-methylacetamide

(IV) Thioester formation: Thioalcohols react with acyl chloride to form thioesters

$$R-\overset{O}{\overset{\|}{C}}-Cl + R'-S-H \longrightarrow H_3C-\overset{O}{\overset{\|}{C}}-S' + HCl$$

$$\underset{\text{acetyl chloride}}{H_3C-\overset{O}{\overset{\|}{C}}-Cl} + \underset{\text{methanethiol}}{H_3C-S-H} \longrightarrow \underset{S\text{-methyl ethanethioate}}{H_3C-\overset{O}{\overset{\|}{C}}-S-CH_3} + HCl$$

(V) Anhydride formation: Acyl halides react with anhydrous sodium salt of carboxylic acid to form acid anhydride

$$\underset{\text{acyl chloride}}{R-\overset{O}{\overset{\|}{C}}-Cl} + \underset{\text{sodium carboxylate}}{NaO-\overset{O}{\overset{\|}{C}}-R} \longrightarrow \underset{\text{acid anhydride}}{R-\overset{O}{\overset{\|}{C}}-O-\overset{O}{\overset{\|}{C}}-R} + NaCl$$

(VI) Peroxide formation: Acyl chloride react with sodium peroxide, $NaO^+ - O^- Na^+$, to form organic peroxide.

$$2\,\underset{\text{acetyl chloride}}{H_3C-\overset{O}{\overset{\|}{C}}-Cl} + \underset{\text{Sodium Peroxide}}{Na-O-O-Na} \longrightarrow \underset{\text{acetic anhydride}}{H_3C-\overset{O}{\overset{\|}{C}}-O-\overset{O}{\overset{\|}{C}}-CH_3} + 2NaCl$$

(VII) Reaction with Organometallic Compounds: Acyl chloride reacts with Grignard reagent in either solution to form ketones

$$R-\overset{O}{\overset{\|}{C}}-Cl + H_3\overset{\sigma^-}{C}-\overset{\sigma^+}{Mg}-Br \xrightarrow{\text{dry ether}} \left[R-\underset{CH_3}{\overset{O^-\,Mg^+Br}{C}}-Cl\right] \longrightarrow R-\overset{O}{\overset{\|}{C}}-CH_3 + MgBrCl$$

However, better yield of ketone is obtained by taking organo cadmium compound in place of Grignard reagent. Thus

$$H_3C-CH_2-Cl + \underset{\text{ethyl cadmium chloride}}{H_3C-CH_2-Cd-Cl} \longrightarrow H_3C-\overset{O}{\overset{\|}{C}}-CH_3 + CdCl_2$$

(VIII) Reaction with Diazomethane: Acyl chlorides readily react with diazomethane to give diazoketones for e.g.

$$\underset{\text{acetyl chloride}}{H_3C-\overset{O}{\overset{\|}{C}}-Cl} + \underset{\text{methanediazonium}}{H_3C-\overset{+}{N}\equiv N} \longrightarrow \underset{\text{diazoketone}}{H_3C-\overset{O}{\overset{\|}{C}}-CH-\overset{+}{N}\equiv N} + CH_3Cl + N_2\uparrow$$

If the diazoketones thus obtained is warmed with water in presence of Ag_2O as catalyst, the next higher homologue of the parent acid of acyl chloride is produced, this method of converting acyl chloride, is referred to as ***Arndt – Eistert Synthesis***

Mechanism

$$H_3C-\overset{O}{\overset{\|}{C}}-CH-\overset{+}{N}\equiv N \longrightarrow H_3C-\overset{O}{\overset{\|}{C}}-OH \xrightarrow[\text{Arrangement}]{\text{Wolff}} O{=}C{=}CH-CH_3 \longrightarrow HO-\overset{O}{\overset{\|}{C}}-CH_2CH_3$$

diazoketone → prop-1-en-1-one → propionic acid

(IX) Reduction:

(a) To aldehydes: Acyl chlorides when reduced with H_2 in presence of palladium (supported over $BaSO_4$) as catalyst from aldehydes. Thus reaction is commonly known as ***Rosenmund's Reaction***

$$H_3C-\overset{O}{\overset{\|}{C}}-Cl + H_2 \xrightarrow[100\text{-}150°C]{Pb\text{-}BaSO_4} H_3C-\overset{O}{\overset{\|}{C}}-H + HCl$$

acetyl chloride → acetaldehyde

The function of barium sulphate is to act as "poison" and reduces the activity of the catalyst to extent the reaction steps at the aldehyde stage.

(b) To alcohols: reduction of acyl chlorides with Lithium aluminium tetrahydrate leads to the formation of alcohols

$$R-\overset{O}{\overset{\|}{C}}-Cl + H_2 \xrightarrow{LiAlH_4} R-\underset{H}{\overset{H}{C}}-OH + HCl$$

acetyl chloride → 1°alcohol

$$R-\overset{O}{\overset{\|}{C}}-Cl + H^- \longrightarrow \left[R-\underset{H}{\overset{O^-}{C}}-Cl\right] \xrightarrow{-Cl^-} H_3C-\overset{O}{\overset{\|}{C}}-H \xrightarrow{+H^-} R-\underset{H}{\overset{H}{C}}-O \xrightarrow[(HCl)]{+H^-} R-\underset{H}{\overset{H}{C}}-OH$$

1° alcohol

The complete reaction involved in the reduction of acyl chloride with lithium aluminium tetrahydride may be represented as.

2RCOCl acyl cchloride	+	$LiAlH_4$ lithium aluminium tetrahydride	→	$(RCH_2O)_2LiAlCl_2$ acyl lithium aluminium dichlroide
$(RCH_2O)_2LiAlAlCl_2$ acyl lithium aluminium dichloride	+	2HCl	→	$2RCH_2OH$ + LiCl + $AlCl_3$ alcohol

(X) Friedel Craft's Reaction: Acyl chlorides after a convenient method for introducing acyl groups into the nucleus of the benzoid compounds in presence of anhydrous aluminium trichloride as catalyst. This reaction called Friedel Craft's Acylation will be discussed in detail in the chapter of Aromatic hydrocarbons.

$$R-\overset{O}{\overset{\|}{C}}-Cl + C_6H_6 \xrightarrow{\text{anhydrous } AlCl_3} C_6H_5-\overset{O}{\overset{\|}{C}}-R + HCl$$

acyl phenol ketone

$$H_3C-\overset{O}{\overset{\|}{C}}-Cl + C_6H_6 \xrightarrow{\text{anhydrous } AlCl_3} C_6H_5-\overset{O}{\overset{\|}{C}}-CH_3 + HCl$$

acetophenone

(XI) Halogenation: (Hell-Volhard Zelensky Reaction): Acyl chloride undergoes chlorination or bromination much readily than aliphatic hydrocarbons and hydrocarbooxylic acid. Further, towards halogenation is presumably due to greater acidity of α-H atoms because of the inductive effect of both the carbonyl group and chlorine atom

$$R-CH_2-\overset{O}{\overset{\|}{C}}-Cl + Cl_2 \rightleftharpoons R-CH(Cl)-\overset{O}{\overset{\|}{C}}-Cl + HCl$$

acyl chloride alpha-chloro-acyl chloride

Mechanism

$$R-CH_2-\overset{O}{\overset{\|}{C}}-Cl \underset{}{\overset{-H^+}{\rightleftharpoons}} H_3C-CH=C(OH)-Cl \longrightarrow H_3C-CH=C(OH)-Cl \rightleftharpoons H_3C-CH(Cl)-\overset{O}{\overset{\|}{C}}-Cl + Cl^-$$

acyl chloride

In actual practice free carboxylic acids, they chlorinated in the presence of a small amount of phosphoric trihalides. The reactions are :

$$\underset{\text{carboxylic acid}}{3R-CH_2-COOH} + PCl_3 \rightarrow \underset{\text{acyl chlroide}}{3R-CH_2-COCl} + H_3PO_3$$

$$\underset{\text{Acyl chloride}}{R-CH_2-COCl} + Cl_2 \rightarrow \underset{\alpha\text{-Chloro acyl chloride}}{R-CHCl-COCl + HCl}$$

$$\underset{\alpha\text{-Chloro acyl chloride}}{R-CHCl-COCl} + \underset{\text{carboxylic acid}}{R-CH_2-COOH} \rightleftharpoons R-CHCl-COOH + \underset{\text{reduced acyl chloride}}{R-CH_2-COCl}$$

Individual Member

FORMYL CHLORIDE

Formula: COHCl

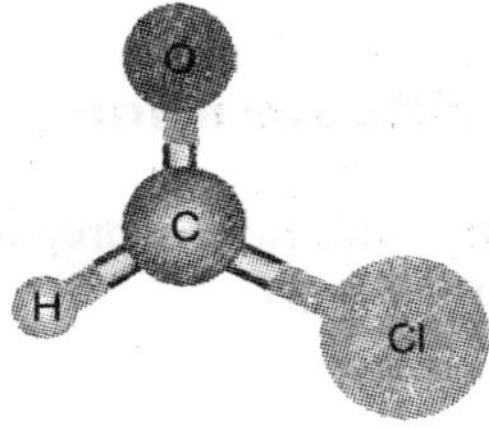

Introduction

Formyl chloride is the first member of the class, it is unstable and is not known to exist in ordinary temperature. There is an evidence of its existence at low temperature (-80°C). A mixture of carbon monoxide and hydrogen chloride, however, behaves as formyl chloride in its reactions.

$$:C{=}\ddot{O} + HCl \longrightarrow H{-}\overset{\overset{O}{\|}}{C}{-}Cl$$

formyl chloride
(unstable)

This mixture id frequently used for introducing a formyl group (formylation) into the benzene ring in presence of anhydrous $AlCl_3$ **(Galterman Kock aldehyde synthesis)**

$$CO + HCl + C_6H_6 \xrightarrow{\text{anhyd } AlCl_3} C_6H_5CHO$$

benzene → benzaldehyde

ACETYL CHLORIDE

Formula: C_2H_3OCl

Introduction

Acetyl chloride is the most important acyl chloride since the abbreviation for the acetyl group (CH_3 – CO —) is Ac acetyl chloride may be written as Ac – C or AcCl for convenience

Laboratory Preparation

Theory: Acetyl chloride is prepared in the lab by the action of phosphorous trichloride on glacial acetic acid. (Fig. 13.3).

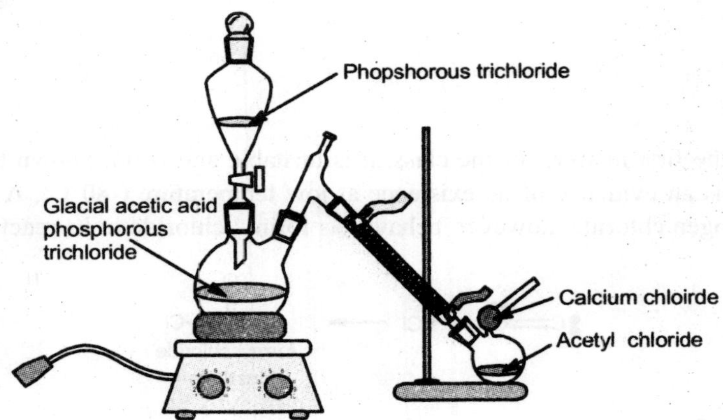

Fig. 13.3. Showing the Lab preparation of Acetyl Chloride

Apparatus	Reagents	Temperature	Pressure
Distillation flask, condenser, mantle stands etc	Glacial acetic acid, phosphorous trichloride, calcium chloride	50°C to 55°C	normal

Reaction Equation:

$3CH_3 - CO - OH$	+	PCl_3	→	$3CH_3 - CO - Cl$	+	H_3PO_3
Acetic acid		phosphorous trichloride		acetyl chloride		metaphosphoric acid

Procedure

Phosphorus trichloride is gradually dropped over glacial acetic acid contained in the distillation flask, which is connected with a receiver. The receiver is protected form atmospheric moisture by fixing a guard tube with calcium chloride to it (see fig. 13.3) hydrogen chloride is expelled from the flask at 50°C and the acetyl chloride is obtained by distillation 55°C. it is purified by re distillation.

Precaution: *The reaction must be carried out in the possible moisture free environment because a viscous white fume will develop while the acetyl chloride is produced.*

Other Methods of Preparation

1. **By the action of Thionyl Chloride on acetic acid:** This reaction can be used for the production of acetyl chloride in large scale:-

$$CH_3COOH + SOCl_2 \rightarrow CH_3COCl + SO_2 + HCl$$

This reaction can contain large amounts of sulphur dioxide dissolved in it.

2. **By the reaction of Carbonyl and Grignard's Reagent:**

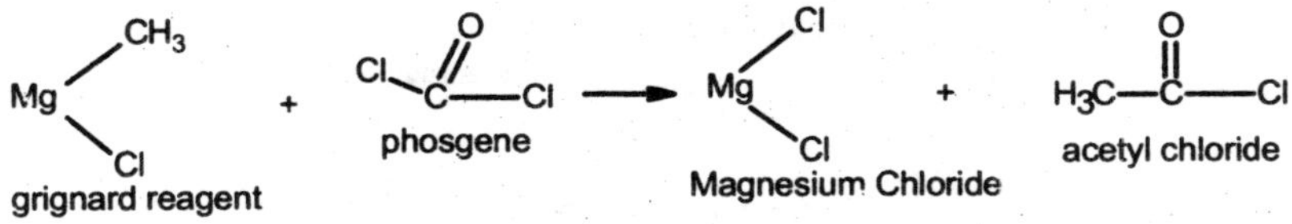

$$CH_3\text{-}Mg\text{-}Cl + Cl\text{-}CO\text{-}Cl \longrightarrow Cl\text{-}Mg\text{-}Cl + H_3C\text{-}CO\text{-}Cl$$

grignard reagent + phosgene → Magnesium Chloride + acetyl chloride

Physical Properties

Molecular Formula	C_2H_3ClO
Formula Weight	78.49762
Composition	C(30.60%) H(3.85%) Cl(45.16%) O(20.38%)
Molar Refractivity	16.19 ± 0.3 cm^3
Molar Volume	70.0 ± 3.0 cm^3
Parachor	155.2 ± 4.0 cm^3
Index of Refraction	1.379 ± 0.02
Surface Tension	24.1 ± 3.0 dyne/cm
Density	1.120 ± 0.06 g/cm^3
Dielectric Constant	Not available
Polarizability	6.41 ± 0.5 $10^{-24} cm^3$
Monoisotopic Mass	77.987242 Da
Nominal Mass	78 Da
Average Mass	78.498044 Da

Chemical Properties

It is quite reactive and readily attacks compounds containing —OH groups. It is important reactions are shown below:

acetyl chloride ($H_3C-C(=O)-Cl$) reacts with:

- + H_2O (Water) → CH_3COOH (Acetic Acid) + HCl
- + C_2H_5OH (Ethyl Alcohol) → $H_3C-C(=O)-NH_2$ (acetamide) + HCl
- + H_3C-SH (methanethiol) → $H_3C-C(=O)-SH$ (ethanethioic S-acid, thioacetic acid) + HCl
- + CH_3MgI (grignard's reagent) → $H_3C-C(=O)-CH_3$ (acetone)
- + $LiAlH_4$ (lithium aluminium tetrahydride) → C_2H_5OH (Ethyl Alcohol) + HCl
- + benzene —($AlCl_3$)→ acetophenone + HCl
- + aniline (NH_2) → $H_3C-C(=O)-N(H)-C_6H_5$ (N-phenylacetamide) + HCl
- + $Na(CH_3COO)_2$ (Sodium Acetate) → $H_3C-C(=O)-O-C(=O)-CH_3$ (acetic anhydride)
- + $H-N(H)-C(=O)-N(H)-H$ (urea) → $H_3C-C(=O)-N(H)-C(=O)-N(H)-H$ (N-(aminocarbonyl)acetamide)

Uses of Acetyl Chloride

It is widely used as an acetylating agent in the synthesis of fine chemicals, agrochemicals and pharmaceuticals. The hydrogen atoms replace oxygen atoms in alcohols or nitrogen atoms in amines, which protects amine groups during amino acid synthesis. It is also used as an intermediate for dyes. It is also used to determine water in organic liquids and in testing cholesterol.

ACID ANHYDRIDES

Formerly speaking acid anhydrides may be derived by the elimination of a molecule of water from two —OH groups of the two molecules of a carboxylic acid.

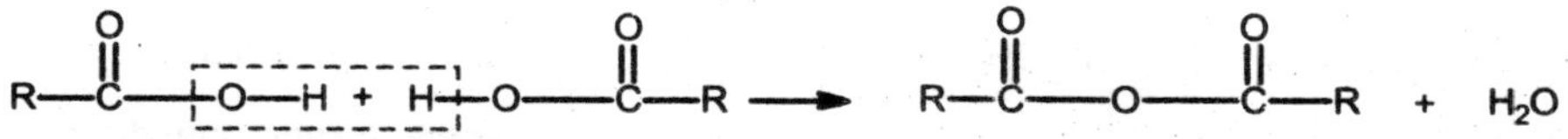

$$R-\overset{O}{\overset{\|}{C}}-O-H + H-O-\overset{O}{\overset{\|}{C}}-R \longrightarrow R-\overset{O}{\overset{\|}{C}}-O-\overset{O}{\overset{\|}{C}}-R + H_2O$$

two molecules of carboxylic acid — acid anhydride

Evidently, acid anhydrides are made of two acyl groups (R – CO —) connected through and oxygen bridge. The functional group —CO – O – CO —, present in anhydrides is called the **anhydride function**.

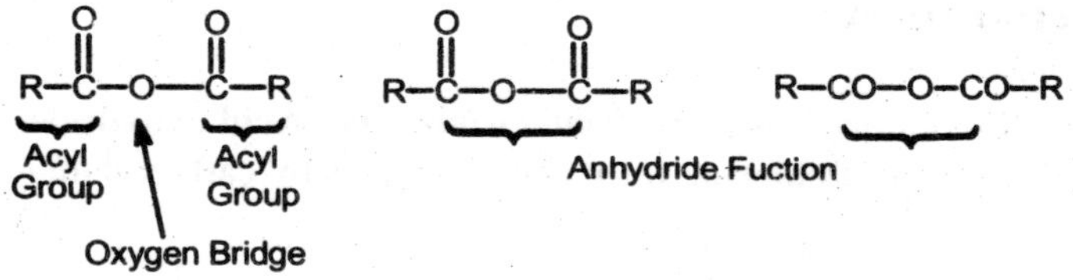

$$R-\overset{O}{\overset{\|}{C}}-O-\overset{O}{\overset{\|}{C}}-R \qquad R-\overset{O}{\overset{\|}{C}}-O-\overset{O}{\overset{\|}{C}}-R \qquad R-CO-O-CO-R$$

Acyl Group, Oxygen Bridge, Acyl Group — Anhydride Fuction

In addition to above general formulas, the acid anhydrides may also be represented as $(RCO)_2O$, to economize space and time.

The acid anhydrides may be **Simple** or **Mixed Anhydrides** according as the two acyl groups attached to the oxygen bridge are same or different.

$$R-\overset{O}{\overset{\|}{C}}-O-\overset{O}{\overset{\|}{C}}-R \qquad R-\overset{O}{\overset{\|}{C}}-O-\overset{O}{\overset{\|}{C}}-R_1$$

Simple anhydride — Mixed anhydride

Almost all the monocarboxylic acids from their corresponding anhydrides, only formed acid, H – CO – OH , is an exception. However, mixed anhydrides of formic acid and other carboxylic acid are known. For example,

$$H-\overset{O}{\overset{\|}{C}}-O-H + H-O-\overset{O}{\overset{\|}{C}}-CH_3 \longrightarrow H-\overset{O}{\overset{\|}{C}}-O-\overset{O}{\overset{\|}{C}}-CH_3 + H_2O$$

Formic acid — Acetic acid — Acetic formic anhydride

Nomenclature

Both in the common and *IUPAC Systems,* acid anhydrides are named by dropping the word 'acid' from the name of the parent acid and adding the word 'anhydride', For naming the mixed anhydrides the first words of the na!lles of the two parent acids are written in alphabetical order and then the word anhydride is added once only. The common names of a few typical anhydrides are listed below.

Formula	Common Name
$CH_3—CO—O—CO—CH_3$	acetic anhydrite
$CH_3CH_2—CO—O—CO—CH_2CH_3$	propionic anhydrite
$CH_3CH_3CH_2—CO—O—CO—CH_2CH_2CH_3$	butryic anhydide
$H—CO—O—CO—CH_3$	acetic formic anhydride
$CH_3—CO—O—CO—CH_2CH_3$	acetic propionic anhydride
$CH_3—CO—O—CO—CH_3CH_2CH_3$	acetic butyric anhydride

Methods of Preparation

(I) *By reaction of an Acid chloride with Sodium carboxylate.* Acid anhydrides are best , prepared by the nucleophilic displacement of chlorine from acid chlorides by carboxylate ion.

$$R—C(=O)—Cl + Na^+\,{}^-O—C(=O)—R \longrightarrow R—C(=O)—O—C(=O)—R + Na^+Cl^-$$

acid chloride; sodium carboxylate; acid anhydride

$$CH_3—C(=O)—Cl + Na^+\,{}^-O—C(=O)—CH_3 \longrightarrow H_3C—C(=O)—O—C(=O)—CH_3 + Na^+Cl^-$$

acetyl chloride; Sodium Acetate; acetic anhydride

Mixed anhydrides can also be prepared by this method when an acid chloride is reacted with sodium salt of another acid. Thus,

$$CH_3—C(=O)—Cl + Na^+\,{}^-O—C(=O)—CH_2—CH_3 \longrightarrow H_3C—C(=O)—O—C(=O)—CH_2—CH_3 + Na^+Cl^-$$

acetyl chloride; Sodium Propionate; acetic propanoic anhydride

This is a very general method for preparing acid anhydrides. The reaction proceeds smoothly and yields are excellent. However, the most serious limitation arises from the fact that unsymmetrical anhydrides are unstable and disproportionate easily.

(2) *By reaction of Carboxylic acids with acetic anhydride.* High-molecular-weight acid anhydrides are prepared by distilling the corresponding carboxylic acid with acetic anhydride.

$$\underset{\substack{\text{acetic anhydride}\\ \text{boiling point 140°C}}}{CH_3-\overset{O}{\overset{\|}{C}}-O-\overset{O}{\overset{\|}{C}}-CH_3} + \underset{\text{Acid}}{2R-\overset{O}{\overset{\|}{C}}-OH} \rightleftharpoons R-\overset{O}{\overset{\|}{C}}-O-\overset{O}{\overset{\|}{C}}-R + \underset{\substack{\text{acetic acid}\\ \text{boiling point 118°C}}}{H_3C-\overset{O}{\overset{\|}{C}}-OH}$$

Acetic acid boils at a lower temperature than the carboxylic acid and acetic anhydride and is the lowest boiling component of the equilibrium mixture. As acetic acid is removed during distiilation, the reaction is forced to the right and proceeds to completion. This method of preparation is applicable to the preparation of simple anhydrides only. It is preferred over the method (1) because, of the low cost and availability of acetic anhydride.

Mechanism This exchange reaction probably occurs through the nucleophilic attack of the carboxylic acid onto the carbonyl carbon of acetic anhydride.

(i)
$$\underset{\text{Acid}}{R-\underset{O-H}{\underset{|}{C}}=O} + \underset{\text{acetic anhydride}}{H_3C-\overset{O}{\overset{\|}{C}}-O-\overset{O}{\overset{\|}{C}}-CH_3} \rightleftharpoons \underset{\text{(intermediate)}}{R-\overset{O}{\overset{\|}{C}}-O-\overset{O}{\overset{\|}{C}}-CH_3} + CH_3COOH$$

(ii)
$$R-\underset{O}{\underset{\|}{C}}-O-\underset{O}{\underset{\|}{C}}-CH_3 + \underset{\substack{\text{acid}\\ \text{(second molecule)}}}{R\text{-}COOH} \rightleftharpoons \underset{\text{higher acid anhydride}}{R-\underset{O}{\underset{\|}{C}}-O-\underset{O}{\underset{\|}{C}}-R} + CH_3COOH$$

Physical Properties (General)

(1) Excepting formic anhydride, which is unstable, the lower members of the series are colourless mobile liquids with a sharp pungent odour. The higher members are colourless crystalline solids.

(2) Their boiling points rise with increase *of* molecular weight as for members of any other class of compounds. Thus acetic anhydride boils at 140°C, propionic anhydride at 168°C and butyric acid at 198°C. relatively boiling points of anhydrides are attributed to the larger size of the anhydride molecule compared to the acid molecule which results in larger hydrogen bonds to an the higher boiling point. The anhydrides, however, do not form hydrogen bonds to an appreciable extent.

(3) They are only slightly soluble in water but dissolve in ethanol and ether. Their insolubility in water is also due to the fact that they do not form hydrogen bonds with water molecules. The large resonance energy (30 Kcal) resulting from the overlap of p orbitals of oxygen and carbon atoms prevents the localization of negative charge on oxygen necessary for hydrogen bonding.

Chemical Properties

Acid anhydrides give reactions analogous to those of acyl chlorides because the latter could by very much regarded as mixed anhydrides of HCl and RCOOH. Thus they react with the same nucleophilic reagents as the acid chlorides but with less strongly electron attracting than the Cl atom in acid chloride R – CO – O — in an anhydride is less strongly electron attracting than the some typical reactions of acid anhydrides are mentioned below. The reactions (1) (2) and (3) are examples of nucleophilic displacement reactions.

1. Hydrolysis: They react with water slowly to produce the corresponding acids.

$$R-\overset{O}{\overset{\|}{C}}-O-\overset{O}{\overset{\|}{C}}-R + H_2O \longrightarrow R-\overset{O}{\overset{\|}{C}}-OH + R-\overset{O}{\overset{\|}{C}}-OH$$

acid anhydride — original acid — (second molecule)

$$H_3C-\overset{O}{\overset{\|}{C}}-O-\overset{O}{\overset{\|}{C}}-CH_3 + H_2O \longrightarrow H_3C-\overset{O}{\overset{\|}{C}}-OH + H_3C-\overset{O}{\overset{\|}{C}}-OH$$

acetic anhydride — acetic acid — (second molecule)

Mechanism

$$R-\underset{O}{\underset{\|}{C}}-O-\underset{O}{\underset{\|}{C}}-R + H-O-H \longrightarrow R-\underset{O}{\underset{\|}{C}}-O + H_3C-\overset{O}{\overset{\|}{C}}-OH_3 \longrightarrow R-\overset{O}{\overset{\|}{C}}-OH + R-\overset{O}{\overset{\|}{C}}-OH$$

two mecules of acid

(2) Alcoholyis: Acid anhydrides react with alcohols to form esters

$$R-\overset{O}{\overset{\|}{C}}-O-\overset{O}{\overset{\|}{C}}-R + R_1-OH \longrightarrow R-\overset{O}{\overset{\|}{C}}-O-R_1 + R-\overset{O}{\overset{\|}{C}}-OH$$

alcohol — ester — acid

$$H_3C-\overset{O}{\overset{\|}{C}}-O-\overset{O}{\overset{\|}{C}}-CH_3 + H_2\overset{CH_3}{C}-OH \longrightarrow H_3C-\overset{O}{\overset{\|}{C}}-O-\overset{CH_3}{CH_2} + H_3C-\overset{O}{\overset{\|}{C}}-OH$$

acetic anhydride — ethanol — ethyl acetate — acetic acid

Similarly, acid anhydrides also react with phenols to give esters

(3) Amonolysis: Acid anhydrides react with concentrated aqueous ammonia to give acid amides

$$R-\overset{O}{\overset{\|}{C}}-O-\overset{O}{\overset{\|}{C}}-R + NH_3 \longrightarrow R-\overset{O}{\overset{\|}{C}}-NH_2 + R-\overset{O}{\overset{\|}{C}}-OH$$

$$H_3C-\overset{O}{\overset{\|}{C}}-O-\overset{O}{\overset{\|}{C}}-CH_3 + NH_3 \longrightarrow H_3C-\overset{O}{\overset{\|}{C}}-NH_2 + H_3C-\overset{O}{\overset{\|}{C}}-OH$$

acetic anhydride — ammonia — acetamide — acetic acid

With primary and secondary amines, acid anhydrides form mono and Di-N – substituted amides respectively. For example.

$$H_3C-\overset{O}{\overset{\|}{C}}-O-\overset{O}{\overset{\|}{C}}-CH_3 + H_2N-CH_3 \longrightarrow H_3C-\overset{O}{\overset{\|}{C}}-NH-CH_3 + H_3C-\overset{O}{\overset{\|}{C}}-OH$$

acetic anhydride, methanamine, N-methylacetamide, acetic acid

$$H_3C-\overset{O}{\overset{\|}{C}}-O-\overset{O}{\overset{\|}{C}}-CH_3 + H_2N-CH_3 \longrightarrow H_3C-\overset{O}{\overset{\|}{C}}-N(CH_3)-CH_3 + H_3C-\overset{O}{\overset{\|}{C}}-OH$$

acetic anhydride, methanamine, N,N-dimethylacetamide, acetic acid

(4) Reaction with dry hydrogen chlo ide: Acid anhydrides react with dry hydrogen chloride to form acyl chlorides

$$R-\overset{O}{\overset{\|}{C}}-O-\overset{O}{\overset{\|}{C}}-R + HCl \longrightarrow R-\overset{O}{\overset{\|}{C}}-Cl + R-\overset{O}{\overset{\|}{C}}-OH$$

acyl chloride

$$H_3C-\overset{O}{\overset{\|}{C}}-O-\overset{O}{\overset{\|}{C}}-CH_3 + HCl \longrightarrow H_3C-\overset{O}{\overset{\|}{C}}-Cl + H_3C-\overset{O}{\overset{\|}{C}}-OH$$

acetic anhydride, acyl chloride, acetic acid

(5) Reaction with Aldehydes: When reacted with aldehydes, acid anhydrides, form alkylidene esters fro example

$$H_3C-\overset{O}{\overset{\|}{C}}-O-\overset{O}{\overset{\|}{C}}-CH_3 + H_3C-\overset{O}{\overset{\|}{C}}H \longrightarrow H_3C-HC\left(O-\overset{O}{\overset{\|}{C}}-CH_3\right)_2$$

acetic anhydride, acetaldehyde, ethane-1,1-diyl diacetate

(6) Reaction with Nitrogen Pentoxide: Acid anhydrides react with nitrogen pentoxide

$$H_3C-\overset{O}{\overset{\|}{C}}-O-\overset{O}{\overset{\|}{C}}-CH_3 + N_2O_5 \longrightarrow 2H_3C-\overset{O}{\overset{\|}{C}}-O-\overset{+}{N}(=O)-O^-$$

acetic anhydride, acetyl nitrate

(7) Reaction with phosphorous Pentachloride: With PCl_5 and anhydrides yield acyl chlorides.

Thus,

$$\underset{\text{acetic anhydride}}{H_3C-\overset{O}{\overset{\|}{C}}-O-\overset{O}{\overset{\|}{C}}-CH_3} + PCl_5 \longrightarrow \underset{\text{acetyl chloride}}{H_3C-\overset{O}{\overset{\|}{C}}-Cl} + POCl_3$$

(8) Friedel Craft's Acylation: Like acyl chlorides, acid anhydrides are used for the introduction of an acyl group into molecule of a benzenoid compound in presence of anhydrous aluminium chloride.

$$\underset{\text{acetic anhydride}}{H_3C-\overset{O}{\overset{\|}{C}}-O-\overset{O}{\overset{\|}{C}}-CH_3} + \underset{\text{benzene}}{C_6H_6} \xrightarrow[\text{(ii) HCl}]{\text{(i) anhydr AlCl}_3} \underset{\text{1-phenylethanone}}{C_6H_5COCH_3} + \underset{\text{acetic acid}}{H_3C-C(=O)OH}$$

(9) Reduction: Acid anhydrides when reduced with $LiAlH_4$ produce alcohols. For example

$$\underset{\text{acetic anhydride}}{H_3C-\overset{O}{\overset{\|}{C}}-O-\overset{O}{\overset{\|}{C}}-CH_3} + 2\,[H] \xrightarrow[\text{in ether at 25 °C}]{LiAlH_4} \underset{\text{ethanol}}{H_3C-CH_2-OH} + H_2O$$

Individual Members

ACETIC ANHYDRIDE

Formula: $C_6H_4O_3$

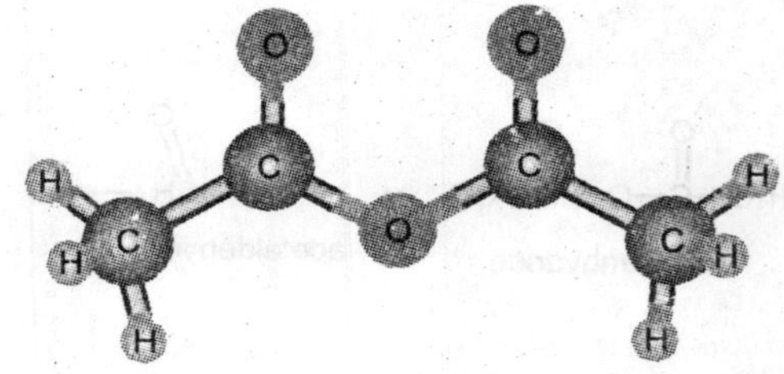

Preparation

Acetic anhydride may be prepared by the following processes

(1) By the pyrolysis of ethylidine diacetate, which is in turn obtained by passing acetylene into glacial acetic acid in presence of mercury salts;

$$2\,H-O-\overset{O}{\overset{\|}{C}}-CH_3 + HC\equiv CH \longrightarrow H_3CHC(O-\overset{O}{\overset{\|}{C}}-CH_3)_2 \xrightarrow{\text{distil}} \underset{\text{acetic anhydride}}{H_3C-\overset{O}{\overset{\|}{C}}-O-\overset{O}{\overset{\|}{C}}-CH_3} + \underset{\text{acetaldehyde}}{H_3C-CH=O}$$

(2) By passing ketone through acetic acid. The ketone required for the process is obtained by the pyrolysis of acetone or acetic acid

$$\underset{\text{acetone}}{H_3C-\overset{\overset{O}{\parallel}}{C}-CH_3} \xrightarrow[700°C]{(C_2H_5)_3\ P=O} \underset{\text{ethylenone ketone}}{H_2C=C=O} + CH_4$$

$$\underset{\text{ethylenone ketone}}{H_2C=C=O} + \underset{\text{acetic acid}}{H_3C-COOH} \longrightarrow \underset{\text{acetic anhydride}}{CH_3-\overset{\overset{O}{\parallel}}{C}-O-\overset{\overset{O}{\parallel}}{C}-CH_3}$$

Laboratory Preparation

Theory: Acetic anhydride is prepared in the lab by the action of anhydrous sodium acetate and acetyl chloride (see fig. 13.4).

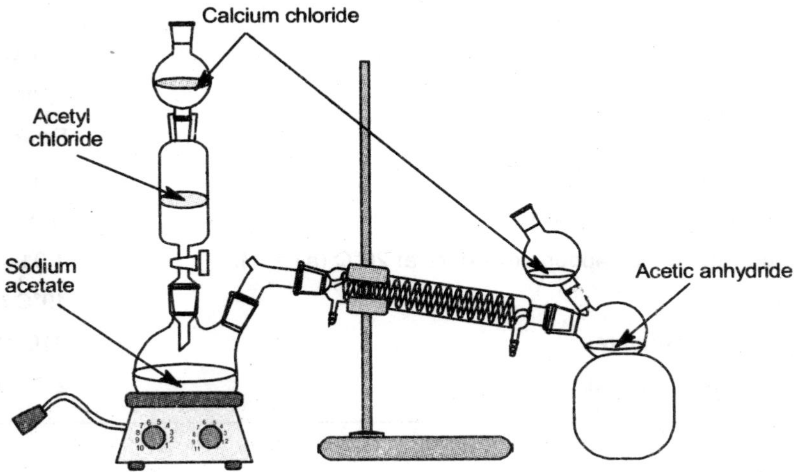

Fig. 13.4. Showing the Lab preparation of Acetic Anhydride

Apparatus	Chemical	Temperature	pressure
Distillation flask, mantle, graham's condenser, reservoir, dropping funnel, clamps, stands etc	Acetyl chloride, sodium acetate, calcium chloride	135-140°C	normal

Reaction Equation:

$$H_3C-C(=O)-Cl + Na-CH_2-C(=O)-OH \longrightarrow H_3C-C(=O)-O-C(=O)-CH_3 + NaCl$$

acetyl chloride + Sodium Acetate → acetic anhydride + NaCl

Procedure:

Acetyl chloride is slowly dropped over anhydrous sodium acetate placed in a distillation flask and a reservoir. After the addition of acetyl chloride, the contents of the distillation flask are thoroughly shaken and distilled 135°C-140°C

Precautions: *The reaction must be carried out in a dry and moisture free condition and the moisture protection for that calcium chloride tube must be used.*

Physical Properties

Boiling point:	139°C
Melting point:	-73°C
Relative density (water = 1):	1.08
Solubility in water:	Reaction
Vapour pressure, kPa at 20°C:	0.5
Relative vapour density (air = 1):	3.5
Relative density of the vapour/air-mixture at 20°C (air = 1):	1.01
Flash point:	49°C c.c.
Auto-ignition temperature:	316°C
Explosive limits, vol% in air:	2.7-10.3

Chemical Properties

Acetic anhydride is a colourless, neutral and pungent liquid; it shows similar reactions like acetyl chloride but much less reactive.

Acetaldehyde Oxidation. The oxidation of liquid acetaldehyde *(q.v.)* with air or oxygen has long been used for the production of acetic acid, the earliest patent on this process having been issued in 1908. The mechanism of this reaction has been studied in many laboratories and can be most simply represented by two equations (1 and 2) involving the intermediate formation of peracetic acid. The latter compound is explosive and its concentration is kept down to a safe limit by the use of catalysts,

$H_3C-CO-O-CO-CH_3$
acetic anhydride

+ H_2O (water) → $H_3C-C(=O)-OH$ (acetic acid)

+ H_3C-CH_2-OH (ethanol) → $H_3C-C(=O)-O-CH_2-CH_3$ (ethyl acetate)

+ NH_3 (ammonia) → $H_3C-C(=O)-NH_2$ (acetamide)

+ HCl (hydrogen chloride) → $H_3C-C(=O)-Cl$ (acetyl chloride) + $H_3C-C(=O)-OH$ (acetic acid)

+ PCl_5 → $H_3C-C(=O)-Cl$ (acetyl chloride) + $POCl_3$

+ N_2O_5 (nitrogen pentoxide) → $H_3C-C(=O)-O-N^+(-O^-)=O$ (acetyl nitrate)

+ benzene → acetophenone + $H_3C-C(=O)-OH$ (acetic acid)

+ $H_3C-C(=O)H$ (acetaldehyde) → $CH_3CH(OCOCH_3)_2$ (Ethylidene diacetate)

+ 2[H] → H_3C-CH_2-OH (ethanol) + H_2O

+ [O] → $H_3C-C(=O)-O-O-C(=O)-CH_3$ (dimethylperoxyanhydride)

especially manganese acetate. Temperatures of 60-80°C. and pressures of 50-100 p.s.i. are generally used.

$$CH_3CHO + O_2 \uparrow CH_3COOOH \quad (1)$$

$$CH_3COOOH + CH_3CHO \uparrow 2CH_3COOH \quad (2)$$

Further study of the reaction showed that acetic anhydride is also formed, according to the equation:

$$CH_3COOOH + CH_3CHO \uparrow (CH_3CO)_2O + H_2O \quad (3)$$

Unless conditions are chosen to favour reaction (3) at the expense of reaction (2) and to separate the anhydride from the water simultaneously formed, all the anhydride formed will be hydrolyzed to acetic acid. In order to increase the yield of anhydride, variations in the catalyst are used, mixtures of cobalt and copper acetates being especially effective. Reduction in temperature and increase in speed of oxidation also aid in giving a higher concentration of anhydride in the final crude oxidation product. A third expedient is to conduct the oxidation in a diluent such as ethyl acetate, which cuts down the rate of hydrolysis of the anhydride. By this means it is possible to obtain conversion to the anhydride of more than two-thirds of the aldehyde oxidized. In order to obtain acetic anhydride from the crude oxidation product, the water must be removed as soon as possible after the oxidation is completed and at as Iowa temperature as possible. Continuous vacuum distillation accomplishes this efficiently, and the diluent used in the oxidation process may assist by lowering the boiling point of the crude or by removing the water as an azeotrope.

This "dual oxidation" process for the simultaneous production of acetic acid and anhydride was discovered and first put into operation by Shawinigan Chemicals Ltd., Canada, but several other plants operating on this principle have been constructed in the United States. In Germany during World War II, a high proportion of the anhydride was made by this process (Fig. 13.5).

Acetic Acid and Acetic Anhydride

Acetic acid is the chemical compound responsible for the characteristic odor and sour taste of vinegar. Typically, vinegar is about 4 to 8% acetic acid. As the defining ingredient of vinegar, acetic acid has been produced and used by humans since before the dawn of recorded history. In fact, its name comes from the Latin for vinegar, *acetum*. Vinegar is formed from dilute solutions of alcohol, such as wine, by the action of certain bacteria in the presence of oxygen. These bacteria require oxygen, and the overall chemical change is the reaction of ethanol with oxygen to form acetic acid and water.

$$CH_3CH_2OH + O_2 \longrightarrow CH_3\overset{\overset{\displaystyle O}{\|}}{C}OH + H_2O$$

The name, vinegar, comes from the French, *vin aigre*, which means "sour wine." Nevertheless, vinegar may also be obtained from other fermented beverages such as malt or cider.

Because vinegar is acidic, it has a variety of properties useful around the house. Mineral deposits left when hard water evaporates, such as those formed on plumbing fixtures and in tea kettles, dissolve in acids, so vinegar can be used to remove them. Because it is acidic, vinegar also inhibits the growth of bacteria, so vinegar is used as a preservative in foods, such as pickled vegetables, and as a mild disinfectant in cleaning. Of course, its sour taste, which is also a result of its acidity, makes it popular as a flavouring in cooking and in salad dressings.

Pure acetic acid was first isolated about 1700 by the distillation of vinegar. When pure, acetic acid is a clear, colourless liquid with a sharp, irritating odor of vinegar. In poorly heated laboratories, the acid was oftentimes found frozen inside its container because its freezing point is only slightly below room temperature at 16.7°C. The term glacial (ice-like) came to be applied to the pure acid in either its solid or liquid state. Glacial acetic acid boils at 118°C, and has a density of 1.049 g/mL at

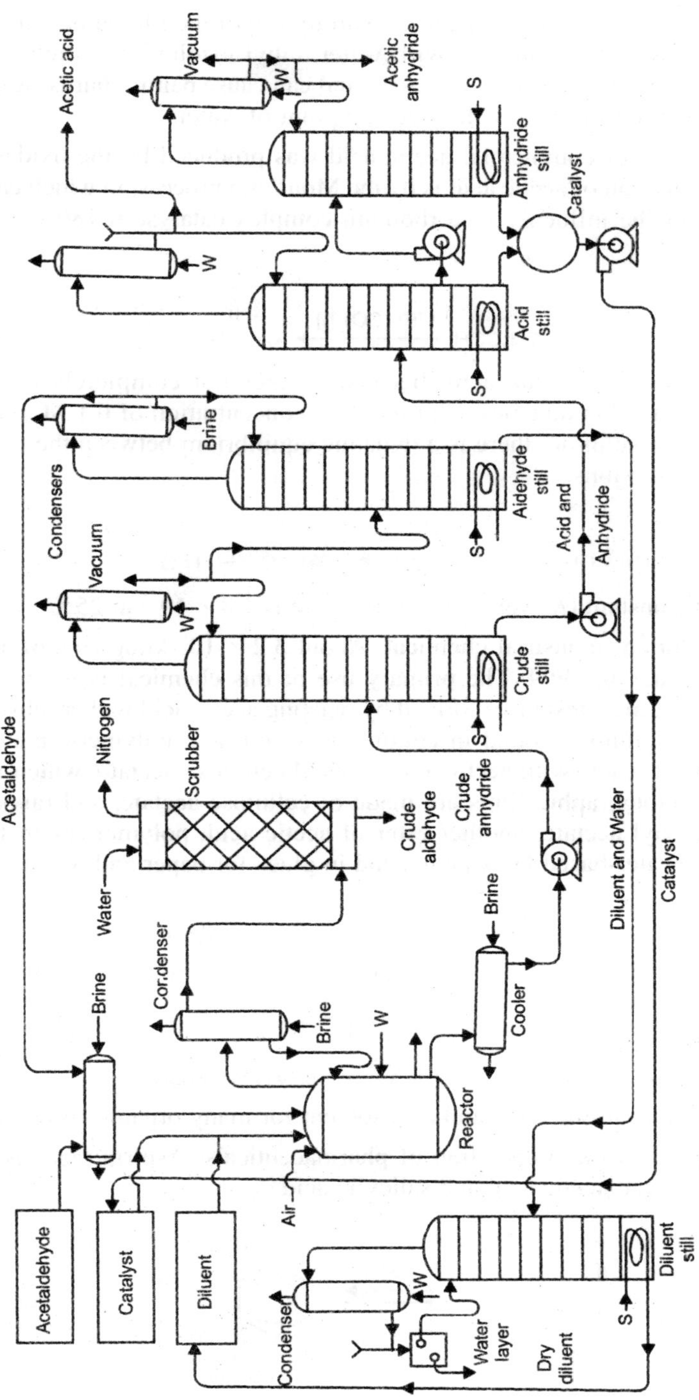

Fig.13.5. Showing the Acetic Anhydride Manufacturing process

25°C. It is flammable with a flash point of 39°C. Through hydrogen-bonding interactions, acetic acid is miscible (mixable) in all proportions with water, ethyl alcohol, and diethyl ether. Pure or concentrated solutions of acetic acid are very corrosive and can cause painful burns. Aqueous solutions of acetic acid freeze at temperatures below the freezing point of water.

For many years, the bulk of commercial acetic acid was produced by the oxidation of ethanol. Today, most industrial production of acetic acid is by the Monsanto process, in which carbon monoxide reacts with methanol under the influence of a rhodium complex catalyst at 180°C and pressures of 30–40 atm.

$$CO + CH_3COH \xrightarrow{[Rh(CO)_2I_2]} CH_3\overset{O}{\overset{\|}{C}}OH$$

Acetic acid is classified as a weak acid, because it does not completely dissociate into its component ions when dissolved in aqueous solution. At a concentration of 0.1 M, only about 1% of the molecules are ionized. In solution, there is a dynamic equilibrium between the neutral molecules and the acetate and hydronium ions.

$$CH_3\overset{O}{\overset{\|}{C}}OH + H_2O \rightleftharpoons CH_3\overset{O}{\overset{\|}{C}}O^- + H_3O^+$$

The acid dissociation constant (K_a value) for acetic acid is 1.8×10^{-5} at 25°C

Acetic acid is an important industrial chemical. About 3.2×10^9 kilograms of acetic acid were produced in the United States in 1999. The primary use of this chemical is in the manufacture of assorted acetate esters. These are substances formed by reacting acetic acid with a substance containing a hydroxyl (–OH) group. Cellulose, found in cotton and wood, is a polymeric material containing multiple hydroxyl groups. It reacts with acetic acid to yield cellulose acetate, which is used to make films and textiles. Some photographic films are made of cellulose acetate, and rayon is made from cellulose acetate fibers. Vinyl acetate, another ester of acetic acid, polymerizes to form poly(vinyl acetate), which is used in water-based latex paints and in glues for paper and wood.

$$CH_2{=}\underset{O-\underset{\overset{\|}{O}}{C}CH_3}{\underset{|}{C}H} \qquad \cdots CH_2-\underset{O-\underset{\overset{\|}{O}}{C}CH_3}{\underset{|}{C}H}-CH_2-\overset{O-\overset{O}{\overset{\|}{C}}CH_3}{\overset{|}{C}H}-CH_2-\underset{O-\underset{\overset{\|}{O}}{C}CH_3}{\underset{|}{C}H}\cdots$$

Vinyl acetate poly (vinyl acetate)

Acetic acid is also used as a fungicide and as a solvent for many organic compound.

Acetic acid is also used in the preparation of pharmaceuticals. Aspirin (acetylsalicylic acid) is formed by the reaction between acetic acid and salicylic acid.

$$C_6H_4(OH)\overset{O}{\overset{\|}{C}}OH + CH_3\overset{O}{\overset{\|}{C}}OH \underset{}{\overset{H_2SO_4}{\rightleftharpoons}} C_6H_4(O-\overset{O}{\overset{\|}{C}}CH_3)\overset{O}{\overset{\|}{C}}OH + H_2O$$

Salicylic acid Aspirin

This esterification reaction is reversible, however, and the presence of water can lead to hydrolysis of the aspirin. Thus, an *anhydrous* ("without water") reagent could lead to better yields of product. This reagent is acetic anhydride. Acetic anhydride is a clear, colourless, mobile (free-flowing) liquid with a sharp odor similar to that of acetic acid. It hydrolyzes in water to yield acetic acid. In its liquid or vapor state, acetic anhydride can irritate body tissues, possibly leading to the death of such tissue. Acetic anhydride melts at –73°C, boils at 139°C, and has a density of 1.080 g/mL at 15°C. It is flammable with a flash point of 54°C. This compound is readily combustible and represents a fire hazard. As the name indicates, an acid anhydride is a compound that is related to an acid by the loss of water. Acetic anhydride can by prepared by the dehydration of acetic acid at 800°C.

$$2\ CH_3\overset{O}{\overset{\|}{C}}OH \longrightarrow CH_3\overset{O}{\overset{\|}{C}}{-}O{-}\overset{O}{\overset{\|}{C}}CH_3 + H_2O$$

Alternatively, the reaction between the acid chloride and a salt of acetic acid (e.g. sodium acetate) yields acetic anhydride and a salt.

$$CH_3\overset{O}{\overset{\|}{C}}Cl + NaO\overset{O}{\overset{\|}{C}}CH_3 \longrightarrow CH_3\overset{O}{\overset{\|}{C}}{-}O{-}\overset{O}{\overset{\|}{C}}CH_3 + NaCl$$

As with acetic acid, the primary use of acetic anhydride is used in the manufacture of cellulose acetate for films and plastic goods; about 75 percent of the acetic anhydride produced annually in the United States is used for this purpose. Approximately 1.5 per cent of the annual acetic anhydride production is used in the synthesis of aspirin. Other uses include the manufacture of industrial chemicals, pharmaceuticals, perfumes, plastics, synthetic fibers, explosives, and weed killers. Because acetic anhydride reacts with water, it is sometimes used as a dehydrating agent in reaction mixtures where removal of water is necessary.

Uses

Acetic anhydride is used in the manufacture of cellulose acetate having the application as a base for magnetic tape and in the manufacture of textile fibres. Also, it is heated with salicylic acid to produce acetylsalicylic acid (aspirin). It is also used in the manufacture of pigments, dyes, cellulose and pesticides etc.

ACID AMIDES

Introduction

In this class of derivatives of carboxylic acid, the (-OH) of the carbonyl group is replaced by an amino group ($-NH_2$)

$$H_3C{-}\overset{O}{\overset{\|}{C}}{-}OH \xrightarrow[+NH_2]{-OH} R{-}\overset{O}{\overset{\|}{C}}{-}NH_2 \text{ OR } R{-}\overset{O}{\overset{\|}{C}}{-}N\begin{smallmatrix}H\\H\end{smallmatrix}$$

These compounds are known as *Acid Amides* of *Amides*

Evidently, the simple amides may be represented by the general formula R – CO – NH_2, where when the amide group – CO – NH_2 acts as the functional group on 'substitution amides'

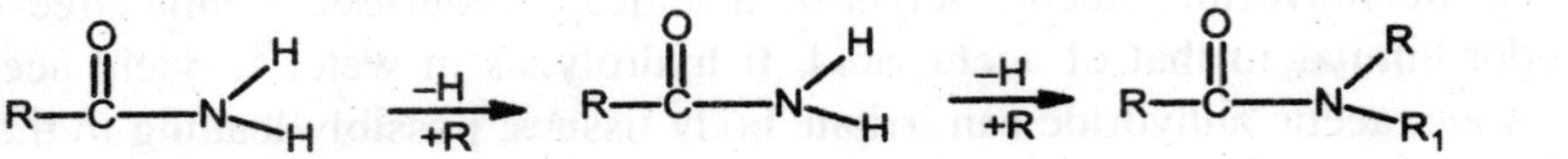

mono N - substututed amide substituted amine

The alkyl amino group – NHR and dialkyl amino group –NR_2 are functional group of mono N – substituted and di – N – substituted respectively. In general, the functional group of amides may be written as

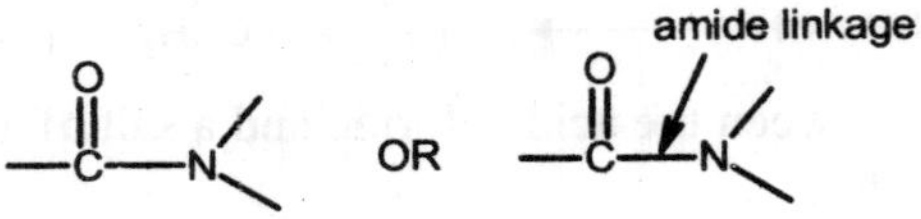

amide linkage

and the carbon-nitrogen bonds is called *amide linkage*. This bond is very stable and is found in the repeating units in protein molecules, nylon and many other industrial polymers.

Alternatively amides could also be conditional as derived ammonia by replacement of one, two or three H atoms by acyl group RCO—. They are designated as primary (1°) amide, secondary (2°) amide and tertiary (3°) amide accordingly as 1, 2 or three acyl groups are attached to the N atom.

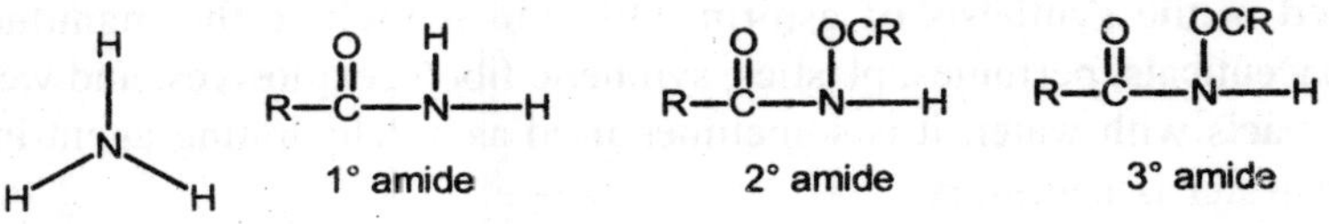

ammonia 1° amide 2° amide 3° amide

The di acyl derivatives of ammonia and primary amines are also called *Imides*

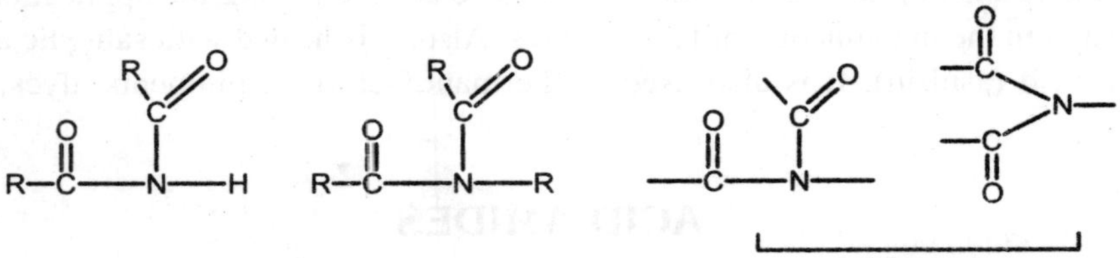

Imide group

Cyclic imides of succinic acid and pthalic acid are important

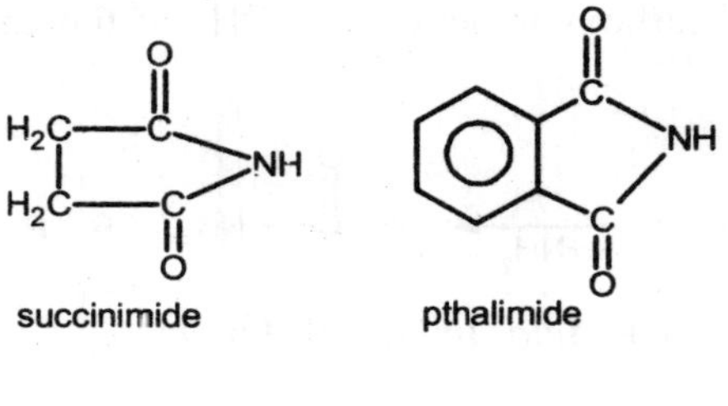

succinimide pthalimide

Under this sub head we are going to discuss only the simple amides of 1° amides R – CO – NH_2 which are important as a class.

Structure of Amides

In amide both the carbonyl carbon and the nitrogen are sp^2 hybridized. The three sp^2 orbitals of carbon overlap with sp^3 orbital of the alkvl group a sp^2 orbital of nitrogen and a p – orbital of oxygen to form three σ-bonds. The remaining two sp^2 orbitals of N overlap with s orbitals of two hydrogens to produce two N – H σ-bonds. Further, the half-filled p orbital or carbon and oxygen being parallel

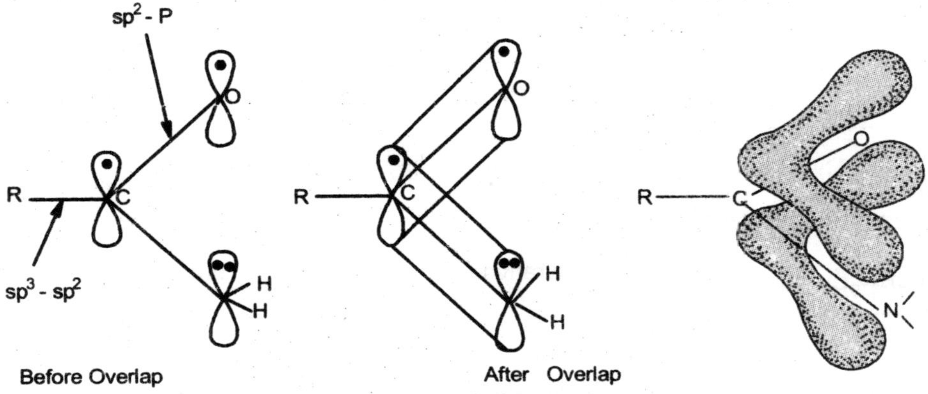

Fig. 13.6. Showing the overlap of the amides

overlap to form α-bond of the carbonyl group. Since the p – orbital N-H containing the lone pair of electrons is also parallel to the p –orbital on carbon, the additional overlap is also possible in view of this additional overlap the molecular orbital pictures of an amide is compared to that of an acyl halide is modified as illustrated in fig. 13.6. In valance bond terms the additional overlap in amide may be visualized as a resonance hybrid of forma I and II.

$$R{-}C(=O){-}NH_2 \;(I) \longleftrightarrow R{-}C(O^-){=}\overset{+}{N}H_2 \;(II) \quad \text{OR} \quad R{-}C(\overset{\delta^-}{O}){-}\overset{\delta^+}{N}H_2 \;(1.32\text{Å})$$

The additional overlap shortens the carbon nitrogen distance in amides to 1.32Å as against the normal carbon nitrogen single bond distance of 1.47Å. This suggests a significant contribution of the dipolar ion II towards the actual structure of amide.

Nomenclature

The *common name* of amides are derived from the trivial name of the parent carboxylic acid by dropping the suffix *–ic acid* and *adding amide*. In the IUPAC system the –oic acid of the name of the corresponding acid is replaced by the word amide. When alkyl groups are attached to the nitrogen atoms, they are so indicated by the capital N. the common and IUPAC names of few amides are listed below for illustration.

Formula	Common Name	IUPAC Name
$H-\overset{O}{\overset{\|\|}{C}}-NH_2$	formamide	methanamide
$CH_3-\overset{O}{\overset{\|\|}{C}}-NH_2$	acetamide	ethanamide
$CH_3CH_2-\overset{O}{\overset{\|\|}{C}}-NH_2$	propionamide	propanamide
$CH_3CH_2CH_2-\overset{O}{\overset{\|\|}{C}}-NH_2$	*n*-butyramide	butanamide
$CH_3-CH(CH_3)-\overset{O}{\overset{\|\|}{C}}-NH_2$	isobutyramide	2-methylpropanamide
$CH_3-\overset{O}{\overset{\|\|}{C}}-N(CH_3)(H)$	*N*-methylacetamide	*N*-methylethanamide
$CH_3-\overset{O}{\overset{\|\|}{C}}-N(CH_3)(CH_3)$	*N, N*-dimethylacetamide	*N, N*-dimethylethanamide

Methods of Preparation

***(i)* By Pyrolysis of ammonium salts carboxylic acid:** The simple amide can be prepared by heating the ammonium salt of the corresponding salt at 170-200°C. The ammonium salt dehydrates to form the amide.

$$\underset{\text{amide carboxylate}}{R-\overset{O}{\overset{\|\|}{C}}-ONH_4} \xrightarrow{170\text{-}200°C} \underset{\text{acetamide}}{H_2N-\overset{O}{\overset{\|\|}{C}}-R} + H_2O$$

Alternatively, ammonium salt of the relevant carboxylic acid can be heated in an inert solvent to get the amide. For example, acetamide is generally made by refluxing a solution of ammonium acetate with glacial acetic acid

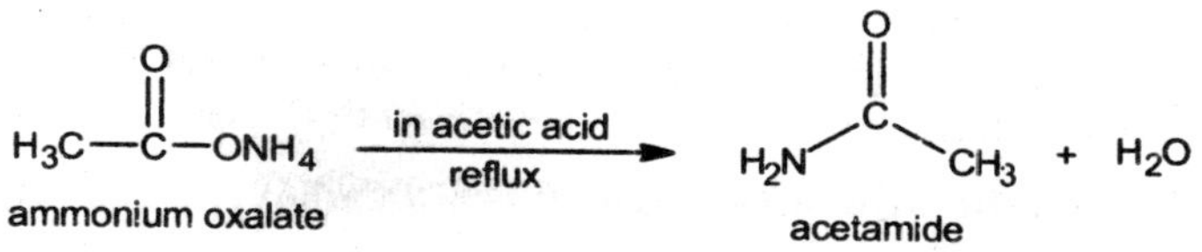

$$H_3C-\overset{O}{\overset{\|}{C}}-ONH_4 \xrightarrow[\text{reflux}]{\text{in acetic acid}} H_2N-\overset{O}{\overset{\|}{C}}-CH_3 + H_2O$$

ammonium oxalate → acetamide

***(ii)* By partial Hydrolysis of Alkyl Nitrites**:-Acid amides are obtained by the controlled or partial hydrolysis of alkyl nitrates.

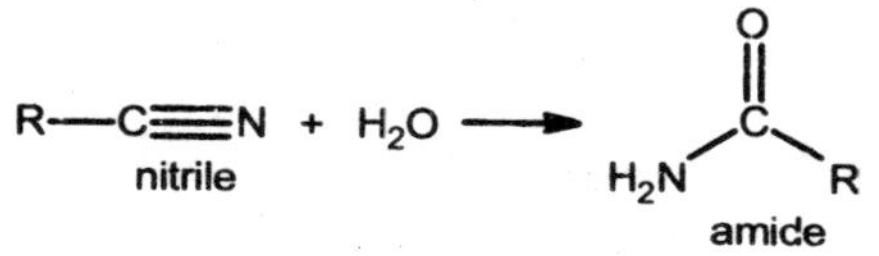

$$R-C\equiv N + H_2O \longrightarrow H_2N-\overset{O}{\overset{\|}{C}}-R$$

nitrile → amide

This cane affected by dissolving the nitrile in cold concentrated sulphuric acid and passing into water, or by shaking with concentrated hydrochloric acid. Formamide may thus be prepared by the partial hydrolysis of hydrogen cyanide.

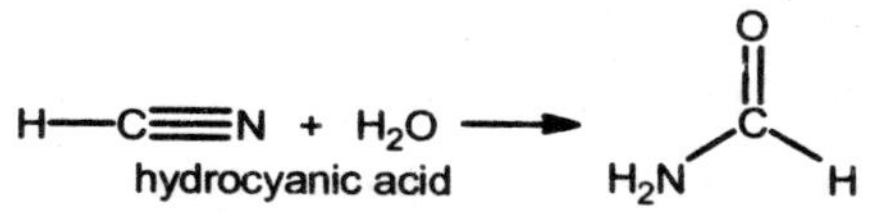

$$H-C\equiv N + H_2O \longrightarrow H_2N-\overset{O}{\overset{\|}{C}}-H$$

hydrocyanic acid

***(iii)* By Amonolysis of Acid chloride, Acid anhydride and esters**:-The most important general method for preparing amides is by treating acid chlorides, anhydrides of esters with a concentrated solution of ammonia

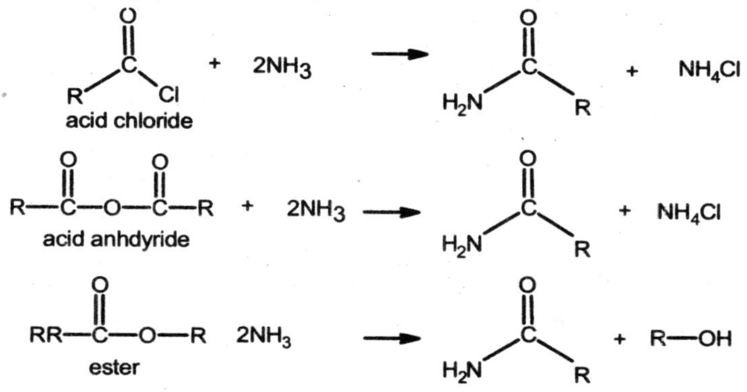

$$R-\overset{O}{\overset{\|}{C}}-Cl + 2NH_3 \longrightarrow H_2N-\overset{O}{\overset{\|}{C}}-R + NH_4Cl$$

acid chloride

$$R-\overset{O}{\overset{\|}{C}}-O-\overset{O}{\overset{\|}{C}}-R + 2NH_3 \longrightarrow H_2N-\overset{O}{\overset{\|}{C}}-R + NH_4Cl$$

acid anhdyride

$$RR-\overset{O}{\overset{\|}{C}}-O-R \quad 2NH_3 \longrightarrow H_2N-\overset{O}{\overset{\|}{C}}-R + R-OH$$

ester

Acid chlorides in general are readily available and are commonly used. For example acetamide is conveniently and quickly prepared by the action of ammonia on acetyl chloride

$$H_3C-\overset{O}{\overset{\|}{C}}-Cl + 2NH_3 \longrightarrow H_2N-\overset{O}{\overset{\|}{C}}-CH_3 + NH_4Cl$$

acetyl chloride → acetamide

The reaction of ammonia on esters is reversible and is of practical value in the case of simple esters only it precedes best with methyl esters.

***(iv)* By heating Carboxylic acid with Urea:-**When a carboxylic acid is heated with urea an exchange reaction takes place to form an acid amide.

$$H_2N-\overset{O}{\overset{\|}{C}}-NH_2 + R-\overset{O}{\overset{\|}{C}}-OH \rightleftharpoons H_2N-\overset{O}{\overset{\|}{C}}-R + H_2N-C-OH$$

urea + carboxylic acid ⇌ amide + carbamic acid

Since the carboxylic acid is decomposed to carbon dioxide and ammonia (both gases) and are thus removed from the reaction mixtures, the above reaction proceeds to completion.

Physical Properties (General)

(*i*) With the exception of formamide which is a liquid at ordinary temperature (mp 2.5°C) all amides are white solids having sharp melting points

(*ii*) Pure amides are colourless. The unpleasant smell associates

(*iii*) The amides have abnormally high boiling points. That is their boiling point are much higher than the acids from which they are derived even through their molecular weight are almost identical for example formic acid (mol weight 46.03) boils at 100°C while formamide (mol wet 45) boils at 211°C. Acetic acid (mol wt 66) boils at 118°C while its amide i.e. acetamide (mol 54) boils at 222°C. The high boiling points of amide are due to the polar character of their molecules and the formation of hydrogen bonds between the oxygen atoms of one molecule and the amine hydrogen atom of the another molecule

$$O=\underset{R}{\underset{|}{C}}-\overset{H}{\overset{|}{N}}-H\cdots O=\underset{R}{\underset{|}{C}}-\overset{H}{\overset{|}{N}}\cdots O=\underset{R}{\underset{|}{C}}-\overset{H}{\overset{|}{N}}-H$$

a portion of the hydrogen bonding of amide $RCONH_2$

Because of salt like dipolar characteristic and complete association in the solid state amides are crystalline solid with melting points higher than that corresponding acids of comparable molecular weight

(*iv*) Lower amides upto those made of five or six carbon atoms are very soluble in water are and are delequescent. All amides are soluble in organic solvents. Amides own their large solubility to associate with water through hydrogen bonding and ionic character of their molecules

O=C(H)—N(H)—H----O(H)—H----O=C(H)—N(H)—H---O(H)—H

amide water amide water

The water solubility goes on decreasing as the molecular weight increases because of the increasing hydrocarbon part (R) of the molecule.

Table of Physical Properties of Some simple amides

Name	Formula	Melting Point (°C)	Boiling Point (°C)	Solubility
Formamide	$HCONH_2$	2.5	210	—
Acetamide	CH_3CONH_2	82	222	very soluble
Propinamide	$CH_3CH_2CH_2CONH_2$	79	213	soluble
Valeramide	$CH_3CH_2CH_2\,CH_2CONH_2$	106	sublimes	soluble
Caproamide	$CH_3(CH_2)_4CONH_2$	100	sublimes	slightly

Chemical Properties (General)

Acid amide gives the following general reactions:-

(a) Basic and acidic properties: Amides are resonance high hybrids of the following ammonical structures

$$H_3C-\overset{O}{\overset{\|}{C}}-\underset{H}{\underset{|}{\ddot{N}}}-H \longleftrightarrow H_3C-\overset{OH}{\overset{|}{C}}=\underset{H}{\underset{|}{\overset{+}{N}}}-H \quad \text{OR} \quad H_3C-\overset{O^{\sigma}}{\overset{|}{C}}\overset{\sigma+}{\equiv}\underset{H}{\underset{|}{N}}-H$$

HYBRID

Because of the delocalization of the non-bonding electron pair of N atom over the system O – C – N it is relatively unavailable for accepting a proton. Hence amides are neutral or weakly basic in character in contest to amines in which the electron pair of N atom is fully available for protonation.

available for protonization

$$R-\underset{H}{\underset{|}{\ddot{N}}}-H$$

strongly basic

relatively unavailable due to delocalization

$$H_3C-\overset{O}{\overset{\|}{C}}-\underset{H}{\underset{|}{\ddot{N}}}-H$$

weakly basic

Further more, as a consequence of the delocalization of the non-bonding electron pair of N-atom O-atom acquires a partial negative and N gets a partial positive charge as shown above in the hybrid structure. Thus similar to carboxylic acids amides will be accepted to evidence acid character. The N – H bond is weakened as a result of delocalization of the electron pair and thereby the release of proton (H^+) is facilitated. Thus acid amides act as very weak acids.

The fact that amides behave both as very weak bases and acids which are only stable in the aqueous solutions

$$H_3C{-}C(=O)NH_2 + HCl \longrightarrow \left[H_3C{-}C(=\overset{+}{O}{-}H)NH_2 \longleftrightarrow H_3C{-}C(O{-}H)=\overset{+}{N}H_2 \right] Cl^-$$

This is an agreement with the fact that amides being resonance hybrids the pronization takes place at the negative O atom rather that at the N atom which carries only a positive charge. This is also supported by NMR studies of amides.

(b) As Weak Acid Amides: React with sodamide [$NaNH_2$] in either solution to form salts of the stronger acid. Here the portion attached to N atom is released in the reaction

$$2R{-}\overset{O}{\overset{\|}{C}}{-}NH_2 + NaNH_2 \longrightarrow \left[2H_3C{-}\overset{O}{\overset{\|}{C}}{-}NH \right] Na + NH_3$$

Amides also form mercury salts with mercury oxides

2. Hydrolysis: On boiling with water, amides are hydrolyzed to give the parent acid and ammonia. The hydrolysis takes place rapidly in a mineral acid (usually hydrochloric acid) and far more rapidly in an alkali such as sodium hydroxide.

$$R{-}\overset{O}{\overset{\|}{C}}{-}NH_2 + H_2O \xrightarrow[(HCl)]{H^+} R{-}\overset{O}{\overset{\|}{C}}{-}OH + NH_4^+$$

$$R{-}\overset{O}{\overset{\|}{C}}{-}NH_2 + H_2O \xrightarrow{OH} R{-}\overset{O}{\overset{\|}{C}}{-}O + NH_3$$

Mechanism

The reaction takes place by the same mechanism as the hydrolysis of esters and acid halides

(a) Acid Hydrolysis

$$R{-}\overset{O}{\overset{\|}{C}}{-}\ddot{N}H_2 \underset{I}{\overset{H^+}{\rightleftharpoons}} R{-}\overset{OH^+}{\overset{\|}{C}}{-}NH_2 \underset{II}{\overset{H_2O}{\rightleftharpoons}} R{-}\underset{H-\overset{+}{O}-H}{\overset{OH}{C}}{-}NH_2 \overset{H^+}{\rightleftharpoons} R{-}\underset{HO}{\overset{HO}{C}}{-}NH_2 \underset{IV}{\overset{H^+}{\rightleftharpoons}} R{-}\underset{O-H}{\overset{OH}{C}}{-}\overset{+}{N}H_3$$

$$\longrightarrow H_3C{-}\overset{OH}{C}{=}O + NH_4^+$$

(b) Base Analysis

$$R-\overset{O}{\overset{\|}{C}}-NH_2 + OH^- \rightleftharpoons R-\underset{OH}{\overset{O^-}{\overset{|}{\underset{|}{C}}}}-NH_2 \longrightarrow R-\underset{O}{\underset{\|}{C}}-O^- + NH_3$$

3. Reaction with Nitrous Acid: Amides when treated with nitrous acid and dilute hydrochloric acid give a qualitative yield of nitrogen and the parent carboxylic acid

$$\underset{\text{acetamide}}{R-\overset{O}{\overset{\|}{C}}-NH_2} + HNO_2 \xrightarrow{HCl} \underset{\text{acetic acid}}{H_3C-\overset{O}{\overset{\|}{C}}-OH} + N_2 + H_2O$$

This reaction of amide is analogous to that given in primary aliphatic amines

4. Reduction: Amides may reduced to primary amines either with sodium and ethanol or with lithium aluminium hydride ($LiAH_4$)

$$\underset{\text{acetamide}}{R-\overset{O}{\overset{\|}{C}}-NH_2} + 4(H) \xrightarrow[\text{or } LiAlH_4]{Na/C_2H_5OH} R-CH_2-NH_2 + H_2O$$

5. Dehydration: Amides when heated with phosphorous pentoxide or thionyl chloride ($SOCl_2$) loose a molecule of water to form the corresponding nitrites

$$\underset{\text{acetamide}}{R-\overset{O}{\overset{\|}{C}}-NH_2} \xrightarrow[\text{or } SOCl_2]{P_2O_5} \underset{1^\circ \text{ nitrile}}{R-C\equiv N} + H_2O$$

Formamide is converted to hydrogen cyanide and water by passing its vapour over thoria (thorium oxide) or pumice stone at 400°C

$$\underset{\text{formamide}}{H-\overset{O}{\overset{\|}{C}}-NH_2} \xrightarrow[400^\circ C]{\text{thoria}} \underset{\text{hydrocyanic acid}}{H-C\equiv N} + H_2O$$

5. Hofmann's Degradation: When an amide is treated with bromine (or chlorine) in alkali solutions it is converted to a primary amine that has one carbon less than starting amide.

$$R-\overset{O}{\overset{\|}{C}}-NH_2 + Br_2 + NaOH \xrightarrow{H_2O} \underset{1^\circ \text{ amine}}{R-NH_2}$$

This reaction which was discovered by Hofmann is 1882 and results in the formation of a primary amine by elimination of carboxyl group (≡ one carbon atom) from an amide, is often referred to *Hofmann Degradation.* The overall reaction of the amide with bromine in the presence of amide with bromine in the presence of three equivalents of alkali may be represented by the equation.

$$R{-}\overset{O}{\overset{\|}{C}}{-}NH_2 + Br_2 + NaOH \xrightarrow{H_2O} \underset{1^\circ\text{ amine}}{R{-}NH_2} + 2NaBr + NaHCO_3 + H_2O$$

Hofmann degradation is particularly useful for stepping down any homologous series by one carbon atom less

Mechanism

The mechanism of Hofmann Degradation involves the following steps:-

(*i*) Base catalyzed bromination of amide produces N – bromoamide

$$R{-}\overset{O}{\overset{\|}{C}}{-}NH_2 + Br_2 + NaOH \xrightarrow{H_2O} R{-}N(H){-}Br + NaBr + H_2O$$

(*ii*) The N – Bromoamide then reacts with alkali (NaOH) begin acyl nitrene in ion

$$H_3C{-}\overset{O}{\overset{\|}{C}}{-}\underset{H}{N}{-}Br + OH^- \longrightarrow \underset{\text{Acyl Nitrene (unstable)}}{H_3C{-}\overset{O}{\overset{\|}{C}}{-}\ddot{N}:} + Br^- + H_2O$$

The nitrogen atom of nitrene is highly unstable.

(*iii*) R group of nitrene then migrates (or rearranges) to form a relatively stable intermediate isocyanate

$$R{-}\overset{O}{\overset{\|}{C}}{-}N \longrightarrow \underset{\text{isocyanate}}{R{-}\ddot{N}{=}C{=}O}$$

(*iv*) The isocyanate finally reacts with water to give unstable carbamic acid which eliminates a molecule of CO_2 to yield the primary amine

$$\underset{\text{isocyanate}}{R{-}\ddot{N}{=}C{=}O} + H_2O \longrightarrow \underset{\text{carbamic acid}}{\left[R{-}\underset{H}{\ddot{N}}{-}\overset{O}{\overset{\|}{C}}{-}OH\right]}$$

$$\longrightarrow \underset{1^\circ\text{ amine}}{R{-}NH_2}$$

Since the key step (iii) in the above mechanism involves rearrangement, the overall reaction resulting in a primary amine is also called Hofmann rearrangement. This reaction is often referred to as Hofmann's Bromoamide reaction because "bromoamide" is an important intermediate as shown in the first step of the mechanism

Individual Member

ACETAMIDE

Formula: C_2H_5NO

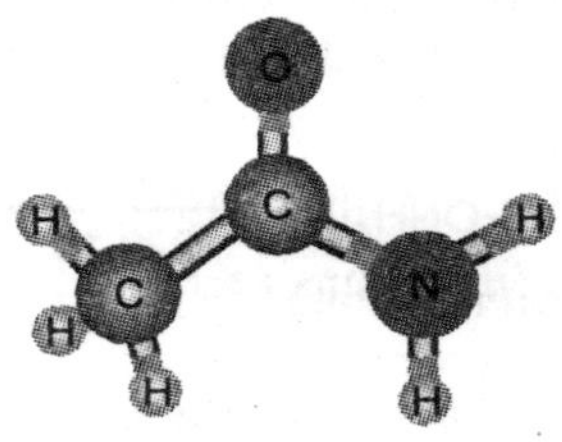

Introduction

Acetamide, the amide of acetic acid, is a white crystalline solid in pure form. It is produced by dehydrating ammonium acetate. It is used as a solvent and plasticizer and in the synthesis of many other organic compounds. Its Mohs scale rating is 1 to 1.5.Acetamide is extremely combustible and releases irritating fumes when ignited. It is toxic by inhalation (of dust), ingestion, skin and eye contact. Skin or eye contact may cause redness and pain.

Preparation

Acetamide may be prepared by the following reactions:-

1. By the action of Ammonia on Acetyl chloride

$$\underset{\text{acetyl chloride}}{H_3C{-}\overset{O}{\overset{\|}{C}}{-}Cl} + NH_3 \longrightarrow \underset{\text{acetamide}}{H_3C{-}\overset{O}{\overset{\|}{C}}{-}NH_2} + HCl$$

2. By the action of Ammonia on acetic anhydride

$$\underset{\text{acetic anhydride}}{H_3C{-}\overset{O}{\overset{\|}{C}}{-}O{-}\overset{O}{\overset{\|}{C}}{-}CH_3} + NH_3 \longrightarrow \underset{\text{acetamide}}{H_3C{-}\overset{O}{\overset{\|}{C}}{-}NH_2} + \underset{\text{acetic acid}}{H_3C{-}\overset{O}{\overset{\|}{C}}{-}OH}$$

3. By the partial hydrolysis of acid nitriles, with sulphuric acid or alkline hydrogen peroxide

$$\underset{\text{acetonitrile}}{H_3C{-}C{\equiv}N} + H_2O \xrightarrow[H_2SO_4]{\text{conc.}} \underset{\text{acetamide}}{H_3C{-}\overset{O}{\overset{\|}{C}}{-}NH_2}$$

4. By the action of ammonia on ethyl acetate

$$\underset{\text{ethyl acetate}}{H_3C{-}\overset{O}{\overset{\|}{C}}{-}O{-}CH_2{-}CH_3} + NH_3 \longrightarrow \underset{\text{acetamide}}{H_3C{-}\overset{O}{\overset{\|}{C}}{-}NH_2} + \underset{\text{ethanol}}{H_3C{-}CH_2{-}OH}$$

Manufacture:-Ethyl acetate is left in contact with strong ammonia until it has dissolved. The solution is distilled and the fraction passing at 200°C is separated. The product on cooling gives crystals of acetamide. The method is used for preparing acetamide in large scale.

5. By partial dehydrating of ammonium acetate

$$\underset{\text{Ammonium Acetate}}{H_3C-\overset{O}{\overset{\|}{C}}-ONH_4} \xrightarrow[220\ ^\circ C]{Heat} \underset{\text{acetamide}}{H_3C-\overset{O}{\overset{\|}{C}}-NH_2} + H_2O$$

a mixture of equal weights of ammonium acetate and glacial acetic acid is heated for a few hours under a reflux condenser. The product is distilled and the fraction collected at 200°C is cooled in a receiver, when acetamide is obtained.

Physical Properties

Physical State	white crystalline solid
Melting Point	78-80°C
Boiling Point	221°C
Specific Gravity	1.16
Solubility In Water	Soluble (decomposes in hot water;
Solvent Solubility	Soluble in alcohol, slightly soluble in ether)
pH	
Vapor Density	
Autoignition	560°C
Nfpa Ratings	Health: 2 Flammability: 1 Reactivity: 0
Refractive Index	1.4274
Flash Point	
Stability	Stable under ordinary conditions

Chemical Properties

1. It is amphoteric in character and reacts both with hydrochloric acid and sodium metal

2. With dilute acids or aikalies: it is hydrolyzed to give acetic acid

$$H_3C-\overset{O}{\overset{\|}{C}}-NH_2 + H_2O \longrightarrow H_3C-\overset{O}{\overset{\|}{C}}-OH + NH_3$$

$$\underset{\text{Acetamide}}{H_3C-\overset{O}{\overset{\|}{C}}-NH_2} + H_2O \longrightarrow \underset{\text{Acetic acid}}{H_3C-\overset{O}{\overset{\|}{C}}-OH} + NH_3$$

3. With Nitrous Acid: It also gives acetic acid

$$\underset{\text{Acetamide}}{H_3C-\overset{O}{\overset{\|}{C}}-NH_2} + HNO_2 \longrightarrow \underset{\text{Acetic acid}}{H_3C-\overset{O}{\overset{\|}{C}}-OH} + N_2 + H_2O$$

4. With Phosphorous Pentoxide: It gives acid nitrile

$$\underset{\text{acetamide}}{H_3C-\overset{O}{\overset{\|}{C}}-NH_2} + P_2O_5 \longrightarrow \underset{\text{acetonitrile}}{H_3C-C\equiv N} + 2HPO_3$$

5. Hofmann's Bromoamide Reaction: On treatment with bromine and aqueous caustic potash, amides yield amines.

$$\underset{\text{acetamide}}{H_3C-\overset{O}{\overset{\|}{C}}-NH_2} + Br_2 + KOH \longrightarrow \underset{\text{N-bromoacetamide}}{H_3C-\overset{O}{\overset{\|}{C}}-NH-Br} + KBr + H_2O$$

$$\underset{\text{N-bromoacetamide}}{H_3C-\overset{O}{\overset{\|}{C}}-NH-Br} + KOH \longrightarrow KBr + CO_2 + \underset{\text{methanamine}}{H_3C-NH_2}$$

$$CO_2 + 2KOH \longrightarrow K_2CO_3 + H_2O$$

The reaction is employed for the preparation of methyl amine and for descending the homologous series

6. With nascent hydrogen (sodium + amyl alcohol): It is reduced to ethyl amine.

$$\underset{\text{acetamide}}{H_3C-\overset{O}{\overset{\|}{C}}-NH_2} + 4H \longrightarrow \underset{\text{ethanamine}}{H_3C-CH_2-NH_2} + H_2O$$

Uses

Acetamide is used as a solvent, plasticizer, and a wetting and penetrating agent and in lacquers, soldering flux and explosives.

Descent of a Homologous Series

(a) Conversion of Acetic Acid into formic acid

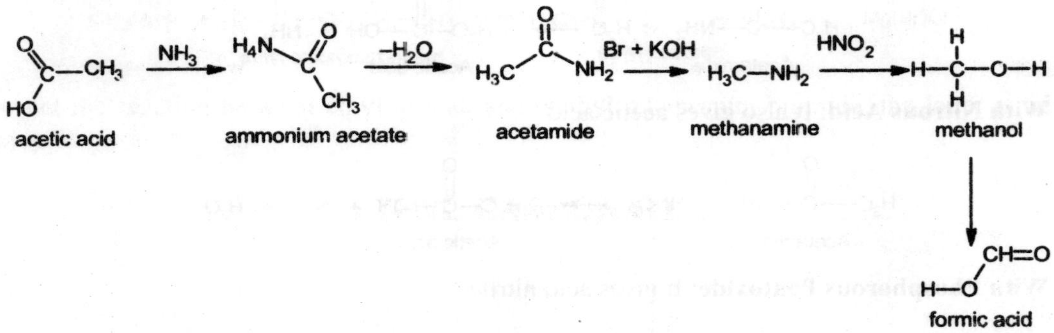

acetic acid $\xrightarrow{NH_3}$ ammonium acetate $\xrightarrow{-H_2O}$ acetamide $\xrightarrow{Br + KOH}$ methanamine (H_3C-NH_2) $\xrightarrow{HNO_2}$ methanol ($H-CH_2-O-H$) → formic acid (H–O–CH=O)

(b) Conversion of Ethyl alcohol

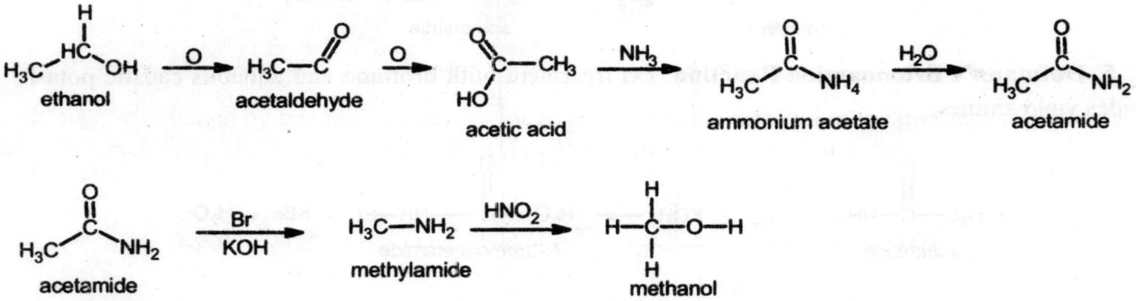

ethanol $\xrightarrow{O}$ acetaldehyde $\xrightarrow{O}$ acetic acid $\xrightarrow{NH_3}$ ammonium acetate $\xrightarrow{H_2O}$ acetamide

acetamide $\xrightarrow[KOH]{Br}$ H_3C-NH_2 (methylamide) $\xrightarrow{HNO_2}$ methanol

Derivatives of Carbamic Acid

Carbamic acid HO—C(=O)—OH, has never been isolated in free state. This is not accepted, since compounds containing two hydroxyl groups attached to the sane carbon atoms are usually unstable. It is believed that the aqueous solution of carbon dioxide containing Carbamic acid in equilibrium with carbon dioxide and water.

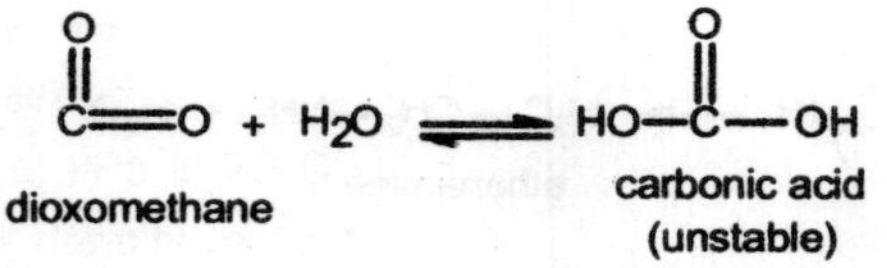

$$O=C=O + H_2O \rightleftharpoons HO-C(=O)-OH$$

dioxomethane; carbonic acid (unstable)

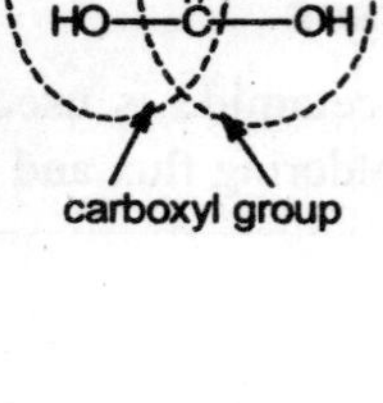

Although carbonic acid itself is unstable, many of its derivatives have considerable stability and are of industrial and biological importance carbonic acid generally behaves like a dibasic acid. This is because it contains two hydroxyl groups and each hydroxyl group can behave like the hydroxyl group of a carboxylic acid

HO—C(=O)—OH

carboxyl group

Carbonic acid forms both monofunctional derivation and difunctional derivatives. The monofunctional derivatives are unstable and are known only as their salts A few of the important derivatives are shown in the table below.

Name	Formula	Melting Point	Boiling point
Carbamic acid	$HO-C(=O)-NH_2$	–6°C	195.55°C
Phosgene (carbonyl chloride)	$Cl-C(=O)-Cl$	–118°C	8.3°C
Urea	$H_2N-C(=O)-NH_2$	132.7°C	—
Ethyl chloroformate	$Cl-C(=O)-O-CH_2-CH_3$	—	—
Ethyl Carbonate	$H_3C-CH_2-O-C(=O)-O-CH_2-CH_3$	50°C	61.5°C

PHOSGENE

Formula: CCl_2O

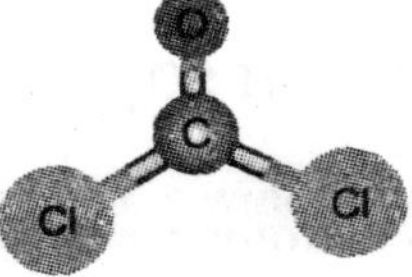

Introduction and History

Phosgene was synthesized by the chemist John Davy (1790-1868) in 1812. It was first used as a weapon by the French, under the direction of French chemist Victor Grignard in 1915. Later, the Germans, under the direction of German chemist Fritz Haber added small quantities of it to chlorine to increase the latter's toxicity. Soon after, use of pure phosgene was begun. Phosgene was responsible for most of the about 100,000 gas-caused deaths during World War I.

The name has derived from Greek Word Phos = light and Latin word generare = generation. Phosgene (also known as carbonyl chloride, $COCl_2$) is a highly toxic gas or refrigerated liquid that was used as a chemical weapon in World War I but is also used to make plastics and pesticides. Phosgene gas has no colour but may appear as a white or yellowish haze when released into air. In lower concentrations its odor resembles recently cut hay or green corn (maize) while at higher concentrations it may be strong and unpleasant.

It is a manufactured chemical, but small amounts occur naturally from the breakdown of chlorinated compounds. Phosgene can also result from the combustion of chlorine-containing organic compounds.

Phosgene is a particularly insidious poison, as phosgene exposure often has no initial symptoms. Symptoms usually appear within 24 hours, but can take up to 72 hours to manifest. The gas combines with water in the tissues of the respiratory tract to form carbon dioxide and hydrochloric acid. The acid then dissolves the membranes in the lungs. Fluid fills the lungs, and death results from a combination of blood loss, shock, and respiratory failure. Phosgene is also an eye and skin irritant. It can also be a contaminant in food or water.

Preparation

Phosgene is usually prepared by the following processes

(1) By passing a mixture of carbon monoxide and chlorine over activated charcoal at 200 – 250 °C (commercial method)

$$CO + CL_2 \xrightarrow[\Delta]{\text{charcol}} COCL_2$$

(2) By the action of fuming sulphuric acid (oleum) on carbon tetrachloride at 78°C

$$CCl_4 + H_2SO_4 + SO_2 \xrightarrow{\Delta} COCl_2 + \underset{\text{chlorosulphuric acid}}{2ClSO_2H}$$

(3) By the oxidation of chloroform with potassium dichromate in the presence of sulphuric acid or by oxygen of the air in the presence of sunlight

$$CHCl_3 + O_2 \xrightarrow{\text{sunlight}} 2COCl_2 + 2HCl$$

Phosgene is also produced when carbon tetrachloride is thrown upon a hot flame, such as that of burning gasoline and for this reason, the use of carbon tetrachloride for extinguishing indoor fires is very dangerous

The use of carbon tetrachloride for extinguishing indoor fires is very dangerous

Physical Properties

Physical

Formula weight	98.9 g/mol
Melting point	155 K (-118°C)
Boiling point	281 K (8°C)
Density	1.43×10^3 kg/m^3 (liquid)
Solubility	slowly decomposes in water
Relative density	1.4
$\Delta_f H^0_{gas}$	-219.1 kJ/mol
$S^0_{gas, 1 bar}$	283.5 J/mol·K

Liquid phase

- Relative liquid density (Water = 1) : 1.4
- Liquid density (1.013 bar at boiling point) : 1403 kg/m^3
- Liquid/gas equivalent (1.013 bar and 15 °C (59 °F)) : 337 vol/vol
- Boiling point (1.013 bar) : 7.5 °C
- Latent heat of vaporization (1.013 bar at boiling point) : 24.39 kJ/kg
- Vapor pressure (at 20 °C or 68 °F) : 1.6 bar

Critical point

- Critical temperature : 182 °C
- Critical pressure : 56.7 bar

Gaseous phase

- Gas density (1.013 bar at boiling point) : 4.35 kg/m^3
- Gas density (15°C and 1 bar) : 4.248 kg/m^3
- Specific gravity (air = 1) (1.013 bar and 21 °C (70 °F)) : 3.4
- Specific volume (1.013 bar and 21 °C (70 °F)) : 0.243 m^3/kg
- Heat capacity at constant pressure (Cp) (1.013 bar and 25 °C (77 °F)) : 0.057 kJ/(mol.K)

Chemical Properties

(1) Hydrolysis: Phosgene is decomposed by water to give hydrogen chloride and carbon dioxide

$$O{=}C(Cl)_2 + H_2O \longrightarrow 2HCl + O{=}C(OH)_2 \longrightarrow CO_2 + H_2O$$

phosgene

(2) Reaction with Alcohols: Phosgene reacts with one mole of an alcohol in an inert solvent (e.g benzene) at 0°C to yield the corresponding alkyl chloroformate

phosgene + ethanol →(Benzene, 0°C) ethyl methyl ether + HCl

Mechanism

The mechanism of this reaction follows the general scheme of nucleophillic substitution and involves the following steps.

Step I: The nucleophillic attacks the carboxyl group carbon of the phosgene molecule.

$$H_3C-C(=O)-Cl + H_3C-CH_2-\ddot{O}-H \rightleftharpoons H_3C-C(=O)(-Cl)-\overset{+}{O}(H)-CH_2-CH_3$$

Step II: One of the chloride ions is eliminated

$$H_3C-C(=O)(-Cl)-\overset{+}{O}(H)-CH_2-CH_3 \rightleftharpoons Cl-C(=O)-\overset{+}{O}(H)-CH_2-CH_3 + Cl^-$$

Step III: A hydrogen ion is lost to give ethyl chlorohydrate

$$H_3C-C(=O)(-Cl)-\overset{+}{O}(H)-CH_2-CH_3 \rightleftharpoons Cl-C(=O)-O-CH_2-CH_3 + H^+$$

ethyl chloroformate

When phosgene is heated with excess of alcohol in the presence of pyridine, the product is an alkyl carbonate

$$O=CCl_2 + 2\,HO-CH_2-CH_3 \xrightarrow[\Delta]{\text{pyridene}} O=C(O-CH_2-CH_3)_2 + 2HCl$$

phosgene + ethanol → diethyl carbonate

Mechanism

The mechanism of this reaction involves the following three steps in addition to the three steps described above

Steps I: The nucleophile attacks the carbonyl carbon of ethyl chloroformate molecule

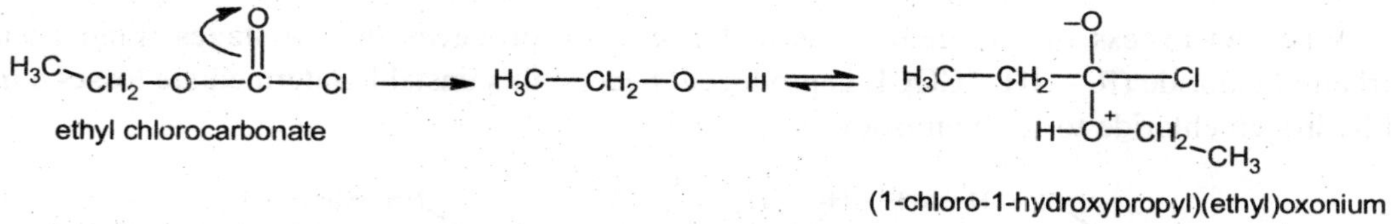

$$H_3C-CH_2-O-C(=O)-Cl \;(\text{ethyl chlorocarbonate}) \longrightarrow H_3C-CH_2-O-H \rightleftharpoons H_3C-CH_2-C(-O^-)(-Cl)-O^+(H)-CH_2-CH_3$$

ethyl chlorocarbonate

(1-chloro-1-hydroxypropyl)(ethyl)oxonium

Step II: Chloride is eliminated

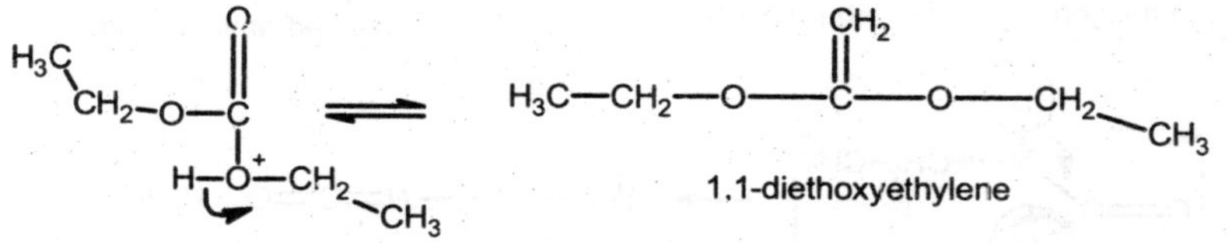

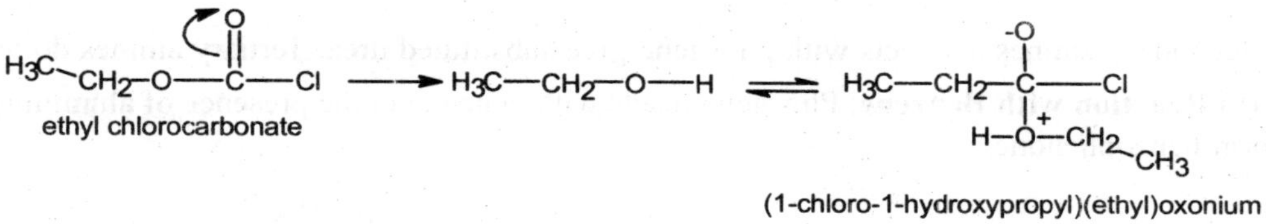

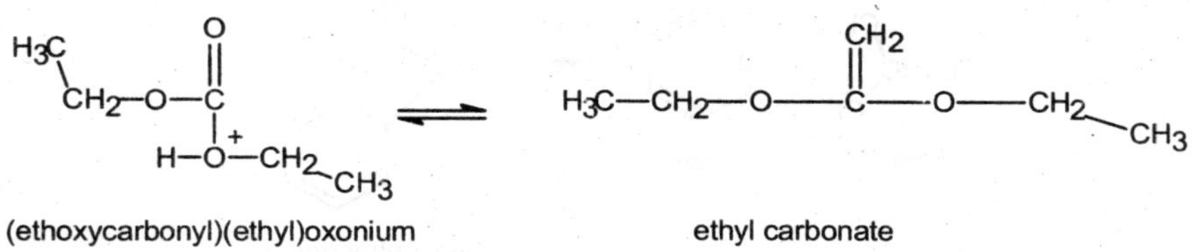

$$H_3C-CH_2-O-C(=O)-O^+(H)-CH_2-CH_3 \rightleftharpoons H_3C-CH_2-O-C(=CH_2)-O-CH_2-CH_3$$

1,1-diethoxyethylene

$$H_3C-CH_2-O-C(=O)-Cl \longrightarrow H_3C-CH_2-O-H \rightleftharpoons H_3C-CH_2-C(-O^-)(-Cl)-O^+(H)-CH_2-CH_3$$

ethyl chlorocarbonate

(1-chloro-1-hydroxypropyl)(ethyl)oxonium

$$H_3C-CH_2-O-C(=O)-O^+(H)-CH_2-CH_3 \rightleftharpoons H_3C-CH_2-O-C(=CH_2)-O-CH_2-CH_3$$

(ethoxycarbonyl)(ethyl)oxonium

ethyl carbonate

(3) Reduction with Ammonia: Phosgene reacts with ammonia to yield urea

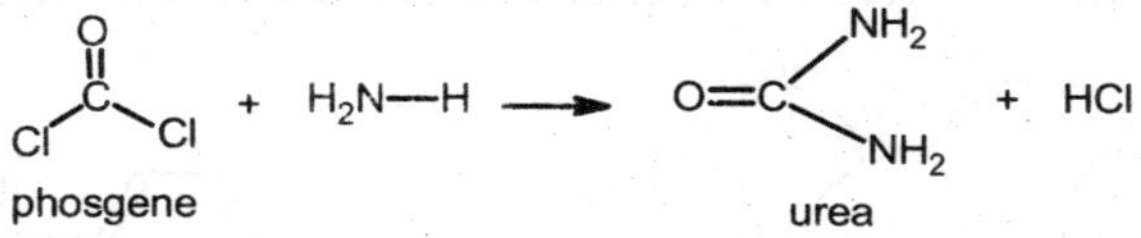

$$Cl-C(=O)-Cl + H_2N-H \longrightarrow O=C(NH_2)_2 + HCl$$

phosgene

urea

$$2HCl + \text{excess } 2NH_3 \longrightarrow 2NH_4Cl$$

(4) Reaction with Amines:-Phosgene reacts with an excess of primary amine to form the corresponding substitutes urea

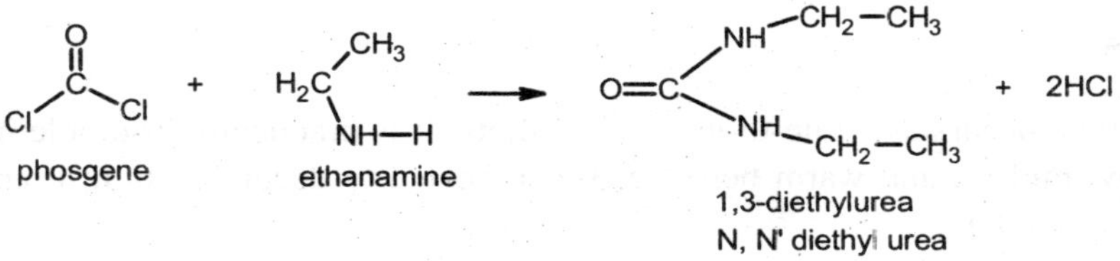

$$Cl-C(=O)-Cl + CH_3-CH_2-NH-H \longrightarrow O=C(NH-CH_2-CH_3)_2 + 2HCl$$

phosgene

ethanamine

1,3-diethylurea

N, N' diethyl urea

The prefix N, N' in the name N, N'-diethyl urea indicator that the ethyl group is attached to the nitrogen atoms

When an excess of phosgene is used the reaction proceeds in two stages. First an unstable carbamyl chloride (R – NH – COCl) is produced. Second this unstable intermediate losses a molecule of hydrogen chloride to form an isocyanate

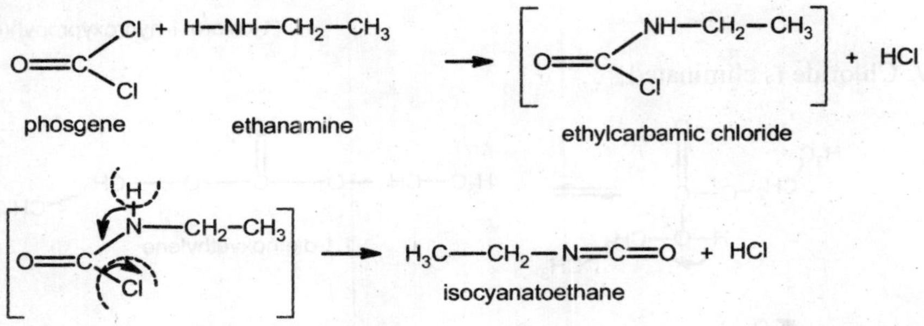

$O{=}CCl_2$ + $H{-}NH{-}CH_2{-}CH_3$ → $[O{=}C(Cl){-}NH{-}CH_2{-}CH_3]$ + HCl

phosgene ethanamine ethylcarbamic chloride

$[O{=}C(Cl){-}N(H){-}CH_2{-}CH_3]$ → $H_3C{-}CH_2{-}N{=}C{=}O$ + HCl

isocyanatoethane

Secondary amines also acts with phosgene give substituted urea. Tertiary amines do not react.

(5) Reaction with Benzene: Phosgene reacts with benzene in the presence of aluminium chloride to form benzophenone.

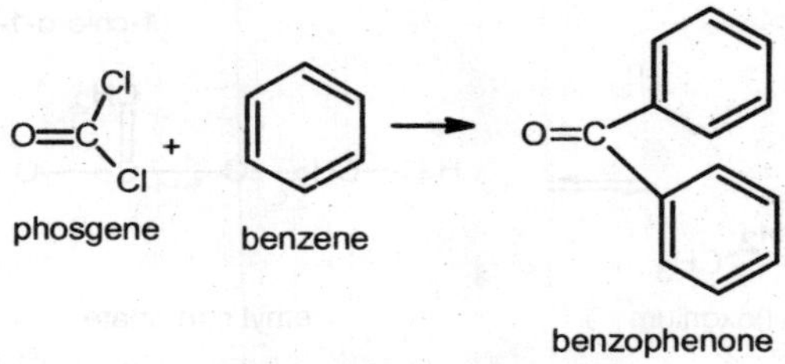

$O{=}CCl_2$ + C_6H_6 → $O{=}C(C_6H_5)_2$

phosgene benzene benzophenone

(6) Reaction with dimethylaniline: Phosgene reacts with dimethylaniline in the presence of zinc chloride to form Michler's Ketone. (Notice that this reaction is the extension of reaction 5).

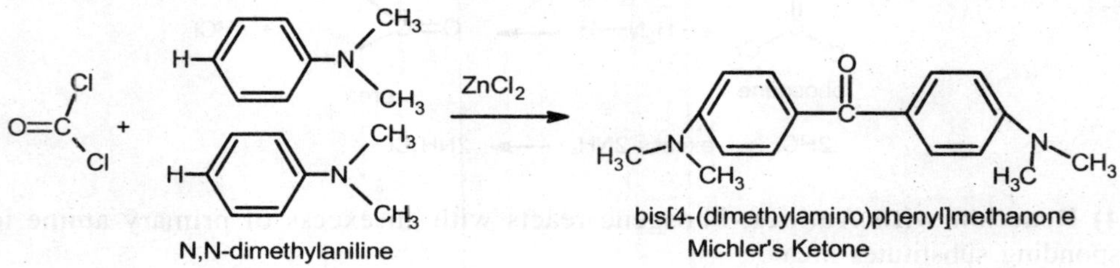

$O{=}CCl_2$ + 2 $H{-}C_6H_4{-}N(CH_3)_2$ $\xrightarrow{ZnCl_2}$ $(CH_3)_2N{-}C_6H_4{-}C(=O){-}C_6H_4{-}N(CH_3)_2$

N,N-dimethylaniline bis[4-(dimethylamino)phenyl]methanone Michler's Ketone

Properties

Michler's ketone occurs as white-to-greenish leaflets. It is practically insoluble in water but soluble in alcohol, pyrimidine, and warm benzene. When heated to decomposition, it emits toxic fumes of nitrogen oxides (NO_x).

Molecular weight:	268.3586
boiling point (°C):	360
Melting point (°C):	173-178

Use

Michler's ketone, 4,4.-(dimethylamino)benzophenone, is a chemical intermediate used in the synthesis of at least 13 dyes and pigments, particularly auramine derivatives. These pigments are used to dye paper, textiles, and leather, and one is used as an antiseptic fungicide

Exposure

The primary routes of potential human exposure to Michler's ketone are inhalation and dermal contact. The risk of potential occupational exposure is greatest for workers in facilities that manufacture the compound or any of the dyestuffs for which Michler's ketone is an intermediate.

Carcinogenicity

There is sufficient evidence for the carcinogenicity of Michler's ketone in experimental animals. When administered orally in the diet, Michler's ketone induced increased incidences of hepatocellular carcinomas in rats of both sexes and in female mice and hemangiosarcomas (primarily subcutaneous) in male mice. In view of an NCI/OTA correlative interpretation, these results can be considered to provide sufficient evidence for the carcinogenicity of Michler's ketone in experimental animals. There were no data available to evaluate the carcinogenicity of Michler's ketone in humans

(7) Reaction with Benzyl Alcohol: Phosgene reacts with benzyl alcohol to produce carbobenzoxy chloride or (benzyl chlorocarbonate)

$$C_6H_5CH_2OH + COCl_2 \longrightarrow C_6H_5CH_2\text{-}O\text{-}CO\text{-}Cl + HCl$$

benzyl alcohol phosgene benzyl chlorocarbonate

carbobenzoxy chloride is an important reagent which is widely used in Peptide Syntheis

The carbobenzoxy protecting group

Max Bergmann and Leonidas Zervas introduced a more general approach in 1932, based on the formal concept of a protecting group. A schematic representation of the idea is shown below. A chemical blocking agent, the protecting group, prevents a reaction from occurring at the modified site. The amino group of the first is temporarily converted into a different, unreactive functional group, such as an amide. Its carboxyl group is activated so that it can react with an unprotected amino group in a compound in which the carboxyl group is protected. (The boxes in the diagram represent the protected amino and carboxyl groups.) After the peptide formation reaction is complete, the protecting groups are removed without affecting any other part of the molecule.

O OH NH$_2$ O OH Ethyl ester NH$_2$

Carbobenzoxy amide

Bergmann and Zervas used carbobenzoxy chloride ($C_6H_5CH_2OCOCl$, formed by the reaction of phosgene [Cl_2CO] and benzyl alcohol [$C_6H_5CH_2OH$]) to react with the amino group. This forms the carbobenzoxy amide derivative of the amino acid. The intermediate is made directly from the amino acid, and the reaction does not disturb the stereochemical array. The free (unprotected) carboxyl group of the carbobenzoxy amino acid (called a Z-amino acid) is activated as a mixed anhydride by reaction with acetic anhydride $(CH_3C{=}O)_2O$, and an amide is formed by reaction with an amino acid in which the carboxyl group is protected by having been converted to an ethyl ester. The carbobenzoxy protecting group can be removed under conditions that do not affect the peptide bond, such as hydrogenation in the presence of a palladium catalyst or specific chemical hydrolysis. Various side chains (especially those with sulfur-or nitrogen-containing substituents) must be protected as well.

CHLOROCARBONIC ACID

Formula: $HCClO_2$

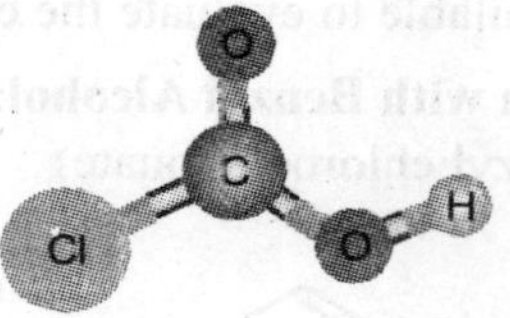

Introduction

Chlorocarbonic acid is the monoacid chloride of carbonic acid. It does not exist in free state but esters well known two if this esters are described below:-

EHTYL CHLOROFORMATE

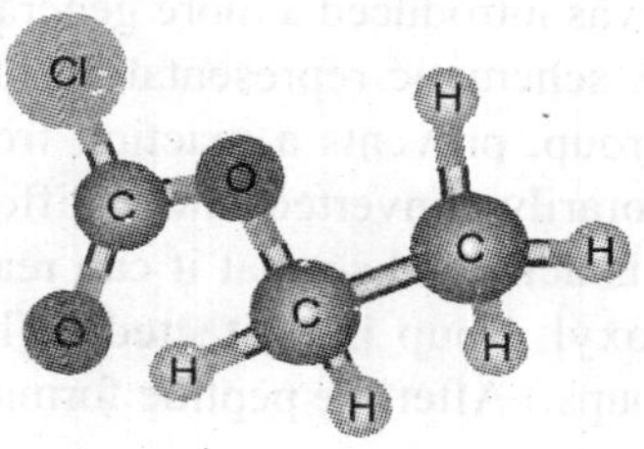

Introduction

It is obtained by treating phosphene with one mole of ethyl alcohol in benzene at 0°C. Ethyl chloroformate is colourless liquid with suffocating odour, having a boiling point at 94°C. it behaves both as an ester and as well as an acid halide. Ethyl chloroformate reacts with compounds containing active hydrogen atoms for example alcohols ammonia and amines

Physical Properties

Physical and Chemical Properties	
Physical State	clear liquid
Melting Point	-50°C
Boiling Point	Decomposes
Specific Gravity	0.985-0.995
Solubility in Water	hydrolysis
pH	acidic
Vapor Density	
Refractive Index	
Flash Point	87°C
Stability	Stable under ordinary conditions

Chemical Properties

$$H_3C-CH_2-OH + Cl-\overset{O}{\overset{\|}{C}}-O-CH_2-CH_3 \longrightarrow H_3C-CH_2-O-\overset{O}{\overset{\|}{C}}-O-CH_2-CH_3 + HCl$$

ethanol   ethyl chlorocarbonate   diethyl carbonate

$$NH_3 + Cl-\overset{O}{\overset{\|}{C}}-O-CH_2-CH_3 \longrightarrow H_2N-\overset{O}{\overset{\|}{C}}-O-CH_2-CH_3 + HCl$$

ethyl chlorocarbonate

$$H_3C-NH_2 + Cl-\overset{O}{\overset{\|}{C}}-O-CH_2-CH_3 \longrightarrow H_3C-NH-\overset{O}{\overset{\|}{C}}-O-CH_2-CH_3 + HCl$$

ethyl chlorocarbonate   ethyl methylcarbamate

As shown above, the ethyl chloroformate may be used in organic synthesis and for the introduction of carboethoxy group on nitrogen.

Ethyl chloroformate also reacts with Grignard's Reagent to produce esters of monocarboxylic acids (for example with methyl magnesium bromide it forms ethyl acetate.

$$CH_3MgBr + Cl-C(=O)-O-CH_2-CH_3 \longrightarrow H_3C-C(=O)-O-CH_2-CH_3 + Mg(Br)Cl$$

ethyl chlorocarbonate; ethyl acetate

to prevent the reaction between ethyl acetate and methyl magnesium bromide, the latter is added gradually

ETHYL CARBONATE

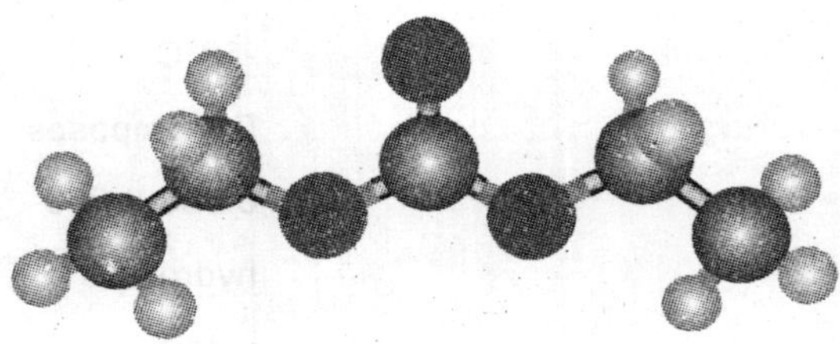

It may be obtained by heating phosgene with an excess of ethyl alcohol in the presence of pyridine. Ethyl carbonate may also be obtained by heating silver carbonate with ethyl iodide.

$$AgNO_3 + 2C_2H_5I \longrightarrow H_3C-CH_2-O-C(=O)-O-CH_2-CH_3 + 2AgI$$

ethyl iodide; diethyl carbonate

Ethyl carbonate is a pleasant smelling liquid with a boiling point of 126°C. It is soluble is water, ethyl carbonate undergoes. Claisen condensation with ketones contain two α-hydrogen atoms to form β-keto esters. For example, with acetone it form ethyl acetoacetate

Physical Properties

Boiling point	126°C
Melting point	-43°C
Relative density (water = 1)	0.98
Solubility in water	none
Vapour pressure, kPa at 20°C	1.1
Relative vapour density (air = 1)	4.07
Flash point	25°C c.c.
Auto-ignition temperature	445°C
Explosive limits, vol% in air	1.4-11.0

$$H_3C-\overset{O}{\overset{\|}{C}}-CH_2-H + H_3C-CH_2-O-\overset{O}{\overset{\|}{C}}-O-CH_2-CH_3 \xrightarrow{C_2H_5ONa} H_3C-\overset{O}{\overset{\|}{C}}-CH_2-\overset{O}{\overset{\|}{C}}-O-CH_2-CH_3$$

acetone; diethyl carbonate; ethyl 3-oxobutanoate (ethyl acetoacetate)

CARBAMIC ACID
OR
AMINOFORMIC ACID

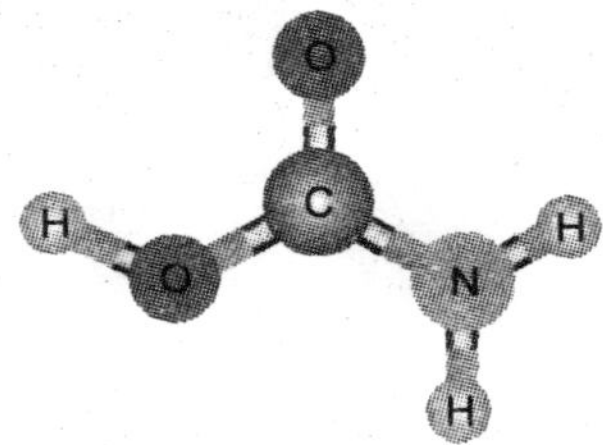

Carbamic acid is the monoacid of carbonic acid. It does not exit in the free state but its salts and esters are important compounds some of these are described below.

AMMONIUM CARBAMATE

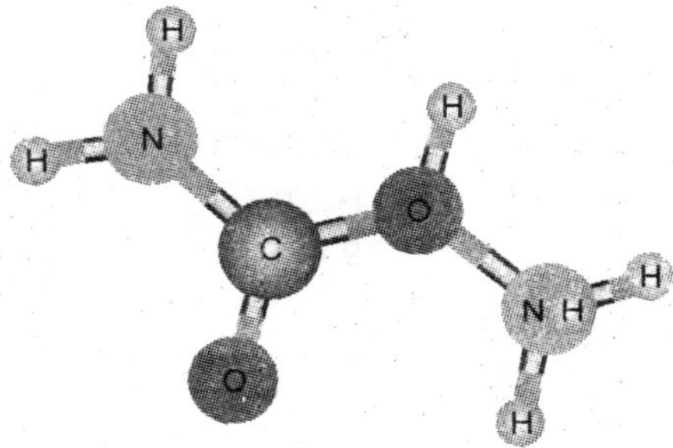

It may be obtained by the treatment of carbon monoxide with ammonia.

$$O{=}C{=}O + NH_3 \longrightarrow H_2N-\overset{O}{\overset{\|}{C}}-\overset{+}{O}H \cdots \overset{+}{N}H_3$$

carbon monoxide; ammonia; ammonium carbamate

Ammonium carbonate is a colourless solid crystal which is very much soluble is water. When an aqueous solution is wormed at 60°C. It is hydrolyzed to ammonium carbonate

$$\underset{\text{ammonium carbamate}}{H_2N{-}C(=O){-}O^{+}H{-}NH_2} + H + H_2O \longrightarrow \underset{\text{ammonium carbonate}}{H_2O^{+}{-}NH_2^{+}{-}C(=O){-}O^{+}H{-}NH_2} + H$$

CARBAMYL CHLORIDE

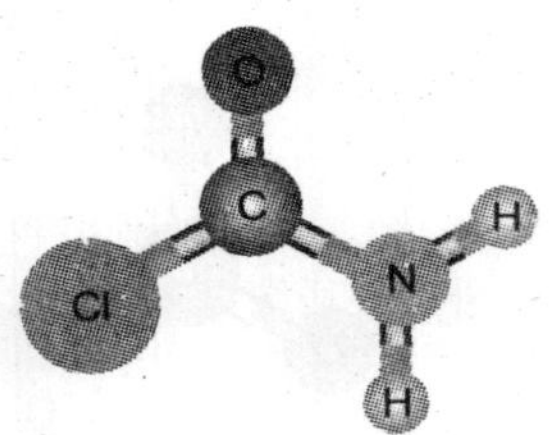

It may be prepared by dry hydrogen chloride over potassium cyanate

$$KOCN + HCl \longrightarrow \underset{\text{carbamic chloride}}{Cl{-}\overset{O}{\overset{\|}{C}}{-}NH_2}$$

Physical Properties

Melting Point	-33°C
Boiling point	165-167°C
Molecular Formula	H_2ClNO
Formula Weight	9.48568
Composition	C(11%) H(2.54%) Cl(44.60%) N(17.62%) O(20.13%)
Molar Refractivity	15.09 ± 0.3 cm^3
Molar Volume	7.2 ± 3.0 cm^3
Parachor	44.4 ± 4.0 cm^3
Index of Refraction	1.440 ± 0.02
Surface Tension	40.6 ± 3.0 dyne/cm
Density	.389 ± 0.06 g/cm^3
Dielectric Constant	Not available
Polarizability	.98 ± 0.5 $10^{-24}cm^3$
Monoisotopic Mass	78.982491 Da
Nominal Mass	9 Da
Average Mass	9.485755 Da

Carbamic acid is a compound of chemical formula H_2NCOOH. But it exists only in the form of carbamate (its salt or ester), carbamide (amide) and carbamoyl (acyl radical). Carbamate structure inhibits cholinesterase and many insecticides and parasiticides contain carbamate fuctional group. It is highly toxic to human also. Heavy exposure to IT can cause carbamate poisoning. Natural carbamide (urea) can be found in protein metabolism in urine. Carbamoyl is the radical NH2CO-, also called carbamyl. It is involved in the biosynthesis of the pyrimidine ring. Carbamoyl compounds, such as salicylamides, are important for the preparation of pharmaceutical products, pesticides, dyes, and biosynthesis researches.

ETHYL CARBAMATE OR URETHANE

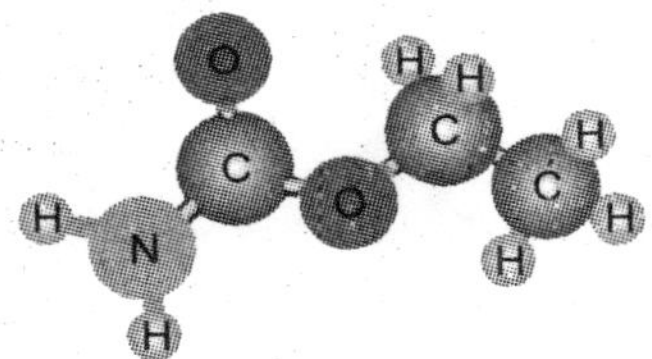

Ester of Carbamic acid as a class are often called urethanes. The ethyl esters is known as urethane or ethyl carbamate. It may be obtained:-

(1) By the action of ammonia on ethyl chloroformate

(2) By heating carbonyl chloride with alcohol

$$\mathrm{Cl{-}C({=}O){-}NH{-}H} + \underset{\text{ethanol}}{\mathrm{HO{-}CH_2{-}CH_3}} \xrightarrow{\Delta} \underset{\text{ethyl carbamate}}{\mathrm{H_2N{-}C({=}O){-}O{-}CH_2{-}CH_3}} + \mathrm{HCl}$$

(3) by heating urea with ethyl alcohol. The reaction prceeds in two steps, *First* urea is converted into ammonia and isocyanic acid which is tautomeric with cyanic acid second isocyanic acid with ethyl alcohols to ethyl carbamate

$$\mathrm{H_2N{-}C({=}O){-}NH_2} \longrightarrow \underset{\text{isocyanic acid}}{\mathrm{H{-}N{=}C{=}O}} \rightleftharpoons \underset{\text{cyanic acid}}{\mathrm{HO{-}C{\equiv}N}} + \mathrm{H{-}NH_2}$$

$$\underset{\text{isocyanic acid}}{\mathrm{H{-}N{=}C{=}O}} + \mathrm{C_2H_5OH} \longrightarrow \underset{\text{ethyl carbamate}}{\mathrm{H_2N{-}C({=}O){-}O{-}CH_2{-}CH_3}}$$

Ethyl carbamate (urethane) is a crystalline solid, mp 50°C it is soluble in water and alcohols esters ethyl carbamate reacts with ammonia to give urea.

ethyl carbamate + NH_3 $\xrightarrow{\Delta}$ urea + ethanol

ethyl carbamate has a mild hypnotic action which is intensified if the ethyl group is replaced by a larger group having a branched chain. It is used in the treatment of cancer and leukemia

UREA OR CARBAMIDE

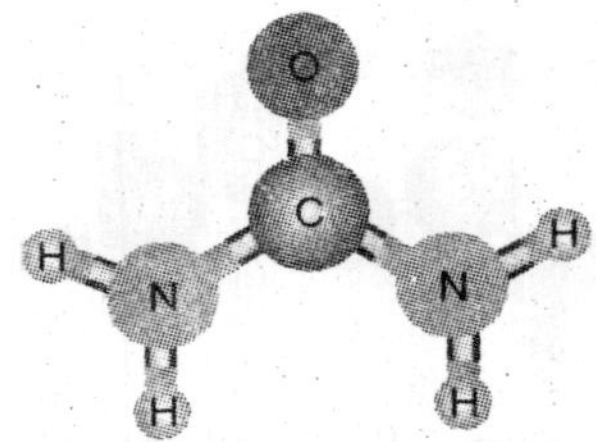

Introduction

Urea was discovered by Hilaire Rouelle in 1773. It was the first organic compound to be artificially synthesized from inorganic starting materials, in 1828 by Friedrich Woehler, who prepared it by the reaction of potassium cyanate with ammonium sulfate. This disproved the theory that the chemicals of living organisms are fundamentally different from inanimate matter and started the discipline of organic chemistry

Physiology

The individual atoms of urea come from carbon dioxide, water, aspartate and ammonia in a metabolic pathway known as the urea cycle, an anabolic process. This expenditure of energy is necessary because ammonia, a common metabolic waste product, is toxic and must be neutralized. Urea production occurs in the liver and is under the regulatory control of *N*-acetylglutamate.Aquatic animals do not produce urea; living in an abundant supply of water, they can simply excrete ammonia immediately as it is produced. Birds, with more severe restrictions on water consumption than most other animals, produce uric acid, a compound even less toxic than urea. Humans produce a little uric acid as a result of purine breakdown. Indeed, excess uric acid production can lead to a type of arthritis known as gout.

Prparation

Urea may be obtained:

(1) By heating an aqueous solution of ammonium cyanate (Wöhler Synthesis)

$$\underset{\text{ammonium cyanate}}{NH_4OCN} \xrightleftharpoons{\Delta} \underset{\text{urea}}{H_2N-CO-NH_2}$$

The reaction is somewhat reversible, for it has been noticed that when silver ion is added to the hot aqueous solution silver cyanate is precipitated

Mechanism

The mechanism of the above reaction involves two steps.

Step I Decomposition of ammonium cyanate gives ammonia and isocyanic acid which is tautomeric with cyanic acid.

$$\underset{\text{ammonium cyanate}}{NH_4OCN} \rightleftharpoons NH_3 + \left[\underset{\text{isocyanic acid}}{H-N=C=O} \rightleftharpoons \underset{\text{cyanic acid}}{HO-C\equiv N}\right]$$

Step II Action of ammonia on isocyanic acid yields urea

$$\underset{\text{isocyanic acid}}{H-N=C=O} + \ddot{N}H_3 \rightleftharpoons H-NH-\underset{\underset{H-\overset{+}{N}H_3}{|}}{C}=O \rightleftharpoons \underset{\text{urea}}{H_2N-\underset{\underset{NH_2}{|}}{C}=O}$$

(2) by the action of ammonia on phosgene (carbonyl chloride)

$$\underset{\text{phosgene}}{Cl-CO-Cl} + \underset{\text{ammonia}}{2\,H-NH_2} \longrightarrow \underset{\text{urea}}{H_2N-\underset{\underset{NH_2}{|}}{C}=O} + 2HCl$$

$$2HCl + \text{excess}NH_3 \longrightarrow \underset{\text{ammonium chloride}}{NH_4Cl}$$

(3) By heating ammonia and carbon dioxide together under pressure at 130-150°C (commercial method)

$$CO_2 + 2NH_3 \xrightarrow[\text{pressure}]{\Delta} H_2N-CO-NH_2$$

Mechanism

The reaction is not as simple as shown as above. It involves the following three steps

Step I Ammonia reacts with carbon dioxide to give an unstable carbamic acid.

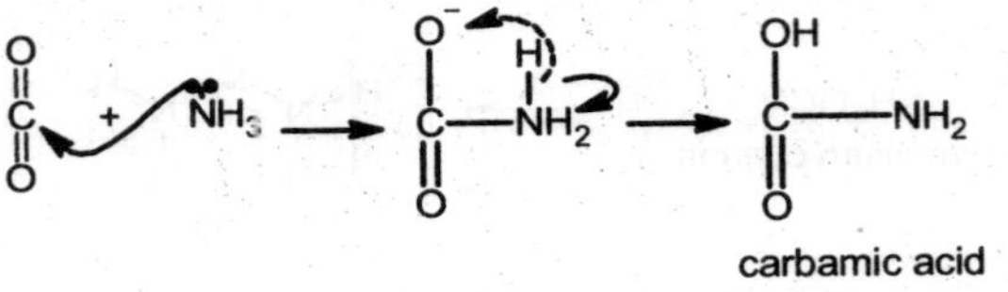

$$O{=}C{=}O + NH_3 \longrightarrow {}^{-}O{-}C(=O){-}\overset{+}{N}H_2{-}H \longrightarrow HO{-}C(=O){-}NH_2$$

carbamic acid

Step II Carbamic acid reacts with ammonia to form ammonium carbamate

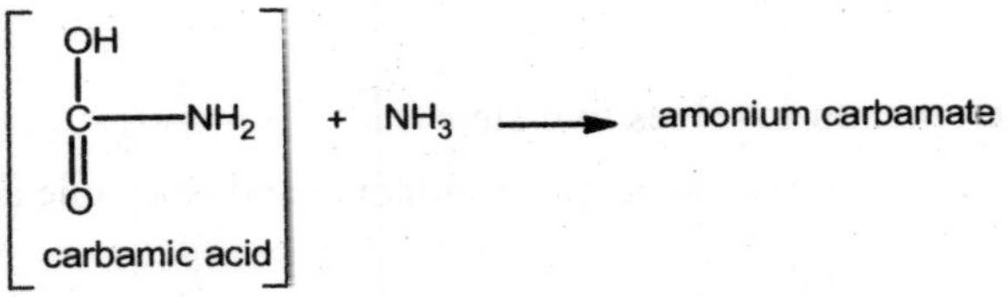

$$[HO{-}C(=O){-}NH_2] + NH_3 \longrightarrow \text{amonium carbamate}$$

carbamic acid

Step III A molecule of water is eliminated from ammonium carbonate to yield urea

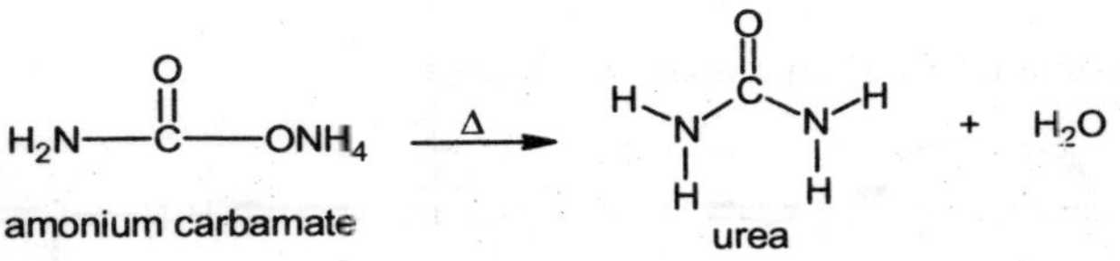

$$H_2N{-}C(=O){-}ONH_4 \xrightarrow{\Delta} H_2N{-}C(=O){-}NH_2 + H_2O$$

amonium carbamate — urea

(4) By the action of ammonia on urethanes and alkyl carbamate

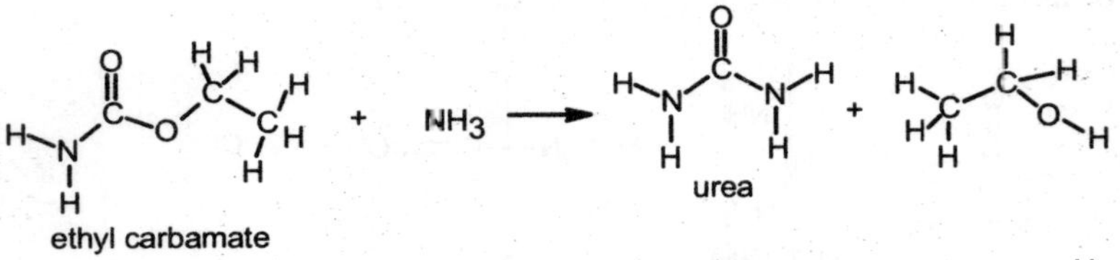

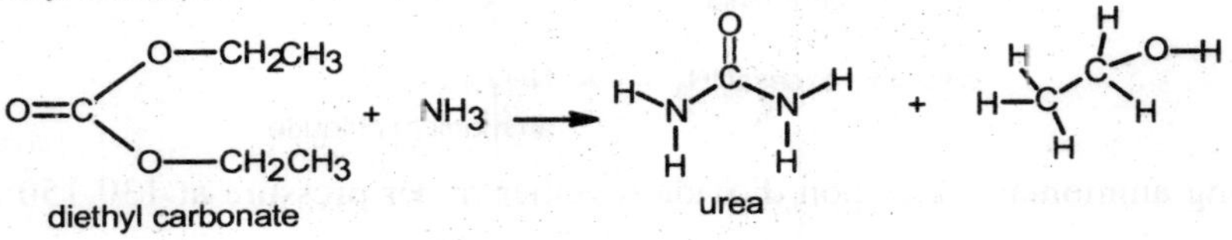

$$H_2N{-}C(=O){-}O{-}CH_2CH_3 + NH_3 \longrightarrow H_2N{-}C(=O){-}NH_2 + CH_3CH_2OH$$

ethyl carbamate — urea

$$O{=}C(OCH_2CH_3)_2 + NH_3 \longrightarrow H_2N{-}C(=O){-}NH_2 + CH_3CH_2OH$$

diethyl carbonate — urea

(5) By the partial hydrolysis of cyanamide with dilute sulphuric acid at 70°C

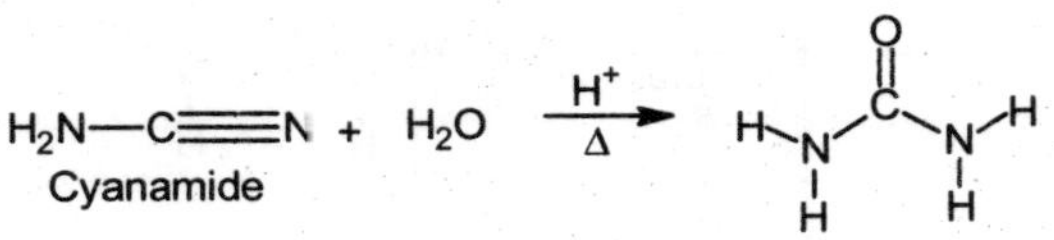

$$H_2N{-}C{\equiv}N + H_2O \xrightarrow[\Delta]{H^+} H_2N{-}C(=O){-}NH_2$$

Cyanamide

(6) By the action of dilute sulphuric acid on calcium cynamide

$$\underset{\text{calcium cynamide}}{Ca{=}N{-}C{\equiv}N} + H_2SO_4 + H_2O \xrightarrow{\text{warm}} H_2N{-}\overset{O}{\overset{\|}{C}}{-}NH_2 + CaSO_4$$

Industrial Preparation of Urea

Hot Gas Rectification Process

In this process(see fig. 13.7) the recycle gas mixture is compressed up to 120 – 130 atmos in five stages in uncooled compressors. Between compression stages air cooling was provided, this however, had to be very carefully controlled so that solid carbamate would not be formed. Recycle of hot gases imposed a greater cooling load on the reaction system than when feeding cold carbon dioxide and liquid ammonia and required the use of a two stage reactor to control the reactor at the desired temperature of 160°C-170°C. Heat was carried off by water boiling inside the tubes under regulated pressure.

The first reactor stage was packed with a number of lead – covered cooling tubes. The recycle compression system also contribution to an obviously high plant investment.

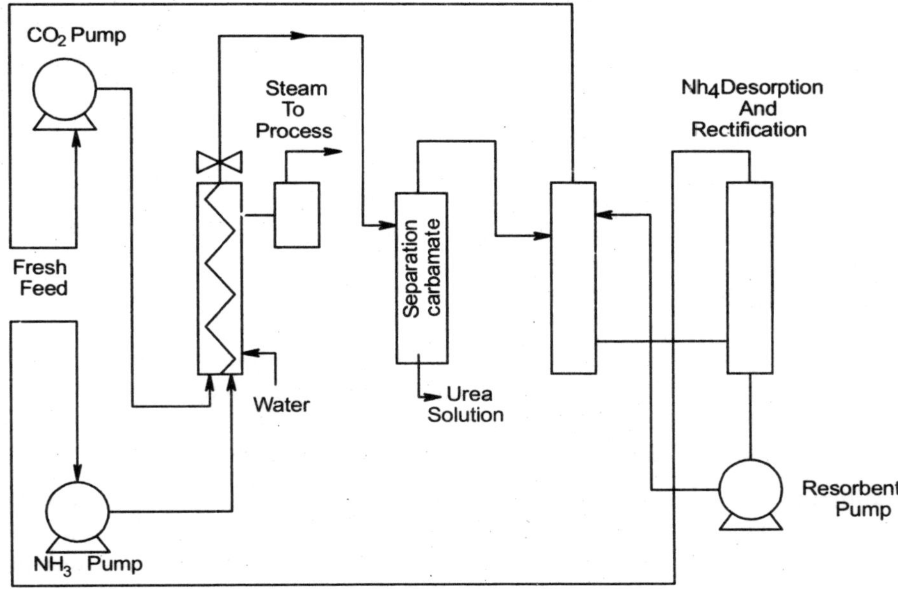

Fig 13.7. Showing the Industrial Preparation of Urea by Hot Recirculation Process

Purification

Crude urea contains bi-uret which is formed to some extent when ever urea is heated on its melting point or above. The final product of will contain varying amounts of bi-uret, depending on the purification process.

Crystal Urea

Crystal urea is the highest purity urea available

Urea USP

It is prepared from crystal urea

Shotted Urea

To provide the greater ease of handling required by some end uses, crude urea is made available in various granular and 'shotted form'

Urea is essentially a waste product: it has no physiological function. It is dissolved in blood (in humans in a concentration of 2.5-7.5 mmol/liter) and excreted by the kidney.

Physical Properties

Melting Point	132.7°C
Density	1.335
Refractive index	1.484 and 1.602
Heat of Fusion	-578 and-600 cal/g
Heat of Combustion	2531 cal/g
Heat of solution in water	-57.8 cal/g
Free energy of formation	-47,120 cal/note
Specific heat at 20°C	0.320 cal/g(°C)

Structure of Urea

Structure of Urea. In urea the two nitrogen atoms, the oxygen atom, and the carbon all *spl* hybridized. These sp^2 hybrid orbitals overlap with each other and with *s* orbitals ur . hydrogen atoms forming C – O of the four hydrogen atoms forming C – O C – N and N – H σ bonds are shown in the fig. 13.8, these σ bonds are lie in one plane and are approximately 120° apart

In addition to the Sp^2 orbitals, each urea also possesses an unhybridized *p* which is perpendicular to the plane of σ bonds. The *p* orbital on each nitrogen atom contains a pair of electrons, the *p* orbital on carbon contains one electron, and the *p* orbital and also contains one electron. These *p* laterally overlap forming a delocalized π-molecular orbital above and below the σ bonds (Fig. 13.8) according the Valence Bond Theory, urea is considered to be a *resonance hybrid* of the following canonical forms.

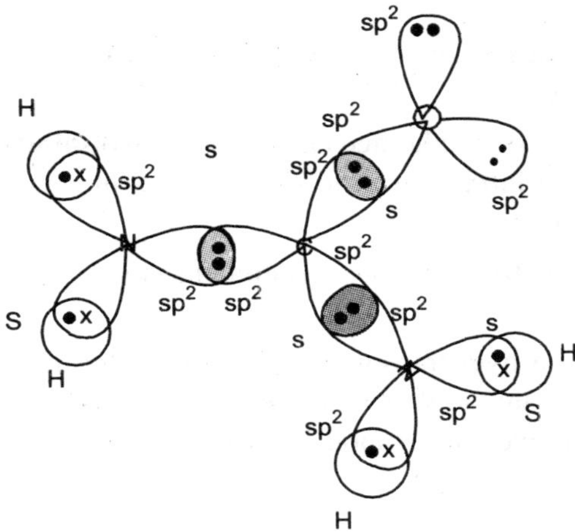

Fig. 13.8.

The Urea Cycle

Urea is the chief nitrogenous waste of mammals.

Most of our nitrogenous waste comes from the breakdown of amino acids.

This occurs by **deamination**.

Deamination of amino acids results in the production of **ammonia (NH_3)**.

Ammonia is an extremely toxic base and its accumulation in the body would quickly be fatal.

However, the **liver** contains a system of carrier molecules and enzymes which quickly converts the ammonia (and carbon dioxide) into urea.

This is called the **urea cycle**.

One turn of the cycle:

- consumes **2** molecules of **ammonia**
- consumes 1 molecule of carbon dioxide
- creates 1 molecule of **urea** ($(NH_2)_2CO$)
- regenerates a molecule of **ornithine** for another turn.

Although our bodies cannot tolerate high concentrations of urea, it is much less poisonous than ammonia.

Urea is removed efficiently by the kidneys.

Link to discussion of the anatomy and physiology of the kidneys.

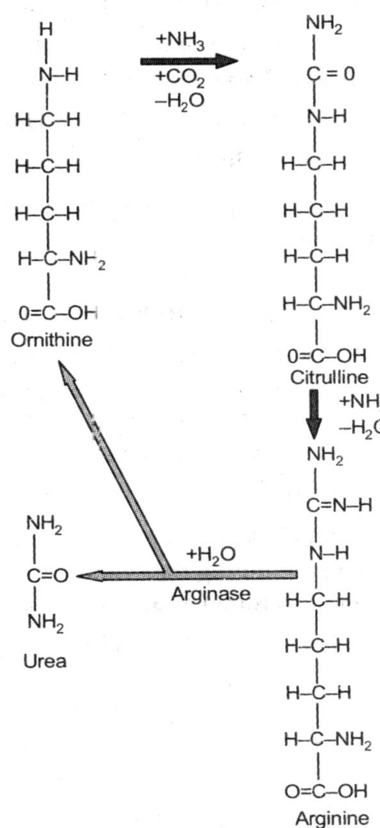

What can go wrong?

There are several inherited diseases of the urea cycle caused by mutations in genes encoding one or another of the necessary enzymes.

The most common of these is an inherited deficiency of **ornithine transcarbamoylase**, an enzyme needed for the conversion of ornithine to citrulline. It results in elevated levels of ammonia that may be so high as to be life-threatening.

It is an X-linked disorder; therefore most commonly seen in males. It can be **cured** by a liver transplant.

It can also be **caused** by a liver transplant! In 1998, an Austrian woman was given a new liver from a male cadaver who — unknown to the surgeons — had a mutation in his single ornithine transcarbamoylase gene. The woman's blood level of ammonia shot up, and she died a few days later.

Uric acid

Humans also excrete a second nitrogenous waste, **uric acid**. It is the product of **nucleic acid**, not protein, metabolism. It is produced within peroxisomes.

Uric acid is only slightly soluble in water and easily precipitates out of solution forming needlelike crystals of sodium urate. These

- contribute to the formation of kidney stones;
- produce the excruciating pain of **gout** when deposited in the joints.

Curiously, our kidneys reclaim most of the uric acid filtered at the glomeruli. Why, if it can cause problems?

- Uric acid is a potent antioxidant and thus can protect cells from damage by reactive oxygen species (**ROS**).
- The concentration of uric acid is 100-times greater in the cytosol than in the extracellular fluid. So when lethally-damaged cells release their contents, crystals of uric acid form in the vicinity. These enhance the ability of nearby dendritic cells to "present" any antigens released at the same time to T cells leading to a stronger immune response.

So the risk of kidney stones and gout may be the price we pay for these protections.

Most mammals have an enzyme — **uricase** — for breaking uric acid down into a soluble product. However, during the evolution of great apes and humans, the gene encoding uricase became inactive. A predisposition to gout is our legacy.

Uric acid is the **chief** nitrogenous waste of

- insects
- lizards and snakes
- birds

(It is the whitish material that birds leave on statues.)

These animals convert the waste products of **protein** metabolism — as well as nucleic acid metabolism — into uric acid.

Because of its lcw solubility in water, these animals are able to eliminate waste nitrogen with little loss of water.

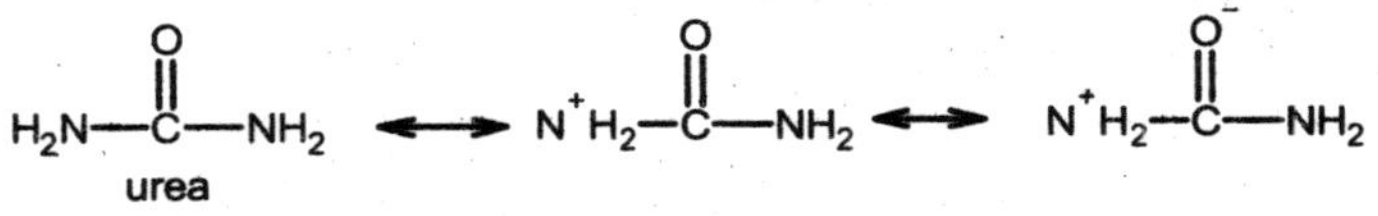

$$H_2N-\overset{O}{\overset{\|}{C}}-NH_2 \longleftrightarrow \overset{+}{N}H_2-\overset{O}{\overset{\|}{C}}-NH_2 \longleftrightarrow \overset{+}{N}H_2-\overset{O^-}{\overset{\|}{C}}-NH_2$$

urea

Chemical Properties

X-ray defragmentation

These studies shows that both C – N bonds in urea are the same length (1.33Å). Thus shorter than the C – N double bond (1.28Å).These values suggests the C – N bond in urea are identical and possess significant double bond character. That is, they confirm the hybrid nature of the urea molecules

Basicity of Urea

The weak basicity of urea

Urea is a weak base. The reason for this is that the loan pair of electrons on nitrogen's are involved in the formation of the delocalized π-molecular orbital. They are therefore not available for the information of a new N – H bond with portion. However, a portion can and does attack the negatively charged oxygen atom confirmed by NMR studies of salt of urea.

The positive ion formed when oxygen takes up a portion is al~o stabilized by resonance

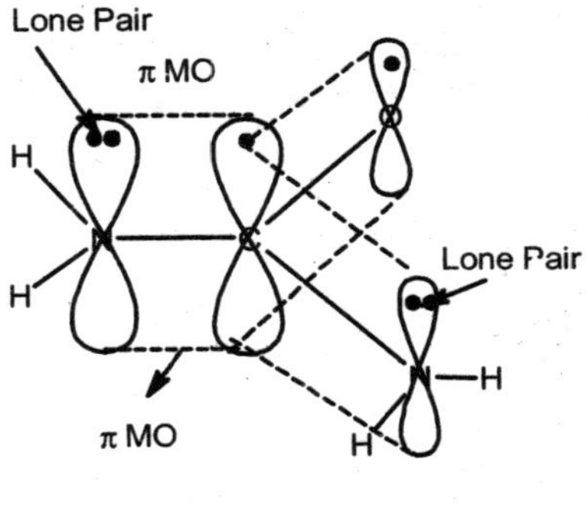

Lone Pair
π MO
H
H
Lone Pair
H
π MO
H

Formation of delocalized π Molecular orbital in urea molecule

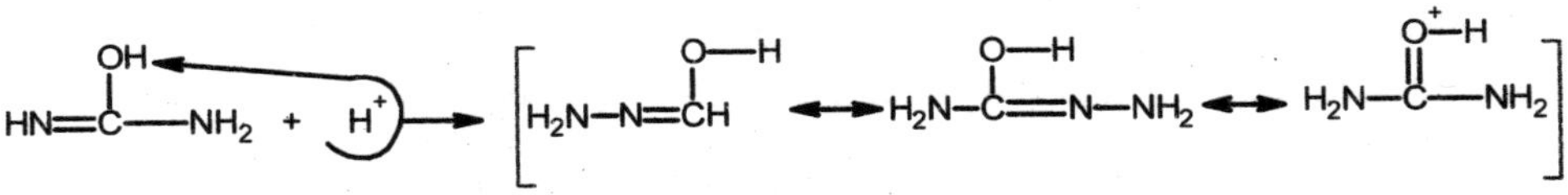

$$HN=\overset{OH}{\overset{|}{C}}-NH_2 + H^+ \longrightarrow \left[H_2N-N=\overset{O-H}{\overset{|}{C}H} \longleftrightarrow H_2N-\overset{O-H}{\overset{|}{C}}=N-NH_2 \longleftrightarrow H_2N-\overset{\overset{+}{O}-H}{\overset{\|}{C}}-NH_2 \right]$$

General Chemical Properties

Urea resembles amides in general chemical behaviours. Some of its important reactions are described below:-

(i) Action of Heat: When urea is heated to a temperature above its melting point (132°C) ammonia is formed and crystalline biuret is also produced.

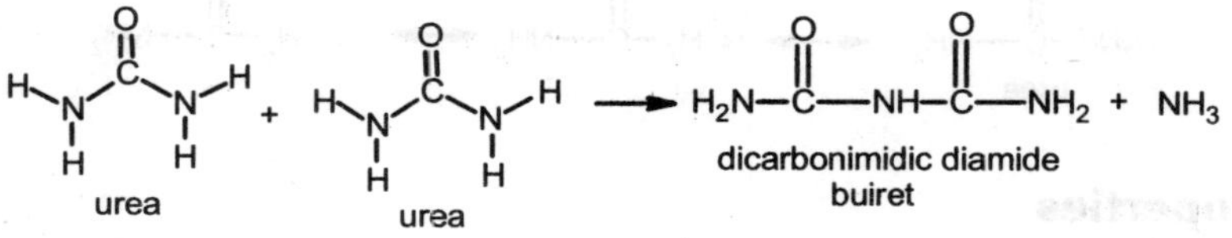

$$H_2N{-}CO{-}NH_2 \;(\text{urea}) + H_2N{-}CO{-}NH_2 \;(\text{urea}) \longrightarrow H_2N{-}CO{-}NH{-}CO{-}NH_2 + NH_3$$

dicarbonimidic diamide
buiret

If an alkaline solution of biuret is treated with copper sulphate solution a violet colour is produced. This proves the test for urea and also for protein peptides and other compounds containing the – CO – NH – linkage.

Mechanism

The mechanism of the above reaction involves two steps

Step (I) Conversion of Urea into Ammonia and isocyanic acid which is tautomeric with cyanic acid

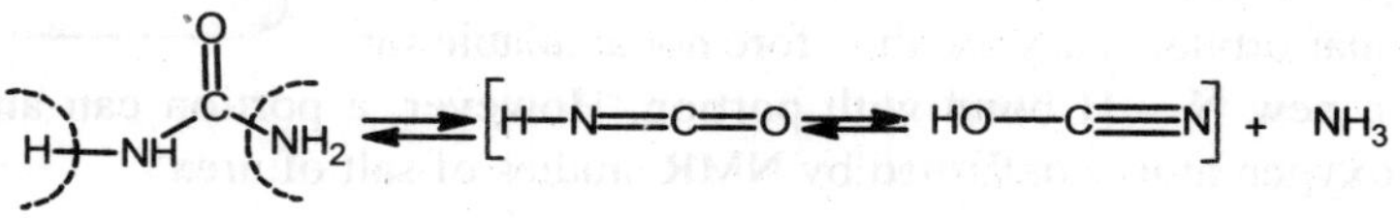

$$H{-}NH{-}CO{-}NH_2 \rightleftharpoons [H{-}N{=}C{=}O \rightleftharpoons HO{-}C{\equiv}N] + NH_3$$

Step (II) Reaction of Isocyanic Acid with Another Molecule of urea gives Biuret

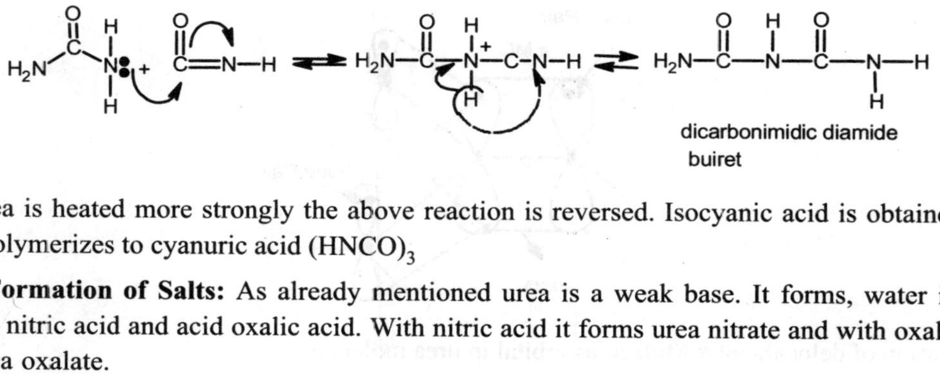

$$H_2N{-}CO{-}NH_2 + O{=}C{=}N{-}H \rightleftharpoons H_2N{-}CO{-}\overset{+}{N}H_2{-}C(O^-){=}N{-}H \rightleftharpoons H_2N{-}CO{-}NH{-}CO{-}NH{-}H$$

dicarbonimidic diamide
buiret

It urea is heated more strongly the above reaction is reversed. Isocyanic acid is obtained which rapidly polymerizes to cyanuric acid $(HNCO)_3$

(ii) Formation of Salts: As already mentioned urea is a weak base. It forms, water insoluble salts with nitric acid and acid oxalic acid. With nitric acid it forms urea nitrate and with oxalic acid it forms urea oxalate.

(iii) Hydrolysis: Urea is hydrolysed on heating on with dilute acids or dilute alkalies to give carbon to give carbon dioxide and ammonia. The hydrolysis can also be brought about by the enzyme urease. This enzyme is found in soil bacteria, water –melon seeds and in soya bean.

$$H_2N{-}CO{-}NH_2 + \begin{matrix} H_2O \\ H_2O \end{matrix} \xrightarrow[\text{or urease}]{H^+ \text{ or } OH^-} 2NH_3 + \left[O{=}C(OH)_2 \right] \longrightarrow 2NH_3 + H_2O$$

with ureane the yield of ammonia and carbon monoxide is quantitative and thus reaction for the acetate estimation of urea in solutions.

***(iv)* Reaction with Thionyl Chloride**:-Urea undergoes dehydration with Thionyl chloride ($SOCl_2$) to yield cynamide

$$H_2N{-}CO{-}NH_2 + SOCl_2 \longrightarrow H_2N{-}C{\equiv}N + SO_2 + 2HCl + H_2O$$

***(v)* Reaction with Nitrous Acid**:-Urea reacts with nitrous acid as ($NaNO_2$ + HCl) to give nitrogen, carbon dioxide and water.

$$\underset{\text{urea}}{H_2N{-}CO{-}NH_2} + \underset{\text{nitrous acid}}{\begin{matrix} O{=}N{-}OH \\ O{=}N{-}OH \end{matrix}} \longrightarrow 2N_2 + 2H_2O + \underset{\text{carbonic acid}}{\left[O{=}C(OH)_2 \right]} \longrightarrow 2N_2 + 3H_2O + CO_2$$

***(vi)* Reaction with Alkaline Solutions of Hypohalites**:-Urea reacts with an alkaline solution of sodium hypobromite (or NaOH + Br_2) to yield hydrazine which is immediately to nitrogen by sodium hydrpobromite.

$$\underset{\text{urea}}{H_2N{-}CO{-}NH_2} + NaOBr + 2NaOH \longrightarrow \underset{\text{hydrazine}}{H_2N{-}NH_2} + NaBr + Na_2CO_3 + H_2O$$

$$\underset{\text{hydrazine}}{H_2N{-}NH_2} + 2NaOBr \longrightarrow 2NaBr + N_2 + H_2O$$

The over all reaction is

$$\underset{\text{urea}}{H_2N{-}CO{-}NH_2} + 3NaOBr + 2NaOH \longrightarrow \underset{\text{hydrazine}}{H_2N{-}NH_2} + 2NaBr + N_2 + 2H_2O + Na_2CO_3$$

The reaction is used for the determination of urea in different substances, this is done by the measuring the volume of nitrogen evolved, when a test sample is introduced to a solution of sodium hypobromite.

(7) Reaction with Sulphuric Acid: Urea reacts with fuming sulphuric acid (oleum) to yield sulphamic acid

$$\underset{\text{urea}}{H_2N{-}CO{-}NH_2} + H_2SO_4 + SO_3 \longrightarrow \underset{\text{sulfamic acid}}{H_2N{-}SO_2{-}OH} + SO_2$$

Sulphamic acid is a colourless crystalline solid with a melting point 205°C. It is a strong acid can be used as a primary standardization of solutions of bases. Two of derivation fine many industrial applications. They are ammonium sulphamate and sodium cyclohexylsulphonate.

AMMONIUM SULPHAMATE

$NH_2SO_3NH_4$

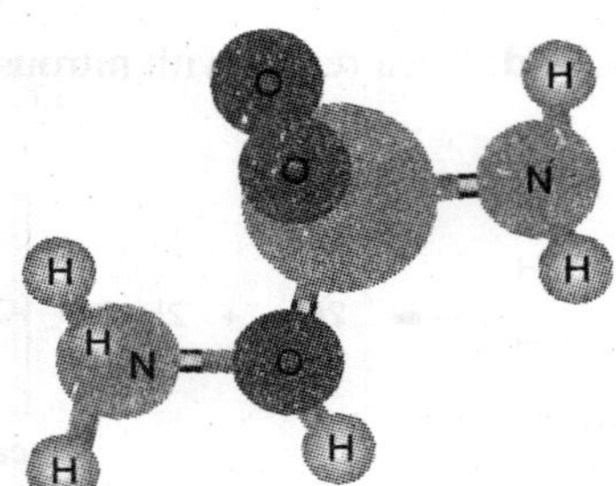

It is used as a flame proofing agent for paper and textiles, it does not cause stiffening of materials and is not removed from fabrics by dry cleaning. It is also effective weed killer and has been used fort he eradication of poison ivy and wild chery.

SODIUM CYCLOHEXYLSULPHONATE OR SUCARYL AND CARBAMATE ($C_6H_{11}NHSO_3NA$)

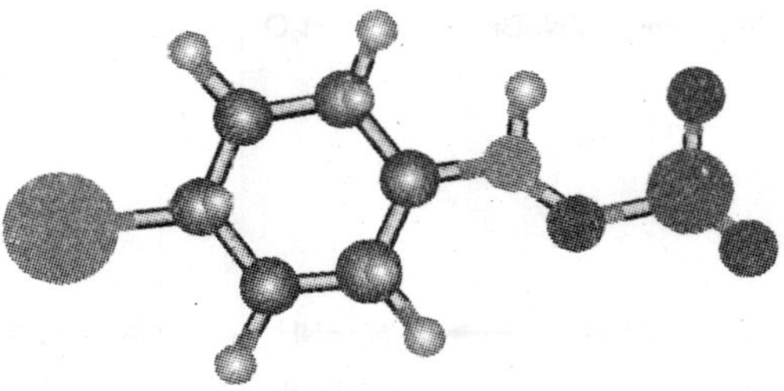

Introduction

Cyclamate: Michael Sveda, while a graduate student at the University of Illinois, discovered cyclamate by smoking a cigarette. While working on the synthesis of anti-pyretic (anti-fever) drugs in the laboratory in 1937, he put his cigarette down on the lab bench. When he put it back in his mouth, he discovered the sweet taste of cyclamate (unsanitary lab technique).

Cyclamate was initially marketed as tablets that were recommended for use as a tabletop sweetener for diabetics.(50 times more sweeter than sugar) In 1958, cyclamates were classified as GRAS (Generally Recognized as Safe). A mixture of cyclamate and saccharin, which had been found to have synergistic sweetening properties and improved taste, was subsequently marketed for use in special dietary foods. In the 1950's diet drinks were introduced using a cyclamate/saccharin blend. The market grew rapidly and soon accounted for about 30% of the soft drink sales.

In 1969, the result of a chronic toxicity study with a mixture of cyclamate and saccharin was interpreted as implicating cyclamate as a bladder carcinogen in rats. Cyclamate was removed from GRAS status and eventually in 1970 banned in the United States from use in foods, beverages and drugs, and is still currently banned. However, many other countries did not act on this incomplete data, and cyclamate continued to be used as a sweetener in those countries. Today over 55 countries have approved the use of cyclamate.

(8) Reaction with Formaldehyde:

(UREA FORMALDEHYDE RESIN)

Urea adds to formaldehyde in the presence of base or an acid to form methylourea and dimethylourea

$$H_2N{-}CO{-}NH_2 + HCHO \longrightarrow H_2N{-}CO{-}NH{-}CH_2{-}OH$$

1-(hydroxymethyl)urea
methylourea

$$H_2N{-}CO{-}NH{-}CH_2{-}OH + HCHO \longrightarrow HO{-}CH_2{-}NH{-}CO{-}NH{-}CH_2{-}OH$$

1-(hydroxymethyl)urea; formaldehyde; 1,3-bis(hydroxymethyl)urea dimethylourea

Under carefully controlled reaction conditions both methylourea and dimethylourea can be isolated. If the reaction is carried further three things happen

(*a*) The dimethylourea molecules can condenser with more urea molecules to form linear polymers.

$$H_2N{-}CO{-}NH_2 + H_2N{-}CO{-}NH{-}CH_2{-}OH + H_2N{-}CO{-}NH_2 + H_2N{-}CO{-}NH{-}CH_2{-}OH \xrightarrow{-n_1H_2O} H_2N{-}CO{-}NH{-}CH_2{-}NH{-}CO{-}NH{-}CH_2{-}NH{-}CO{-}NH{-}CH_2{-}NH{-}CO{-}NH_2$$

1-{[(aminocarbonyl)amino]methyl}-3-({[({[(aminocarbonyl)amino]methyl}amino)carbonyl]amino}methyl)urea

(*b*) The linear polymers can condense with more formaldehyde molecules to form cross-linked polymers

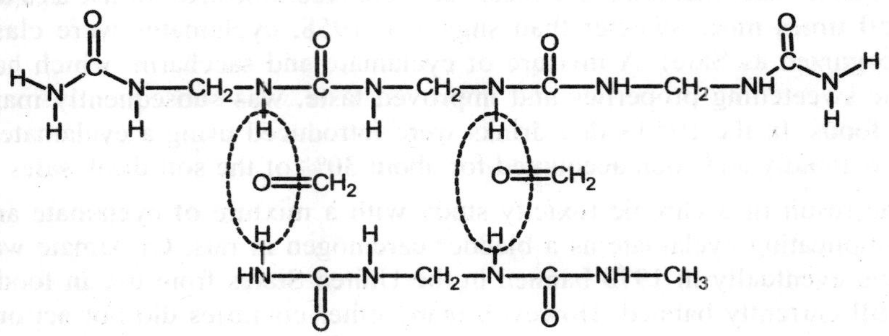

(*c*) The dimethylourea molecules can condense with each other to form linear polymers

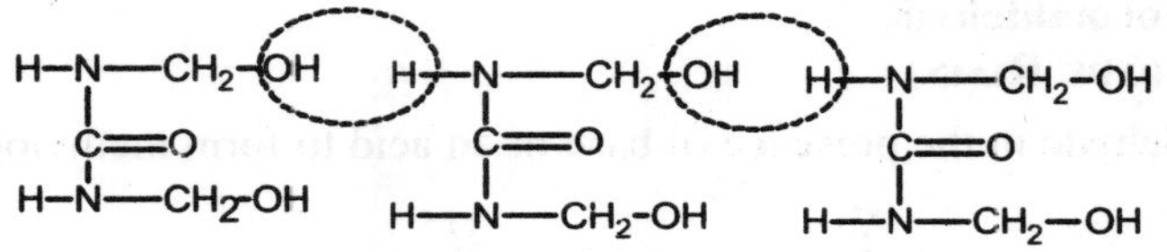

This gives rise to a complex mixture of linear and cross linked condensation polymers known as urea formaldehyde resin.

UREA FORMALDEHYDE RESIN

Introduction

The resin is the general term used technically for solid, hard to soft, organic, non-crystalline products having a broader or narrower molecular mass distribution. They normally have a melting or softening range, are brittle in the solid state. Resins are in general raw materials, for example for binders, curable molding compositions adhesives and coatings. Resin can be divided as natural resins and synthetic resins.

The most important synthetic resins are phenol-formaldehyde resins, urea-formaldehyde resins, melamine-formaldehyde resins, polyesters resins, silicone resins and ketone-aldehyde resins.

They are usually thermoset type polymers. Thermoset resins are usually prepared as pre-polymer and than in the second stage they are cross-linked by the effect of catalyst, heat and pressure. The possible reaction of urea with formaldehyde and formation of urea-formaldehyde resin are summarized in the following scheme.

Urea-formaldehyde resins are usually synthesized by condensation in aqueous, basic media. Depending on the ultimate application of the product, a 1.5-to 2-fold excess of formaldehyde is employed. Suitable catalysts are all basic compounds, provided they are water-soluble to some extent. The most widely used catalysts are the alkali hydroxides. The pH present of the reaction mixture

should not exceed 8-9 to prevent occurrence of the Cannizzaro reaction of formaldehyde. The pH must be kept constant by buffering or by repeated addition of base.

The reaction is 10-20 minutes at 50-60°C. Catalytic crosslinking is carried out with free acids (acid hardening). Large amount acid (pH=2) cause the crosslinking to occur at room temperature (cold cusing).

Experimental Procedures

Urea-formaldehyde Condensation

In a 500ml three-necked flask equipped with stirrer, thermometer, and reflux condenser 243 g of a 37 % aqueous formaldehyde solution (3 moles) are warmed to 50° C, and 120 g (2 moles) of urea

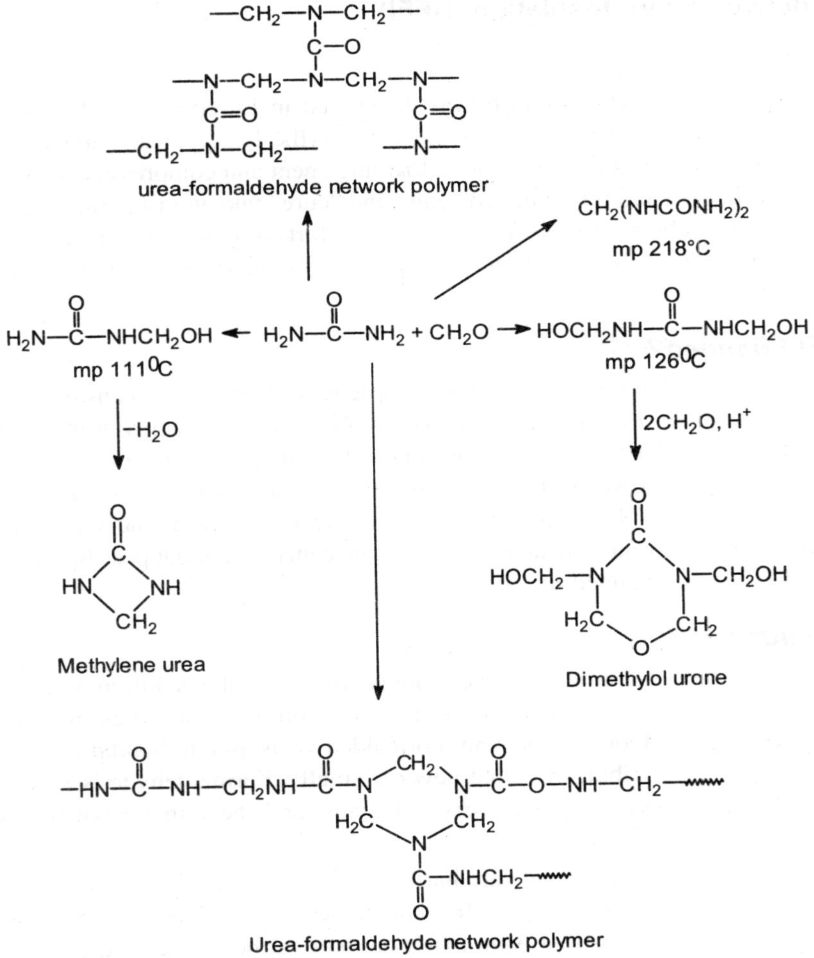

urea-formaldehyde network polymer

$CH_2(NHCONH_2)_2$
mp 218°C

$H_2N{-}CO{-}NHCH_2OH \leftarrow H_2N{-}CO{-}NH_2 + CH_2O \rightarrow HOCH_2NH{-}CO{-}NHCH_2OH$

mp 111°C   mp 126°C

$-H_2O$ $2CH_2O, H^+$

Methylene urea

Dimethylol urone

Urea-formaldehyde network polymer

are then dissolved in it. After addition of 10 ml of conc. ammonia the solution is heated to 85°C within about 20 minutes the reaction mixture becomes turbid, and the viscosity increases while the pH declines to a value of about 5. After a total of one hour the heat is removed, and a small sample is taken to check whether the condensate is still water soluble.

Preparation of a Urea-formaldehyde Resin

To a 500 ml, three-necked flask equipped with a stirrer and after condenser is charged 130 g of 37 % aqueous formaldehyde (1.6 moles) which is brought to a pH about 7.5 by the addition of 10 % sodium hydroxide solution. Then, 60 g (1 mole) urea is added and the mixture gently refluxed and stirred for 2 hr. The mixture is than concentrated to 30 % solids by distillation of 40 ml of water under water aspirator pressure. The resulting syrup, after acidifying with acetic acid, can be heated further at 100°C for several hours to effect gelation.

Urea-Formaldehyde Foam Insulation (UFFI)

What Is UFFI?

Urea-formaldehyde foam insulation (UFFI) was developed in Europe in the 1950s as an improved means of insulating difficult-to-reach cavities in house walls. It is typically made at a construction site from a mixture of urea-formaldehyde resin, a foaming agent and compressed air. When the mixture is injected into the wall, urea and formaldehyde unite and "cure" into an insulating foam plastic. During the 1970s, when concerns about energy efficiency led to efforts to improve home insulation in Canada, UFFI became an important insulation product for existing houses. Most installations occurred between 1977 and its ban in Canada in 1980.

Why Was UFFI Banned?

In the insulating process, a slight excess of formaldehyde was often added to ensure complete "curing" with the urea to produce the urea-formaldehyde foam. That excess was given off during the curing, almost entirely within a day or two of injection. Properly installed, UFFI might not have resulted in any problem. Unfortunately, however, UFFI was sometimes improperly installed or used in locations where it should not have been. Enough complaints were received, particularly from people living in small, well-sealed homes, that Canadian authorities became concerned about possible health implications. The further use of UFFI was banned in 1980.

What Is Formaldehyde?

Formaldehyde is a pungent, colourless gas commonly used in water solution as a preservative and disinfectant. It is also a basis for major plastics, including durable adhesives. It occurs naturally in the human body and in the outdoor environment. Formaldehyde is used to bond plywood, particleboard, carpets and fabrics, and it contributes to "that new car smell." Formaldehyde is also a by product of combustion; it is found in tobacco smoke, vehicle exhaust and the fumes from furnaces, fireplaces and wood stoves.

While small amounts of formaldehyde are harmless, it is an irritating and toxic gas in significant concentrations. Symptoms of overexposure to formaldehyde include irritation to eyes, nose and throat; persistent cough and respiratory distress; skin irritation; nausea; headache; and dizziness.

Health Canada has determined that 0.1 parts per million (ppm) is a safe level of formaldehyde in the home. Sensitivity to this level may vary based on individual age and health.

Should You Be Concerned About UFFI Today?

Tests show that UFFI is not a source of over-exposure to formaldehyde after the initial curing and release of excess gas. As it was last installed in 1980, it would certainly not be causing excess indoor formaldehyde today. Houses with UFFI show no higher formaldehyde levels than those without it. However, if UFFI comes in contact with water or moisture, it could begin to break down. Wet or deteriorating UFFI should be removed by a specialist and the source of the moisture problem should be repaired. In new or other well-sealed houses, significant indoor formaldehyde levels may still occur when new carpets or wood composite materials, such as plywood, particleboard and wafer board, are used in home construction, cabinetry and furnishings. These are the most likely sources of high formaldehyde levels in the home today.

If you are asked for a UFFI declaration

Since 1993, a UFFI declaration has not been required for mortgage insurance under the National Housing Act. However, a UFFI declaration may still be requested as part of a real estate listing or an agreement of purchase and sale. Even though UFFI should not be a cause for concern, you may, depending on where you live in Canada, be asked to declare whether or not it is in your home. Some home inspectors will have the training or experience to identify UFFI. You can make a physical check of the home yourself. Look for a series of small patched holes, 1.2 to 2 cm (1/2 to 3/4 in.) across, at regular intervals on exterior or interior walls. Foam may be obvious where floor joists meet the exterior walls of the basement or around electrical outlets or switch plates. These indicators do not necessarily mean that UFFI is present, but they may alert you to the possibility.

(9) Reaction with Ethylene Oxide: Urea reacts with ethylene oxide to form bis (hydroxyethyl urea)

$$\underset{\text{urea}}{H_2N{-}CO{-}NH_2} + \underset{\text{oxirane}}{H_2C{-}CH_2{-}O \text{ (ring)}} \longrightarrow \underset{\text{1,3-bis(2-hydroxyethyl)urea}}{O{=}C(NH{-}CH_2{-}CH_2{-}OH)_2}$$

Bis (hydroxyethyl) urea is an important intermediate in the manufacture of non-ionic detergents

(10) Reaction with Hydrazine: Urea reacts with hydrazine in the presence of aqueous amyl alcohol as solvent at 100°C to yield semicarbazide.

$$\underset{\text{urea}}{H_2N{-}CO{-}NH_2} + \underset{\text{urea}}{H_2N{-}CO{-}NH_2} \longrightarrow \underset{\text{hydrazinecarboxamide}}{H_2N{-}\overset{O}{\overset{\|}{C}}{-}\overset{H}{\overset{|}{N}}{-}NH_2} + NH_3$$

Semicarbazide:-is an important reagent for the identification of aldehydes and ketones

(11) Reaction with Acid Chloride:-Urea reacts with acid halides to form acyl derivatives or ureids. For example, acetyl chloride reacts with urea to give N-acyl urea

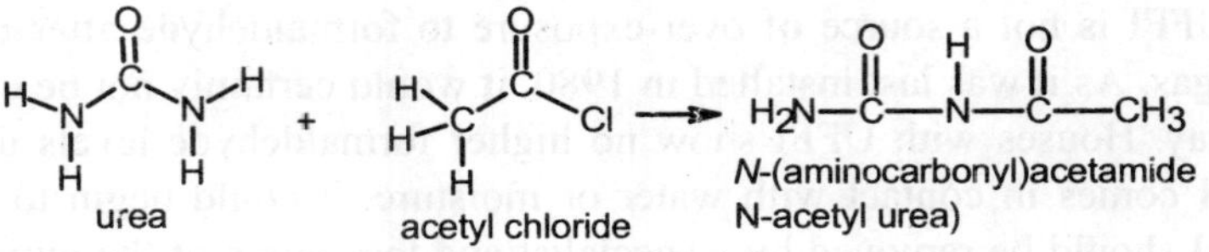

urea + acetyl chloride → *N*-(aminocarbonyl)acetamide N-acetyl urea)

(12) Reaction with Malonic acid and Malonic Ester:-Urea reacts with Malonic acid in the presence of phosphorous oxychloride ($POCl_3$) to yield a tautomeric mixture of barbituric acid (malonylurea). Barbituric acid can also be obtained by heating urea with Malonic ester in the presence of sodium ethoxide.

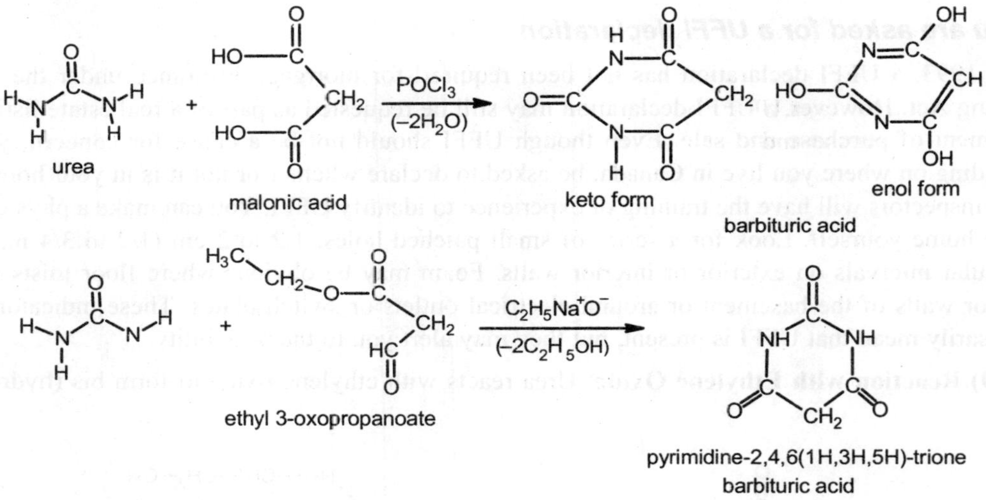

urea + malonic acid $\xrightarrow[(-2H_2O)]{POCl_3}$ keto form ⇌ enol form

barbituric acid

urea + ethyl 3-oxopropanoate $\xrightarrow[(-2C_2H_5OH)]{C_2H_5Na^+O^-}$ pyrimidine-2,4,6(1H,3H,5H)-trione

barbituric acid

BARBITURIC ACID

Formula: $C_4H_4N_2O_3$.

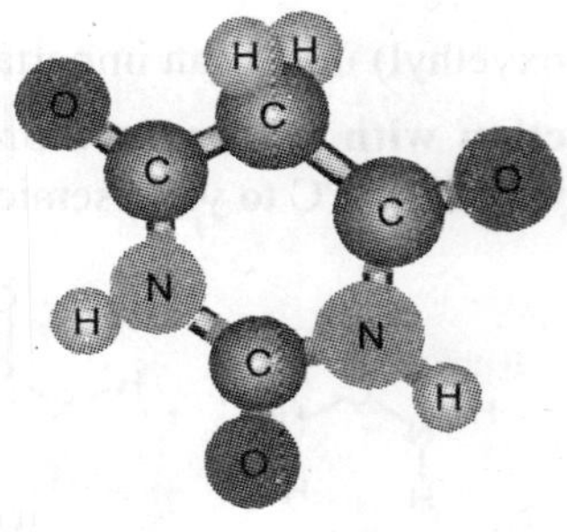

Introduction

Barbituric acid or malonylurea or 4-hydroxyuracyl is an organic compound based on a pyrimidine heterocyclic skeleton. It is an odorless powder soluble in hot water. Barbituric acid is the parent compound of a large class of barbiturates that have central nervous system depressant properties although barbituric acid itself is not pharmacologically active. The compound was discovered by the German chemist Adolf von Baeyer in 1864 by combining urea and malonic acid in a condensation reaction. Malonic acid has since been replaced by diethyl malonate.

$$\text{HOOC-CH}_2\text{-COOH} + (H_2N)_2C{=}O \longrightarrow \text{barbituric acid} + 2H_2O$$

General Description & Applications

Barbituric acid, chemically 2,4,6-trioxohexahydropyrimidine, a cyclic amide used as the parent compound to produce barbiturates that act as central nervous system depressants. Barbituric acid itself does not give sedative and hypnotic effects but the substituted derivatives with alkyl or aryl group at position 5 provide effects. Thiobarbituric acid, replaced oxygen atom of the urea component by sulfur, the parent compound of the thiobarbiturates which resembles the barbiturates in its effects. Barbiturates are drugs that acts as sedative-hypnotic agents. The short-acting barbiturates such as thiopental are used as intravenous anesthetics. The long-acting barbiturate such as phenobarbital is an anticonvulsant used in the treatment of epilepsy. They are used for the suppression of anxiety, the induction of sleep, and the control of seizures. Commercially available barbiturates are;

- Amobarbital
- Aprobarbital
- Butabarbital
- Butalbital
- Hexobarbital
- Mephobarbital
- Pentobarbital
- Phenobarbital
- Secobarbital
- Talbutal
- Thiobarbital
- Thiopental

Physical Properties

Physical State	white to off-white crystalline powder
Melting Point	250-255 C (Decomposws)
Boiling Point	260 °C
Specific Gravity	
Solubility In Water	142 g/l at 20 C
pH	2-3 (50 g/l Aq. solution)
Refractive Index	
Stability	Stable under ordinary conditions. Decomposed if heated

The table below shows the accepted barbiturates in the use of sedatives

Accepted barbituric acid derivatives

Formula	Chemical Name	Trate Name	Melting Point
$C_8H_{12}N_2O_3$	5, 5 diethyl barbituric acid	Barbital	147°C
$C_9H_{14}N_2O_3$	5, ethyl 5 iso propyl barbituric acid	Probarbital (ipral)	205°C
$C_{10}H_{12}N_2O_3$	5, 5 diallyl barbituric acid	Dial	146°C
$C_{10}H_{13}BrN_2O_3$	5 isopropyl – 5 – (β-bromoallyl) barbituric acid	Nostal	179°C
$C_{10}H_{14}N_2O_3$	5 – ethyl – 5 – n – butyl barbituric acid	Neonal	127°C
$C_{11}H_{15}BrN_2O_3$	5 – sec – 5 – (β-bromo barbituric acid)	Pernoston	133°C
$C_{11}H_{16}N_2O_3$	5 – ally – 5 – isobutyl barbituric acid)	Sadoptal	139°C
$C_{11}H_{16}N_2O_3$	5 – allyl – 5 – isobutyl barbituric acid	Vinibarbital (delivinal)	160°C
$C_{11}H_{16}N_2O_3Na$	5 – ethyl – 5 – (1 – methyl butyl sodium barbiturate)	**Pentobarbital sodium salt neobarbital**	130°C
$C_{11}H_{18}N_2O_3$	Ethyl – 5 – isoamyl barbituric acid	Amytal	155°C
$C_{12}H_{12}N_2O_3$	5 – ethyl 5 pentyl barbituric acid	Phenol barbital, Luminal	178°C
$C_{12}H_{16}N_2O_3$	5 – ethyl – 5 phenyl cyclohexyl barbituric acid	Phamodorn	179°C

Chemical Properties

Barbituric acid, itself, is not pharmacologically active, but chemists immediately began making a great variety of derivatives for potential use as drugs. The medical value of these substances was not realised, however, until 1903 when two other German chemists, Emil Fischer and Joseph von Mering, discovered that one of these compounds, diethylbarbituric acid, was very effective in putting dogs to sleep. It is said that von Mering proposed that the new substance be called '*Veronal*', because the most peaceful place he knew on Earth was the Italian city of Verona.

When this discovery was announced, chemists rapidly produced many new medically active barbiturates, including the sleeping pill phenobarbital (sold under the tradename *Luminal*), and pentobarbitone (*Nembutal*) which is often used to induce sleepiness before full pre-operative anesthesia. By 1955 the production of barbiturates in the US alone amounted to enough to provide 10 million adults with a sleeping pill every night of the year!

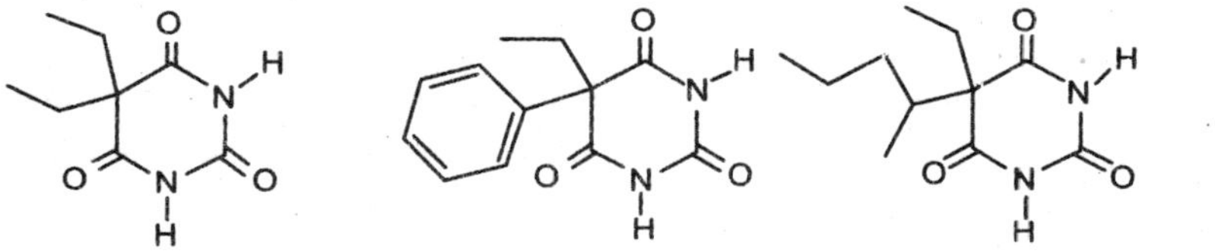

Barbital (Veronal) Phenobarbital (Luminal) Pentobarbital (Nembutal)

Various barbiturates. The differences from barbituric acid are shown in red.

We still do not know how barbiturates work, although it is believed that they act by depressing the central nervous system and inhibiting certain nerve response centres. We do know, however, that their physiological effects are very structure dependent. The basic structure is a ring composed of 4 carbon atoms and 2 nitrogens. Certain sidechains added to the ring can dramatically alter the potency of the drug-indeed, in some cases the addition of a single carbon atom transforms an inactive form of the compound into an active one.

Nothing but the Truth...

One class of barbiturate derivatives is called **thiobarbiturates**, which contain a sulphur molecule in place of one of the oxygens. The most important member of this class is **sodium pentothal**, which is used as an intravenous anesthetic from which recovery is rapid and which has the advantage of having very little or no side effects.

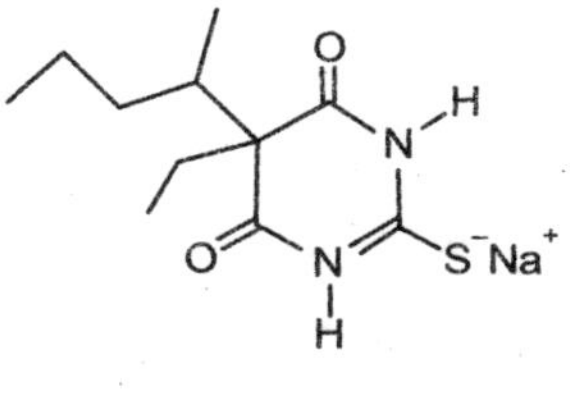

Sodium pentothal-**the 'truth serum'. The changes from barbituric acid are shown in red.**

Barbiturates, particularly sodium pentothal, also have interesting applications in the field of psychology, since they can provide access to levels of consciousness not normally attainable except in dreams or trances. This allowed them to be used for therapy of traumatised GIs in WW2 who could relive and verbalise battle experiences which were buried beyond voluntary recollection. The inhibition-relieving action of these drugs has also been employed by the police (and other agencies), allowing prisoners that have been placed into a pentothal-induced dreamy trance to volunteer information they otherwise would have suppressed. This has earned sodium pentothal the nickname of a 'truth serum', even though in actuality it is neither a serum, nor does it guarantee the truth.

A Habit Forming Danger

Barbiturates present problems because patients develop tolerance for them and even higher doses are then required to produce the same effect. They are also habit forming, and the fatal dose is not much more than the normal therapeutic dose. Thus they are truly dangerous drugs and are widely used by suicides. They also present the danger of automatism, that is the patient may take a dose, leave the bottle on the bedside table, and then, when half asleep, forget that the dose has already been taken and take some more. This is one of the theories for the death of Marilyn Monroe. Another major problem is that barbiturates and alcohol are eliminated by the same system of enzymes in the liver. This means that consumption of the 2 chemicals at the same time can lead to a build up of the toxic drug in the liver, resulting in liver failure and death.

(13) Formation of Inclusion Complexes: It has been noticed that certain compounds from liquid media have the ability to trap other molecules in the crystal lattice. The crystalline compound which result are often referred to as the inclusion complexes.

Urea forms such inclusion compounds with a number of straight chain aliphatic compounds that form inclusion complex are hydrocarbon, alkyl halides, alcohols, ketones, acids and esters. The procedure for making such compounds consists of mixing a saturated methenol solution of urea with saturated solution of other compounds. A crystalline precipitate result from which the components can be separated by extracting urea with water or the trapped compound with suitable organic solvents. The process can be used for removing unbranched chain compounds form branched compounds.

Uses of Urea

General Uses

Pharmaceutical

Urea and malonic acid react to form barbituric acid.

$$H_2NCONH_2 + HOOCCH_2COOH \xrightarrow{-2H_2O} \text{Barbituric acid}$$

Barbituric acid

This was discovered by **Adolf Bayer** in *1864*. But the barbiturates were not exploited as hypnotics until the early 1900's.Urea is also used in the production of various acylureas and urethanes for use as sedatives and hypnotics.

Resins

Melamine is formed by the dehydration of urea.

$$6H_2NCNH_2 \text{ (urea, C=O)} \longrightarrow \text{melamine} + 6NH_3 + 3CO_2$$

melamine

Almost 100 years elapsed between **Liebig's** discovery in *1834* and a commercial process being developed. Melamine is primarily used in the production of melamine-formaldehyde resins which have much greater hardness and stain resistance than urea-formaldehyde resins. Both melamine-formaldehyde and urea-formaldehyde have very varied uses including adhesives, laminates, moulding compounds, coatings and textile finishes.

Industrial Uses

Urea has the ability to form 'loose compounds' with many organic compounds. The organic compounds are held in channels formed by interpenetrating helices comprising of hydrogen bonded urea molecules. This behaviour can be used to separate mixtures and has been used in the production of aviation fuel and lubricating oils. As the helices are interconnected all helices in a crystal must have the same 'handedness'. This is determined when the crystal is nucleated and can thus be forced by seeding. This property has been used to separate racemic mixtures.

Agricultural Uses

Fertilizer Urea

Urea, a white crystalline solid containing 46% nitrogen, is widely used in the agricultural industry as an animal feed additive and fertilizer Here we discuss it only as a nitrogen fertilizer.

Physical Forms of Urea

Commercially, fertilizer urea can be purchased as prills or as a granulated material. In the past, it was usually produced by dropping liquid urea from a "prilling tower" while drying the product. The prills formed a smaller and softer substance than other materials commonly used in fertilizer blends. Today, though, considerable urea is manufactured as granules. Granules are larger, harder, and more resistant to moisture. As a result, granulated urea has become a more suitable material for fertilizer blends.

Advantages of Fertilizer Urea

- Urea can be applied to soil as a solid or solution or to certain crops as a foliar spray.
- Urea usage involves little or no fire or explosion hazard.

- Urea's high analysis, 46% N, helps reduce handling, storage and transportation costs over other dry N forms.
- Urea manufacture releases few pollutants to the environment.
- Urea, when properly applied, results in crop yield increases equal to other forms of nitrogen.

Incorporate Urea for Best Use

Nitrogen from urea can be lost to the atmosphere if fertilizer urea remains on the soil surface for extended periods of time during warm weather. The key to the most efficient use of urea is to incorporate it into the soil during a tillage operation. It may also be blended into the soil with irrigation water. A rainfall of as little as 0.25 inches is sufficient to blend urea into the soil to a depth at which ammonia losses will not occur.

Urea Losses to the Air

Urea breakdown begins as soon as it is applied to the soil. If the soil is totally dry, no reaction happens. But with the enzyme urease, plus any small amount of soil moisture, urea normally hydrolizes and converts to ammonium and carbon dioxide. This can occur in 2 to 4 days and happens quicker on high pH soils. Unless it rains, urea must be incorporated during this time to avoid ammonia loss. Losses might be quite low in the spring if the soil temperature is cold. The chemical reaction is as follows:

$$\underset{\text{(urea)}}{CO(NH_2)_2} + H_2O + \text{urease} \rightarrow 2NH_3 + CO_2$$

The problem is the NH_3, because it's a gas, but if incorporated the NH_3, acts the same as incorporated anhydrous ammonia. Also, half of 28% liquid N is urea and the same thing happens with this half as with regular urea.

Urea Losses Related to Soil Temperature and pH

The volatility of urea depends to a great extent on soil temperature and soil pH. Tables 1 and 2 show that after a few days warm temperatures or high pH would cause losses.

Table 13.1. Per cent of surface-added urea volatilized as ammonia at different temperatures and days on the surface

Days	Temperature (F) 45 degrees	60 degrees	75 degrees	90 degrees
	(% of added N volatilized)			
0	0	0	0	0
2	0	0	1	2
4	2	2	4	5
6	5	6	7	10
8	5	7	12	19
10	6	10	14	20

Data abstracted from curves in SSSP 24, pages 87-90, 1960. Urea was added on a silt loam soil at 100 lbs N.

Table 13.2. Per cent of surface-added urea volatilized as ammonia at various soil pH levels and days on the surface.

Days	Soil pH 5.0	5.5	6.0	6.5	7.0	7.5
	(% of added N volatilized)					
0	0	0	0	0	0	0
2	0	0	0	0	1	5
4	1	2	5	10	18	20
6	4	5	7	11	23	30
8	8	9	12	18	30	33
10	8	10	13	22	40	44

Data from SSSP 24, pages 87-90, 1960. Urea added on silt loam soil at 100 lb. N.

Fall Application Comparisons

Urea can be readily nitrified—that is, converted to nitrate (NO_3)— even when applied late in the fall, and can be quite susceptible to denitrification or leaching the following spring. Anhydrous ammonia (AA) applied in the fall does not nitrify as quickly, due to the stunting of microorganisms in the AA application band. A two-year study conducted at Waseca compared late-October applications of both AA and urea for continuous corn (Table 13.3). These data show a 6 bu/A advantage for AA over urea when applied in the fall without a nitrification inhibitor. But when N-Serve was added, a 16 bu/A advantage was shown with AA. This indicates that the inhibitor has a better degree of contact with the AA mix than is possible with urea.

Table 13.3. Corn yield as influenced by N source, time of application, and nitrification inhibitor at Waseca

N Source*	1981 - 82 Avg. Fall	Spring
	- - - Yield (bu/A) - - -	
AA (82% N)	162	168
AA + N-serve	170	172
Urea (45% N)	156	164
Urea + N-serve	154	162
*150 lb N/A	Malzer & Randall	

Studies with continuous use of urea have been conducted at Lamberton since 1960. Corn yields over a 24-year period averaged 5 to 6 bushels per acre higher with spring application of urea compared to the fall plowed-down application (Table 13.4).

Table 13.4. Corn yield as influenced by fall and spring applications of urea at Lamberton.

Time/method of Urea Application*	24-yearAvg. Yield
	bu/A
Fall, plowed-down	97
Spring, top-dressed	102
Spring, side-dressed	103

*80 lb N/A

Urea applied in the fall has generally not been as effective as AA. This is especially true in south-central Minnesota and Iowa. When fall soil-moisture conditions are dry, there is little difference between AA and urea. But when soil-moisture content is high, fall applications of urea haven't performed as well as AA. Applications of urea-ammonium nitrate (UAN) in the fall are not recommended due to rapid nitrification and a high potential for loss.

Soil Application and Placement of Urea

If properly applied, urea and fertilizers containing urea are excellent sources of nitrogen for crop production. After application to the soil, urea undergoes chemical changes and ammonium (NH_4 +) ions form. Soil moisture determines how rapidly this conversion takes place. When a urea particle dissolves, the area around it becomes a zone of high pH and ammonia concentration. This zone can be quite toxic for a few hours. Seed and seedling roots within this zone can be killed by the free ammonia that has formed. Fortunately, this toxic zone becomes neutralized in most soils as the ammonia converts to ammonium. Usually it's just a few days before plants can effectively use the nitrogen. Although urea imparts an alkaline reaction when first applied to the soil, the net effect is to produce an acid reaction. Urea or materials containing urea should, in general, be broadcast and immediately incorporated into the soil. Urea-based fertilizer applied in a band should be separated from the seed by at least two inches of soil. **Under no circumstances should urea or urea-based fertilizer be seed-placed with corn.** With small grains, 10 lb. of nitrogen as urea can generally be applied with the grain drill at seeding time even under dry conditions. Under good moisture conditions, 20 lb. of nitrogen as urea can be applied with the grain drill. Research results at North Dakota State University indicate that under dry conditions, urea at the rate of more than 20 lb. nitrogen per acre, applied with a grain drill in a 6-inch spacing, can reduce wheat stands more than 50% (Table 13.5) Research at the University of Wisconsin indicates that seed-placed urea with corn, even at low rates of nitrogen, is very toxic to the seed and greatly reduces yields (Table 13.6). When urea was side-placed as a 2" x 2" starter, however, little if any damage was noted (Table 13.7). In Minnesota, good crop production usually requires an application of more than 20 lb. of nitrogen per acre. Farmers can avoid damage from urea by broadcasting most of the urea nitrogen fertilizer ahead of seeding. Data in Table 13.8 indicate that urea broadcast prior to seeding is equal to or more effective than similar ammonium nitrate treatments.

Table 13.5. Seed-placed ammonium nitrate (AN) and urea comparisons on seedling damage to spring wheat under limited moisture conditions. North Dakota, 1975.

Treatments N(lb./A)	N Source	Seedlings per 40 ft. of Row Location		
		Absaraka	Williston	Casselton
0	-	600	270	760
20	AN	570	220	600
30	AN	590	240	690
40	AN	590	260	660
20	Urea	400	200	550
30	Urea	280	110	430
40	Urea	220	70	220

Table 13.6. Effect of urea and ammonium nitrate placed with seed on corn grain yield. Wisconsin, 1973

lb. N/A*	Yield, bu/A	
	Urea	Ammonium Nitrate
0	137	137
5	60	142
10	36	143
20	33	92

*Sufficient N broadcast prior to planting.
Source: Liegel & Walsh Plainfield Sand, Hancock, Wisconsin

Table 13.7. Effect of urea and ammonium nitrate side-placed on corn grain yield. Wisconsin, 1973.

lb. N/A*	Yield, bu/A	
	Urea	Ammonium Nitrate
0	142	142
25	145	145
50	146	146
100	150	141

*Sufficient N broadcast prior to planting.
Source: Liegel & Walsh Plainfield Sand, Hancock, Wisconsin.

Table 13.8. Effect of source and placement of urea and ammonium nitrate (AN) on corn yields. Lamberton Minnesota Experiment Station, 1960-84.

lb. N/A	Treatment	Source	Av. Yield bu/A
0	—		62
40	Plow-down—fall	AN	79
40	Plow-down—fall	Urea	86
40	Surface—fall	AN	82
40	Surface—fall	Urea	85
80	Plow-down—fall	AN	98
80	Plow-down—fall	Urea	97
160	Plow-down—fall	AN	104
160	Plow-down—fall	Urea	105
40	Topbress—spring	AN	89
40	Topdress—spring	Urea	88
80	Topdress—spring	AN	100
80	Topdress—spring	Urea	102

Source: MacGregor, Malzer and Nelson, University of Minnesota

Spreading of Urea

Urea can be bulk-spread, either alone or blended with most other fertilizers. It is recommended that the spreading width not exceed 50 feet when combined with other fertilizer materials. Urea often has a lower density than other fertilizers with which it is blended. This lack of "weight" produces a shorter "distance-of-throw" when the fertilizer is applied with spinner-type equipment. In extreme cases this will result in uneven crop growth and "wavy" or "streaky" fields.

Blending Urea with Other Fertilizers

Urea and fertilizers containing urea can be blended quite readily with monoammonium phosphate (11-52-0) or diammonium phosphate (18-46-0). Urea should not be blended with superphosphates unless applied shortly after mixing. Urea will react with superphosphates, releasing water molecules and resulting in a damp material which is difficult to store and apply.

Fluid Urea

Uniformity of particle size is important with dry solid urea, whether applied directly or in blended formulations. Some imported urea appears to be below U.S. quality standards on granule uniformity. Dissolving urea and marketing the liquid solution is an attempt to overcome this lack of uniformity and still take advantage of the favourable urea price. The liquid mix of urea and ammonium nitrate (UAN 28% N) has been on the market for a long time. The characteristics of this solution, however, are not the same as when urea alone is dissolved in water. A solution of 50% urea by weight results in 23-0-0 and has a salting-out temperature of 60 degrees F. In order to store and handle liquid urea during cooler temperatures, the nitrogen concentration must be lowered to reduce salting problems. There are several possible formulations that can be used for this, such as adding small amounts of ammonium nitrate, ammonium sulfate, or anhydrous ammonia. Research, particularly on liquid urea, is very limited. Generally, where dry urea functions successfully, the fluid urea should perform equally well and may have the advantage of better uniformity over some dry urea sources.

Biuret in Urea

Biuret in urea can cause agronomic problems if placed near the seed. or even if added preplant in bands where seeds will later be planted. Most U.S. manufacturers of urea keep biuret content low by keeping high temperatures to a minimum. Biuret content is typically around 0.3%, although urea of foreign origin appears to be higher. High heat is normal during the manufacture of urea. If heat exceeds 200 degrees F there is a slight conversion of urea to biuret, but this takes place only during the manufacturing process. No such conversion happens in storage or in the soil. Biuret converts to ammonia, but conversion is much slower than for urea. Since biuret remains in the soil for several weeks, the potential for seed damage continues beyond the brief period of conversion of urea to ammonia. The major damage of biuret is to germinating seeds. There is little damage through plant absorption, although some citrus crops have been affected.

Application of Urea to Growing Crops

Urea can be applied to sod crops, winter wheat. or other small grains. This application, however, should be made during cool seasons. During warm periods (60 degrees F or above), urea in contact with vegetative material will tend to give off ammonia. If urea must be applied on grass pastures in the summer, apply when there is a high probability of rainfall.

Foliar Application of Urea

Urea can also be applied as a foliar spray on some crops, such as potatoes, wheat, vegetables, and soybeans. Urea is highly watersoluble. At normal atmospheric temperatures, approximately 1 lb. Of urea can be dissolved in 1 lb. of water. Research data indicate that urea should contain no more than 0.25% biuret for use in foliar sprays. For many crops the quantity of nitrogen applied at one time should not exceed 20 lb. of nitrogen per acre.

Urea Storage

Urea is neither combustible nor explosive. It can be stored safely with no loss of quality under normal circumstances. Small or fast-moving augers should not be used to move granular urea. Urea particles are generally soft and abrasion can break the granules. Belt conveyers should be used whenever possible. Urea should not be stored with ammonium nitrate. These materials, when in contact, rapidly absorb water when the relative humidity is above 18%. Table 13.9 indicates the relative humidity at which urea and ammonium nitrate absorb moisture from the air.

Table 13.9. Critical relative humidities (CRH) of urea, ammonium nitrate, and a mixture of the two

Material	CRH%
Urea	75.2
Ammonium Nitrate (A.N.)	59.4
Urea + ammonium nitrate	18.1

Slow Release of Urea

Urea fertilizer can be coated with certain materials, such as sulfur, to reduce the rate at which the nitrogen becomes available to plants. Under certain conditions these slow-release materials result in more efficient use by growing plants. Urea in a slow-release form is popular for use on golf courses, parks, and other special lawn situations.

Urea Do's and Don'ts

- Store separately from ammonium nitrate.
- Do not use small, fast-moving augers to move the urea.
- Do not exceed a spreading width of 50 feet when urea is applied.
- Do not place in direct contact with corn seed.
- Keep rates of nitrogen applied together with small grain in drill to 10 1b. on dry soils, 20 lb. when soil is moist.
- Apply urea on sod crops when atmospheric temperature is below 60 degrees F.
- When urea is broadcast on soils of high pH (above 7.5), the material should be incorporated into the soil as soon as possible.

Estimation of Urea

The estimation of urea in any sample can be done by the following method

Hypobromite Method: In this method urea id decomposed by alkaline hypobromite solution (see reaction 6 of urea) and the volume of the nitrogen evolved is measured. The estimation is carried out in an apparatus called ureometer see fig 65. The large central tube is filled with hypobromite solution (10 g of sodium hydroxide + 25ml of water + 25 ml of Bromine). While the sample is placed in the side tube.

By opening the tap 1 ml sample is carefully introduced into the main central tube (with the hypobromite solution) evolution of nitrogen occurs and the urea is directly estimated from the graduation.

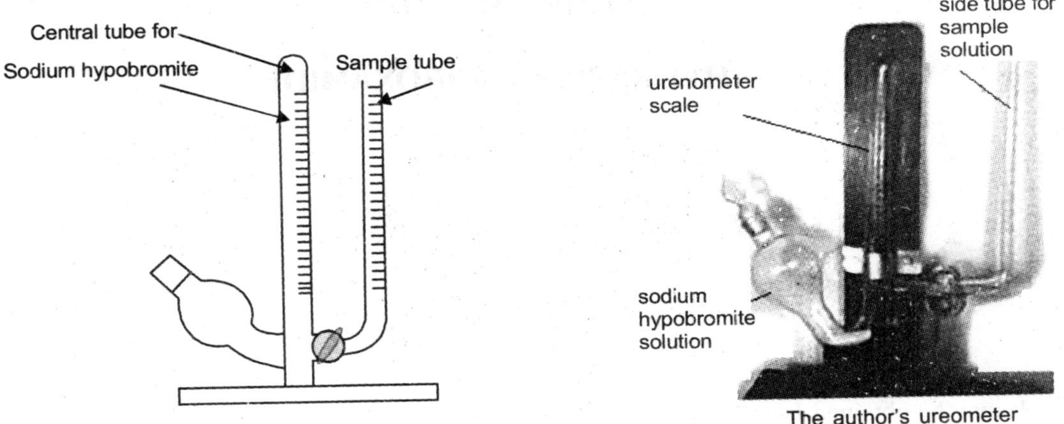

Fig. 13.9. Showing the Doremus Ureometer

Compounds of Urea

NITRO UREA

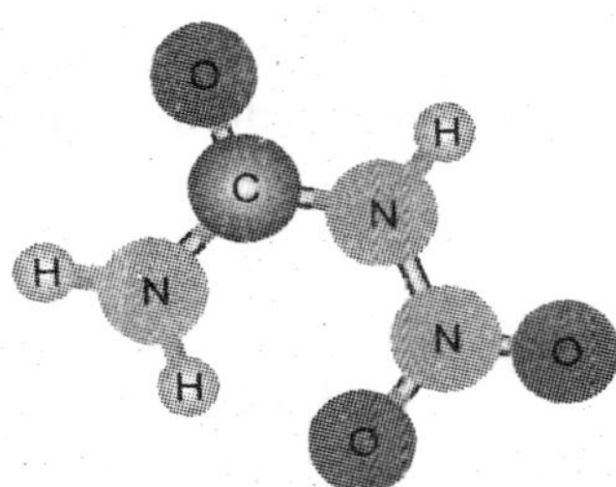

It may be obtained by treating urea nitrate with cold conc. sulphuric acid

$$\left(H_2N-\underset{|}{\overset{O-H}{CH}}-NH_2\right)^{+}ONO_2^{-} \xrightarrow[0°C]{H_2SO_4} H-\underset{H}{N}-\overset{O}{\overset{\|}{C}}-\overset{H}{N}-\overset{+}{N}(=O)-O^{-} + H_2O$$

urea nitrate — 1-nitrourea

Nitrourea is a crystalline solid with a melting point of 152°C. It is a soluble in cold water but unstable in hot water. Nitrourea is reduced by zinc and acetic or better electrolytically to yield semicarbazide.

$$H_2N-\overset{O}{\overset{\|}{C}}-NH-\overset{+}{N}(=O)-O^{-} + 6[H] \longrightarrow H_2N-\overset{O}{\overset{\|}{C}}-NH-NH_2 + 2H_2O$$

1-nitrourea — Hydrazinecarboxamide semicarbazide

SEMICARBAZIDE OR HYDRAZIDE CARBOXAMIDE

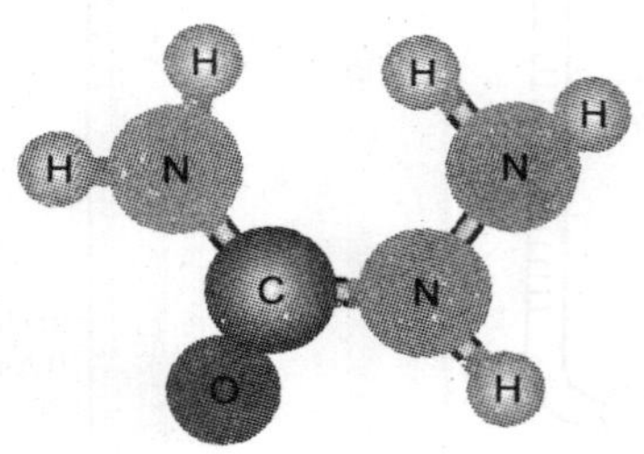

It may be obtained from Nitrourea or by heating urea with hydrazine in the presence of aqueous amyl alcohol at 100°C. It can also be obtained by treating hydrazine sulphate with potassium cyanate.

$$H_2N{-}NH{-}O{-}S(=O)_2{-}OH + 2KOCN \longrightarrow 2H_2N{-}C(=O){-}NH{-}NH_2 + K_2SO_4$$

(sulfooxy)hydrazine — hydrazinecarboxamide

Semicarbazide is a crystalline solid with a melting point of 96°C

Physical Properties

Molecular Formula	$H_4N_2O_4S$
Formula Weight	128.10776
Composition	H(3.15%) N(21.87%) O(49.96%) S(25.03%)
Molar Refractivity	21.40 ± 0.4 cm^3
Molar Volume	65.9 ± 3.0 cm^3
Parachor	208.7 ± 4.0 cm^3
Index of Refraction	1.562 ± 0.02
Surface Tension	100.4 ± 3.0 dyne/cm
Density	1.943 ± 0.06 g/cm^3
Dielectric Constant	Not available
Polarizability	8.48 ± 0.5 $10^{-24}cm^3$
Monoisotopic Mass	127.989177 Da
Nominal Mass	128 Da
Average Mass	128.106956 Da

It is basic in nature and forms crystalline salts with strong acids. Its hydrochloride is frequently used for the identification of aldehydes and ketones, with it forms sparingly soluble semicarbazones. Semicarbazide is hydrolysed on heating with dilute acids or alkalies. It gives hydrazine ammonia and carbon dioxide.

GUANIDINE

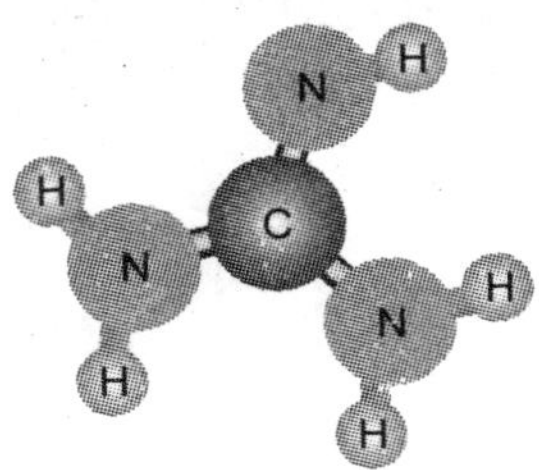

It is the imide corresponding to urea. Guanidine occurs in beet juice and is on of the degradation products of the purines. Several derivatives of guanidine are of biological importance e.g. creatine, creatinine and agrinine. Guanidine is colourless solid with melting point at 50°C, is very hydroscopic and difficult it isolate in pure form, it is isolated in its hydrochloride or nitrite form. Gunidine hydrochloride is obtained heating cyanmide (or dicyandiamide) with ammonium chloride solution at 100°C under pressure.

$$H_2N{-}C{\equiv}N + NH_4Cl \longrightarrow H_2N{-}\underset{}{\overset{\overset{H_2N-Cl}{\|}}{C}}{-}NH_2$$

cyanamide — guanidine hydrchloride

Guanidine nitrate is prepared by treating dicyandiamide with ammonium nitrate

Physical Properties

Molecular Formula	CH_5N_3
Formula Weight	59.0705
Composition	C(20.33%) H(8.53%) N(71.14%)
Melting Point	148°C
Boiling Point	170°C (decomposes)
Molar Refractivity	12.89 ± 0.5 cm^3
Molar Volume	37.9 ± 7.0 cm^3
Parachor	114.4 ± 8.0 cm^3
Index of Refraction	1.595 ± 0.05
Surface Tension	82.9 ± 7.0 dyne/cm
Density	1.55 ± 0.1 g/cm^3
Dielectric Constant	Not available
Polarizability	5.11 ± 0.5 $10^{-24}cm^3$
Monoisotopic Mass	59.048347 Da
Nominal Mass	59 Da
Average Mass	**59.071086 Da**

Guanidine is a very strong base. It forms stable salts giving each with weak acid. The basic strength of guanidine is due to the stability of the positive gunidinium ion, which is stabilized by resonance. Guanidine is hydrolyzed on warming with dilute alkalies to give urea and ammonia. When fused with potassium hydroxide it yields cynamide.

NITROGUANIDINE

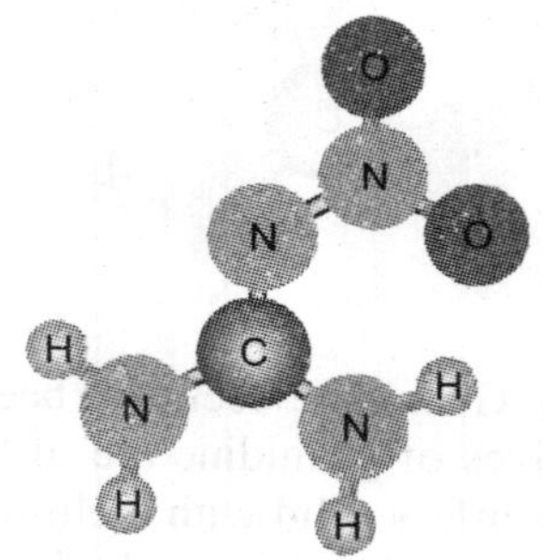

It is produced by the action of cold concentrated sulphuric acid on guanidine nitrate.

$$\underset{\text{guanidine nitrate}}{H_2N{-}C({=}NH_2^{+}\cdot{}^{-}ONO_2){-}NH_2} \xrightarrow[0^\circ C]{H_2SO_4} \underset{\text{2-nitroguanidine}}{H_2N{-}C({=}N{-}NO_2){-}NH_2} + H_2O$$

EXPLOSIVE

Nitroguanidine is used as a component of some explosive. It is as powerful as TNT and explodes without flash.

Physical Properties

MW	104
colour	white
crystal density	1.72 g/cm^3
crystal form	ortho-rhombic
melting point	245°C(dec)
solubility in water at 20°C	0.4g/100g
oxygen balance, % to CO_2	-31
heat of formation	0.950 kJ/g

heat of combustion	8.35 kJ/g
heat of detonation	3.01 kJ/g
gas volume	1,077 cm^3/g @ STP
collision constant	7.5 log.sub.10.sub./S
detonation velocity	8.16 km/s @ D. 1.72g/cc
detonation pressure	27.3 GPa @ D. 1.78g/cc
explosion temperature (at 5 s)	275°C (dec)
effect of prolonged storage	negligible
relative impact test value	200% of TNT
frict'on pendulum	no explosion
lead block	100
ballistic mortar	105
sand test	75
plate dent test	95
activation energy	87.5 kJ/mol

THIOUREA

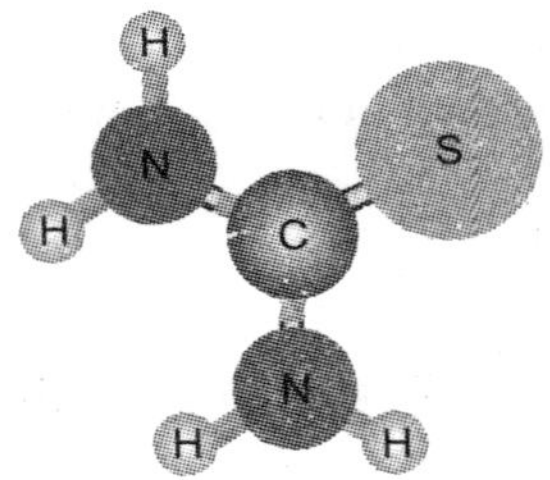

Thiourea; IUPAC name 2-thiourea; also known as thiocarbamide, sulfourea) is a white crystalline solid. Thiourea (CH_4N_2S) occurs in two tautomeric forms:

$$H_2N-C(=S)-NH_2 \rightleftharpoons H_2N-C(-SH)=NH$$

thiourea   Isothiourea

and thus has three functional groups: amino, imino, and thiol (BUA, 1995).

Other Methods of Preparation

By treating cynamide wiih hydrogen sulphide in the presence of ammonia under pressure

$$H_2N{-}C{\equiv}N + H_2S \xrightarrow[\text{pressure}]{NH_3} (H_2N)_2C{=}S$$

cyanamide → thiourea

By heating calcium cynamide with hydrogen sulphide at 160°C - 180°C(commercial method)

$$Ca{=}N{-}C{\equiv}N + 2H_2S \xrightarrow{\Delta} (H_2N)_2C{=}S + CaS$$

calcium cynamide → thiourea

The substance has no sharp melting point, as rearrangement to ammonium thiocyanate (NH_4SCN) occurs at temperatures above about 135 °C (Mertschenk et al., 1995). Data on melting between 167 and 182 °C are reported in the literature (BUA, 1995). Information on the boiling point is not available, as decomposition occurs. The temperature of decomposition is not known.

Thiourea is soluble in water (137 g/litre at 20 °C), soluble in polar protic and aprotic organic solvents, and insoluble in non-polar solvents (BUA, 1995). A UV absorption maximum at 238 nm was measured in water at pH 7.4 (Weast & Astle, 1979). A significant pH dependence of the *n*-octanol/water partition coefficient (log K_{ow}) was not detected (Govers et al., 1986).

Additional physicochemical properties for thiourea are presented in Table 1 and in the International Chemical Safety Card (ICSC 0680) reproduced in this document.

Chemical Reaction of Thiourea

Thiourea behaves as a monoacidic base and like urea forms well defined salts. It forms inclusion complexes with branched chain and cyclic aliphatic compounds and with straight chain compounds containing more through fourteen carbon atoms

Some of the important reaction of thiourea is described below:

(1) Hydrolysis: Thiourea is decomposed on boiling with dilute alkalies to give ammonia carbon dioxide and hydrogen sulphide

$$(H_2N)_2C{=}S + H_2O \xrightarrow[\Delta]{OH^-} 2NH_3 + CO_2 + H_2S$$

(2) Oxidation: Thio urea is oxidized on treatment with an alkaline solution of potassium permanganate or an alkaline solution of hydrogen peroxide to yield urea

$$H_2N-C(=S)-NH_2 + [O] \xrightarrow[\text{or } H_2O_2/OH^-]{KMnO_4/OH^-} H_2N-C(=O)-NH_2 + S$$

(3) Reaction with Metal Oxides: Thio urea reacts with oxides of mercury, silver or lead to form cynamide

$$\underset{\text{thiourea}}{H_2N-C(=S)-NH_2} + HgO \longrightarrow \underset{\text{cyanamide}}{H_2N-C\equiv N} + HgS + H_2O$$

(4) Reaction with alkyl Halides: Thiourea reacts with alkyl halides to form S-alkyl thiouronium halides (the prefix S- indicates that the alkyl group is attached to the sulphur atom)®

$$\underset{\text{thiourea}}{H_2N-C(=S)-NH_2} \rightleftharpoons \underset{\substack{\text{diaminomethanethiol} \\ \text{isothiourea}}}{H_2N-CH(SH)-NH_2} + R-Cl \longrightarrow \left(\underset{\text{s- alkyl thiourea}}{H_2N-C(SR)=NH} + HCl\right) \longrightarrow \underset{\text{s-alkyl thiouronium chloride}}{H_2N-C(SR)=NH_2Cl}$$

Table Physicochemical properties of thiourea

Property	Value
Relative molecular mass	76.1
Density (g/cm^3)	1.405
Vapour pressure (kPa) at 20 °C	9.98 x 10^{-9}
n-Octanol/water partition coefficient (log K_{ow}) (measured)	–1.61 to –0.92
Water solubility (g/litre)	95 at 10 °C137 at 20 °C
Henry's law constant (Pa·m^3/mol) at 20 °C	5.6 x 10^{-9}

The s-alkyl isothiouronium halides are stable, crystalline salts they are used to characterize sulphamic acids with which they form insoluble salts. They are used to prepare the alcohols by alkaline hydrolysis.

$$\underset{\text{s-alkyl thiouronium chloride}}{H_2N-C(SR)=NH_2Cl} + Na-OH \longrightarrow R-SH + H_2N-C\equiv N + NaCl + H_2O$$

Like Urea thiourea also reacts with formaldehyde to produce resins.

®Note it is the iso form of thiourea–which takes part in the reaction.

3. Analytical Methods

The determination of thiourea in workplace air can be carried out by adsorption on a glass fibre filter, filter elution with water in an ultrasonic bath, C_{18} reversed-phase HPLC with water as the mobile phase, and UV detection at 245 nm. The detection limit is 0.4 μg thiourea/litre sample solution; a recovery rate of 106 ± 6% is given (BUA, 1995).

This method can also be applied to the detection of thiourea in water. The detection limit is 0.1 mg/litre water. Thiourea concentrations above 10 mg/litre have to be diluted before analysis; solutions with very low concentrations of the chemical can be concentrated in a rotary evaporator (sample solution 2.1 μg/litre; BUA, 1995).

In soil, thiourea can be determined by HPLC, but with a cationic exchange resin as the separating phase and under salting-out conditions (with an aqueous solution of ammonium sulfate as the mobile phase). Detection is carried out by UV absorption at 240 nm. The method works especially well at a column temperature of 60°C. At a substance concentration of 160 μg/litre, a recovery rate of 99.3 ± 2.7% is given. The detection limit is 2.7 ng absolute. For the detection of thiourea in biological material, reversed-phase HPLC with methanol/water as the mobile phase and UV detection (240 nm) is applied.

4. Sources of Human and Environmental Exposure

Natural sources

Thiourea has been detected but not quantified in laburnum shrubs (*Laburnum anagyroides*) and is a natural metabolite of the fungi *Verticillium alboatrum* and *Bortrylius cinerea* (IARC, 1974).

Anthropogenic sources

Thiourea is industrially produced by the reaction between technical-grade calcium cyanamide ($CaCN_2$) and hydrogen sulfide (H_2S) or one of its precursors in aqueous solution — e.g., ammonium sulfide ($(NH_4)_2S$) or calcium hydrogen sulfide ($Ca(HS)_2$). Calcium cyanamide must not contain calcium carbide, as explosive acetylene can be liberated with water or hydrogen disulfide. In Germany, thiourea is produced by a continuous process in a closed reaction vessel (BUA, 1995; Mertschenk et al., 1995). In 1993, the global annual production of thiourea was about 10 000 tonnes (BUA, 1995). Of this, about 40% (4000 tonnes) was produced by the German manufacturer, which is the sole manufacturer in Western Europe; 20% (2000 tonnes) was contributed by a Japanese manufacturer; and another 40% (4000 tonnes) was contributed by at least seven Chinese companies. A more recent global production figure is not available.

Uses

Table 2 gives use patterns derived from 1993 global data (BUA, 1995), but the use pattern may vary widely between countries.

Table 2. Estimated global use pattern of thiourea.[a]

Use
• Ore leaching (e.g., gold and silver extraction from minerals)
• Auxiliary agent (diazo paper)
• Isomerization catalyst (conversion of maleic to fumaric acid)
• Additive (slurry explosives)
• Metal refinement (copper)
• Metal cleaning (including silver polish)
• Other (e.g., drilling auxiliary in petroleum industry, fertilizer)
• Processing
• Production of thiourea dioxide
• Modification of resins
• Various chemical intermediates

[a] From BUA (1995).

In the USA, thiourea is used in animal hide glue, which contains thiourea at a concentration of 10–20% as a liquefying agent. Reports indicate its use in the production of flame retardant resins and as a vulcanization accelerator (NTP, 2000). In Germany, thiourea is not used in the leaching of ore mines and not processed to thiourea dioxide. Instead, the following use pattern is reported (BUA, 1995): auxiliary agent in diazo paper (light-sensitive photocopy paper) and almost all other types of copy paper (19%); metal cleaning, including silver polish (4%); precipitation of heavy metals (3%); additive in slurry explosives (3%); electroplating/electroforming (1%); corrosion inhibitor (1%); processing to organic intermediates (41%); mercaptosilanes (6.5%); vulcanization accelerators (0.5%); resin modification (4.5%); and chemicals industry and miscellaneous (16.5%) (BUA, 1995). In Japan, thiourea is added to fertilizers to inhibit the nitrification process (Hashimoto, 1979; Kubota & Asami, 1985). Data on the quantities used are not available. Thiourea is emitted by manufacturers of electronic components and accessories and manufacturers of aircraft and aircraft parts (CARB, 1997). Organic thiourea derivatives are used as vulcanization accelerators, pharmaceuticals (antiseptic, thyrotherapeutic, narcotic, and tuberculostatic agents), and plant protection agents and pesticides (e.g., chloromethiuron, diafenthiuron, thiophanate, and thiophanate-methyl) (Mertschenk et al., 1995).

CHAPTER

14

Esters

Introduction

Esters are the derivatives of carboxylic acids in which the –OH of carboxyl group is replaced by alkoxy group (-OR')

$$R-C(=O)-OH \xrightarrow[+OR]{-OH} \underset{\text{ester}}{R-C(=O)-O-R_1}$$

The general formula of esters id represented as

$$R-C(=O)-O-R_1 \text{ or } R-CO-R' \text{ or } R-COOR$$

The functional group of esters which determines their characteristics behaviour may be written as

$$R-C(=O)-O-R_1 \text{ - CO – O – R' -or simply -COOR'}$$

Esters are the most important class of acid derivatives and are widely distributed in nature. The active ingredient responsible for pleasant flavours and colour of a fruit and flower is often an ester. For example amyl acetate is present in bananas, octyl acetate in oranges, amyl propionate in apricots, ethyl butyrates in peaches, n – butyl butyrate in pineapples and isoamyl isovalerate in apples. Animal and vegetable fats and oils and esters derived from glycerol and higher acids.

Nomenclature

To name an ester you treat it as a **derivative of an alcohol and an acid**, which it is. First you name the part from the alcohol and then the part from the acid using an **-ate** ending. The resulting name should be written as two words. This diagram shows an ester that is made from **methyl alcohol** and **acetic acid**. So its name is **methyl acetate**. Methyl from the methyl alcohol. Acetate from the acetic acid.

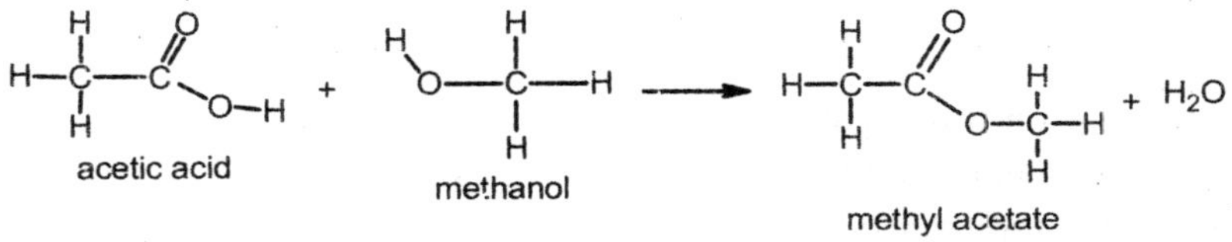

acetic acid + methanol → methyl acetate + H_2O

Notice that as far as naming is concerned, it is somewhat similar to naming a salt. In **sodium acetate**, we have a sodium ion bonded to the acetate ion that is the remainder of the compound. In **methyl acetate**, we have a methyl group bonded to the acetate group that is the remainder of the compound.

acetic acid + methanol → acetate / methyl + H_2O

methyl acetate

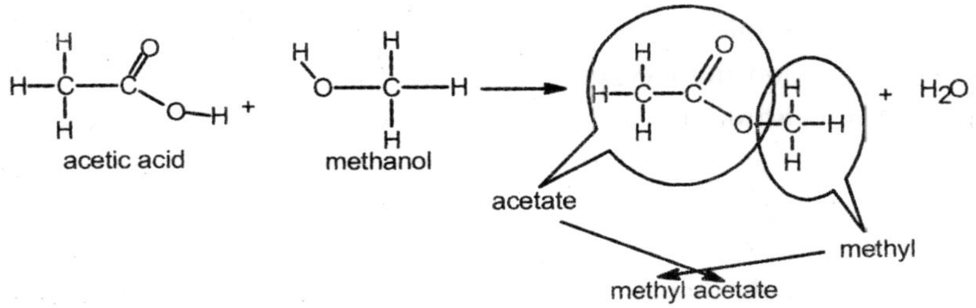

In both cases the acetate is bonded to the other part. In the case of a salt, however, it is an **ionic** bond between the acetate and the sodium. In the case of the ester it is a **covalent** bond between the acetate and the methyl. So there is quite a difference in that you are dealing with a molecular compound instead of an ionic compound. But as far as naming esters, you can approach the naming of esters the same way you approach naming a salt. You name the group that is attached to the oxygen of what was the acid. Then you name what remains of the acid by using the **-ate** ending.

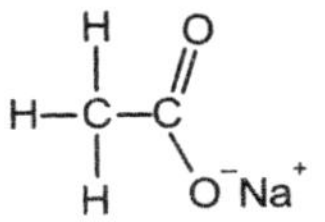

O^-Na^+ — Sodium acetate (a salt)

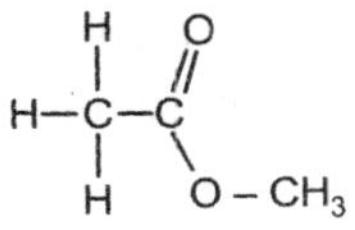

$O-CH_3$ — Methyl acetate (a ester)

There is one disadvantage to the naming of esters as though they were similar to salts, and that is that it may lead you to the feeling that the oxygen between the two groups originally started out with the acid. This is not true. In forming esters, the oxygen that was originally bonded to the carboxylic acid becomes part of the water. The oxygen that was in the alcohol ends up being the oxygen that is the link between the methyl and the acetate in methyl acetate. The general reaction is shown below.

$$\underset{\text{carboxylic acid}}{R-\overset{O}{\overset{\|}{C}}-OH} + \underset{\text{alcohol}}{R'OH} \xrightarrow{H^+} \underset{\text{ester}}{R-\overset{O}{\overset{\|}{C}}-O-R'} + HOH$$

Guidelines

Before turning you loose on the names and formulas of esters, let me point out that it is very important to make sure that you get correct which part of the ester came from the alcohol and which part came from the acid, because the **-ate** ending goes with the part that was part of the acid. Two examples are shown here, one in the left series of boxes and one in the right. The simplest way to identify the **-ate** part is to find the **oxygen atom that is part of the continuous chain** (second box). It is bonded to carbon on both sides. On one side of that oxygen there will be a **carbonyl group** (third box). On the other side there won't be. Instead there will be an alkyl group. The side of the molecule that has the carbonyl group is the side that came from the acid (fourth box) and that part will have an **-ate** ending (fifth box). The other part, the alkyl group without the carbonyl, will simply have a -yl ending.As shown above, the formulas for these molecules can be written in either direction. That is, the part from the acid may be on the left or it may be on the right. You have to look at the structural formula to determine which part is from the acid (and gets the **-ate** ending) and which part is from the alcohol (and gets the **-yl** ending).

$C-C-\overset{O}{\overset{\|}{C}}-O-C-C$	$C-C-\overset{O}{\overset{\|}{C}}-O-C-C$
$C-C-\overset{O}{\overset{\|}{C}}-O-C-C$	$C-C-\overset{O}{\overset{\|}{C}}-O-C-C$
$C-C-\overset{O}{\overset{\|}{C}}-O-C-C$	$C-C-\overset{O}{\overset{\|}{C}}-O-C-C$
$C-C-\overset{O}{\overset{\|}{C}}-O-C-C$	$C-C-\overset{O}{\overset{\|}{C}}-O-C-C$

If there are side groups on either portion of the molecule, you will, of course, need to use numbers to show where they are attached. Number each portion starting from the carbon atom that is attached to the oxygen in the carboxylate group. In a sense this means numbering the atoms from the middle of the molecule out to the ends. However, if you think about an ester being made from a carboxylic acid and an alcohol, the numbering goes from the oxygen ends of those molecules. The subscripts here are to number the carbon atoms rather than indicate how many there are. In the example above, a methyl group is attached to the first carbon atom of the two-carbon ethyl group chain (this can also be called an isopropyl group) and a methyl group is also attached to the second carbon atom of the three-carbon propanoate chain.

$$C_2-C_1-O-\underset{C}{\underset{|}{\overset{O}{\overset{\|}{C_1}}}}-\underset{C}{\underset{|}{C_2}}-C_3$$

1-methylethyl 2-methylpropanoate
or
isopropyl 2-methylpropanoate

Name of the esters and their smells

methyl butanoate	smells of pineapple or apple
methyl salicylate (oil of wintergreen)	smells of the ointments called Germolene, iodex
methyl benzoate	smells of fruity-ylang ylang
ethyl methanoate	smells of rum
ethyl butanoate	smells of pineapple or strawberry
ethyl salicylate	smells of oil of wintergreen
ethyl heptanoate	smells of grape
butyl ethanoate	smells of raspberry
pentyl ethanoate	smells of banana
pentyl pentanoate	smells of apple
pentyl butanoate	smells of pear or apricot
octyl ethanoate	smells of fruity-orange
benzyl ethanoate	smells slightly of jasmine
isopentyl ethanoate	smells of pear or banana (it is used as the flavouring in the manufacturing of old fashioned Pear Drops)
isopentyl acetate	smells (banana oil)
n-propyl 2-methylpropanoate	smells (rum)
isobutyl methanoate	smells (raspberries)
octyl acetate	smells (oranges)
n-butyl butanoate	smells (pineapple)

Preparation

Esters can be formed by the **intermolecular dehydration** reaction of a **carboxylic acid with an alcohol**. Note that this is a reversible reaction. Note also that water is formed in this reaction. This water molecule is a result of an **intermolecular dehydration reaction**. Also known as **Fischer – Speier Esterification reaction**

$$R-\overset{O}{\overset{\|}{C}}-OH + R-OH \underset{}{\overset{H^+}{\rightleftharpoons}} H_3C-\overset{O}{\overset{\|}{C}}-O-R_1 + H_2O$$

$$\underset{\text{propionic acid}}{H_3C-CH_2-\overset{O}{\overset{\|}{C}}-OH} + \underset{\text{ethanol}}{CH_3-CH_2-OH} \overset{H^+}{\rightleftharpoons} \underset{\text{ethyl propionate}}{H_3C-CH_2-\overset{O}{\overset{\|}{C}}-O-CH_2-CH_3} + H_2O$$

There is an -OH group from the carboxylic acid combining with the hydrogen from the alcohol to form water. So we can say that water is being split out between these two compounds. These intermolecular dehydration reactions are very common types of reactions and very important ones, biologically. By the way, intermolecular dehydration reactions are sometimes called **condensation reactions**.

Other Methods

Esters are also obtained when a mixture of the vapour of an acid and an alcohol is passed over thoria.

$$R-\overset{O}{\overset{\|}{C}}-OH + R_1-OH \xrightarrow[300°C]{ThO_2} H_3C-\overset{O}{\overset{\|}{C}}-OH + H_2O$$

(1) Alcoholysis of acid Chlorides and acid anhydrides

Esters are conveniently prepared by the action of alcohols on acid chlorides or acid anhydrides.

(*a*) $R - CO - Cl + R' - OH \rightarrow R - CO - O - R' + HCl$

e.g $CH_3 - CO - Cl + C_2H_5 - OH \rightarrow CH_3 - CO - CO - C_2H_5 + HCl$
acetyl chloride ethyl alcohol ethyl acetate

(*b*) $(R-CO)_2O + H_3C-OH \rightarrow H_3C-\overset{O}{\overset{\|}{C}}-O-R_1 + H_3C-C(=O)OH$

Acid anhydrides, however, react a little slowly than acid chlorides which are preferred on a laboratory scale.

(2) Silver Salt Method

Esters may be prepared by treating ethanoic solution of an ethyl halide which silver salt of an appropriate carboxylic acid.

$CH_3CH_2 - CO - OAg^+ + C_2H_5Br \rightarrow CH_3CH_2 - CO - O - C_2H_5 + AgBr$
Silver propionate ethyl bromide

Although sodium salts of carboxylic acids also give satisfactory yields, the use of silver salts gives improved yields. Sodium or potassium salts of carboxylic acids. However, produce excellent yields of esters when treated with a dialkyl sulphate in alkaline solution.

$$H_3C-\overset{O}{\overset{\|}{C}}-O-K + (H_3C-O)_2SO_2 \rightarrow H_3C-\overset{O}{\overset{\|}{C}}-O-CH_3 + H_3C-O-SO_2-O-K$$

(acetate)potassium + dimethyl sulfate → methyl acetate + methyl potassium sulphate

(3) Addition of an acid to Alkenes

When the solution of an alkenes in the preparation of esters of testing alcohols

$$H_3C-\underset{\underset{CH_3}{|}}{C}=CH_2 + H_3C-\overset{\overset{O}{||}}{C}-OH \xrightleftharpoons{H_2SO_4} H_3C-\overset{\overset{O}{||}}{C}-O-\underset{\underset{CH_3}{|}}{\overset{\overset{CH_3}{|}}{C}}-CH_3$$

2-methylprop-1-ene acetic acid tert-butyl acetate

(4) By the addition of Diazomethane to Carboxylic acid

Methyl esters can be prepared conveniently when ethereal solution of diazomethane is added to solution of a carboxylic acid in ester.

$$H_3C-\overset{\overset{O}{||}}{C}-OH + H_3C-\overset{+}{N}\equiv N \longrightarrow H_3C-\overset{\overset{O}{||}}{C}-O-CH_3 + N_2$$

acetic acid methanediazonium methyl acetate

This reaction proceeds even below 0°C and gives quantitative yield of esters. This method is however suitable only for laboratory preparation because of the difficulty enunciated in preparing and handling the reagent.

(4) Tischenko Reaction

Synthesis of complex esters by disproportionation of aldehydes.

$$2RC\begin{matrix}{}^{\nearrow O}\\ {}_{\searrow H}\end{matrix} \xrightarrow{Al(OC_2H_5)_3} RC\begin{matrix}{}^{\nearrow O}\\ {}_{\searrow OCH_2R}\end{matrix}$$

(~50%)

R' = Alk

*All $(OR')_3$ + $AlCl_3$ or $HgCl_2$, $Mg(OR')_2$, $Zn(OR')_2$

e.g

$$2CH_3CH_2CHO \xrightarrow{Al(OC_2H_5)_3} \underset{\text{propyl propionate}}{CH_3CH_2-CO-O-CH_2CH_2CH_3}$$

The reaction proceeds with aliphatic, aromatic and heterocyclic compounds.

Physical Properties (General)

1. Lower members are volatile liquids, lighter than water and having pleasant smell and are usually described as fruity. The higher esters are colourless wax solids.

2. The boiling points of the methyl and the ethyl esters are lower than those of parent acids because association through hydrogen bonding does not occur in esters (hydroxyl groups are absent) Ester with branched chain isomers.

The inability of esters to form hydrogen bonds is evident from that their boiling points are unpairable acid chlorides of similar molecular weights.

Compound	Formula	Molecular weight	Boiling point
Butyl Chloride	$H_3C-CH_2-CH_2-C(=O)-Cl$	106	101°C
Butyl Formate	$H-C(=O)-O-CH_2-CH_2-CH_2-CH_3$	102	107°C
Propyl Acetate	$H_3C-C(=O)-C-CH_2-CH_2-CH_3$	102	102°C

3. Low Molecular weight esters are fairly soluble in water. For example, methyl formate dissolve up to the extent of 30 g/100g of water at 25°C. Water solubility of esters decrease with increase hydrogen chain. On the other hand, all esters are soluble in most organic solvents (ethanol, esters benzene etc)

Chemical Properties

Esters give nucleophillic substitution (S_N) reactions anhydrous to the acid halides and anhydrides since the carbonyl carbon in their molecule is electron deficient and carries a partial positive charge.

$$R-C(=O)-O-R + :Nu \longrightarrow H_3C-C(OH)(Nu)-O-R \longrightarrow H_3C-C(=O)-Nu\ :\bar{O}R + \text{Alkoxy ion (leaving group)}$$

The esters undergo S_N reactions considerably less readily than halides and anhydride. This is due to two reasons:

(*i*) Esters are resonance halides of two cannonical* forms

$$R-C(=O)-\ddot{O}-R \longleftrightarrow H_3C-C(O^-)=C^+-H$$

As a result of this resonance, the carbonyl carbon is not as electron deficient (as positive) as in acid and anhydrides this makes esters less prone to nucleophillic attacks.

(*ii*) Alkoxide ions (-OR) are much stronger bases than halides ions and carbohydrate ions ($RCOO^-$), and hence there are poor leaving groups relative to X^- ions and $RCOO^-$ ions.

Nevertheless, esters undergo a good number of S_N reactions. As few more reactions will make the understanding more clear, which are discussed below:

***(1) Hydrolysis*:** Esters are hydrolyzed by boiling with mineral acids or alkalies

An ester can be hydrolyzed to give the parent carboxylic acid and alcohol by using mineral acid (sulphuric or hydrochloric) a catalyst.

* The simplest form of a matrix specially the form of a square matrix that the zero elements every except at the principle diagonal.

$$R-\overset{O}{\overset{\|}{C}}-O-R + H_2O \underset{}{\overset{H^+}{\rightleftharpoons}} H_3C-\overset{O}{\overset{\|}{C}}-OH + R_1-OH$$

ester — acetic acid — alcohol

$$H_3C-\overset{O}{\overset{\|}{C}}-O-CH_2-CH_3 + H_2O \overset{H^+}{\rightleftharpoons} H_3C-\overset{O}{\overset{\|}{C}}-OH + H_3C-CH_2OH$$

ethyl acetate — acetic acid — ethanol

The above reaction is reversible and is exactly opposite of Esterification. In order to hydrolyze most of the ester, the equilibrium must be pushed to the right by giving excess of water (one of the reactants)

(2) Alkaline Hydrolysis (Saponification)

What is saponification?

The term saponification is the name given to the chemical reaction that occurs when a vegetable oil or animal fat is mixed with a strong alkali. The products of the reaction are two: soap and glycerin. Water is also present, but it does not enter into the chemical reaction. The water is only a vehicle for the alkali, which is otherwise a dry powder. The name saponification literally means "soap making". The root word, "sapo", is Latin for soap. The Italian word for soap is sapone. Soap making as an art has its origins in ancient Babylon around 2500 - 2800 BC.The oils used in modern handmade soap are carefully chosen by the soap maker for the character they impart to the final soap. Coconut oil creates lots of glycerin, makes big bubbly lather, and is very stable. Olive oil has natural antioxidants and its soap makes a creamier lather. Tallow, or rendered beef fat, makes a white, stately bar that is firm and creates abundant lather. Many other oils can be used, each one for a specific reason. Your soap maker will be glad to tell you which oils are used to make her or his soap.The alkali used in modern soap is either potassium hydroxide, which is used to make soft soap or liquid soap because of its greater solubility, or sodium hydroxide, which is used to make bar soap. The common term for the alkali became simply "lye", which curiously is not short for alkali, but originated in the Anglo-Saxon language. Soap made in cottages and on farms in earlier American times became known as "lye soap". That term is now pejorative and derogatory and denotes a harsh soap that would irritate your skin. The old soap got a bad name because it had an excessive amount of caustic. Weighing and measuring techniques were crude, and knowledge of soap chemistry was elementary or non-existent.

The true fact is that modern handcrafted soap, though necessarily made with lye to get true soap, has no lye in the final product. It has all been reacted with the oils to form soap and glycerin.

A curious fact about modern soap is that most common soap found in the grocery store made in mass-produced factories does have a small amount of excess alkali in it. Also, it has had all of its naturally-occurring glycerin removed so it can be sold as a separate commodity. Why? Greater profit. An important difference between most commercial soap and our Real Handmade soap is that the glycerin is left in Real Handmade Soap and thus it retains its natural moisturizing property.

CH_2-OOC-R - CH-OOC-R - CH_2-OOC-R (fat) + 3 NaOH (or KOH)

both heated→

CH_2–OH–CH–OH–CH_2–OH (glycerol) + 3 R–CO_2–Na (soap) R=$(CH_2)_{14}CH_3$ (for example)

Mechanism

Mechanism of the Base Hydrolysis of Esters

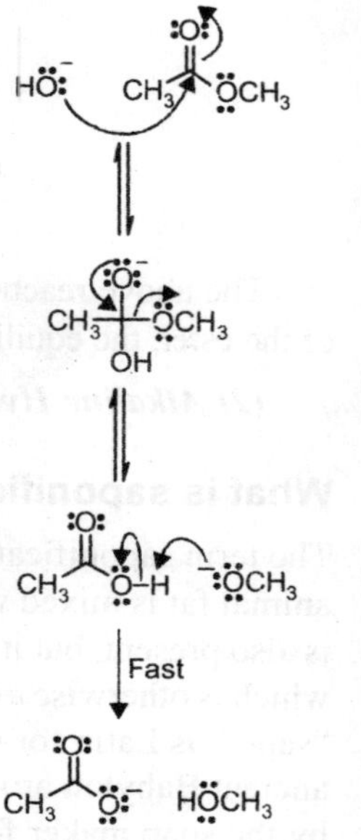

Step 1: The hydroxide nucleophiles attacks at the electrophilic C of the ester C=O, breaking the p bond and creating the tetrahedral intermediate

Step 2: The intermediate collapses, reforming the C=O results in the loss of the leaving group the alkoxide, leading to the carboxylic acid.

Step 3: An acid / base reaction. A very rapid equilibrium where the alkoxide functions as a base deprotonating the carboxylic acid (an acidic work up would allow the carboxylic acid to be obtained from the reaction).

Reaction under ACIDIC conditions

- Note that the acid catalysed mechanism is the reverse of the **Fischer esterification.**
- The mechanism shown below also leads to *acyl-oxygen* cleavage (see **step 5**).
- The mechanism is an example of the **less reactive system type.**

HÖ: CH3 ÖCH3; CH3 ÖCH3 ÖH; CH3 Ö-H ⁻:ÖCH3; Fast; CH3 O: HÖCH3

Mechanism of the Acid Catalysed Hydrolysis of Esters

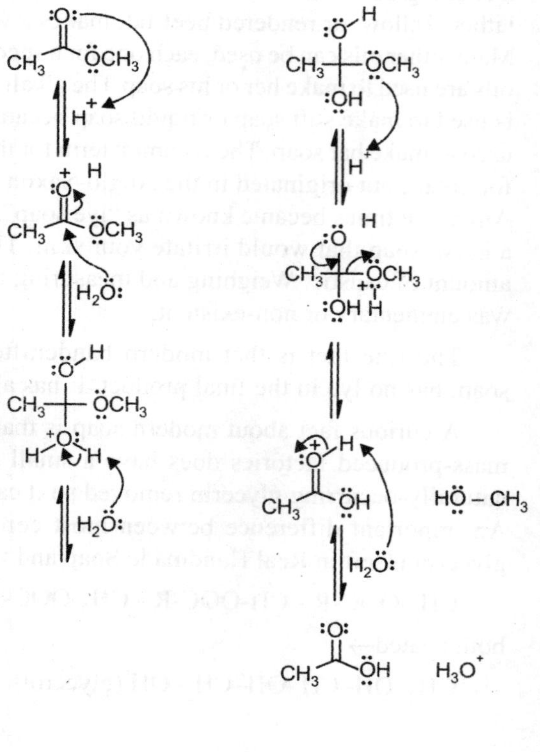

Need to activate the ester. Protonation of the ester carbonyl makes it more electrophilic

Step 1: An acid/base reaction. Since we only have a weak nucleophile and a poor electrophile

Step 2: The water O functions as the nucleophile attacking the electrophilic C in the C=O, with the electrons moving towards the oxonium ion, creating the tetrahedral intermediate.

Step 3: An acid/base reaction. Deprotonate the oxygen that came from the water molecule.

Step 4: An acid/base reaction. Need to make the $-OCH_3$ leave, but need to convert it into a good leaving group first by protonation.

Step 5: Use the electrons of an adjacent oxygen to help "push out" the leaving group, a neutral methanol molecule.

CH3 ÖCH3 H+; CH3 ÖCH3 H2Ö:; CH3 ÖCH3 H2Ö:; CH3 ÖCH3 ÖH; H; CH3 ÖCH3 ÖH H; CH3 ÖH HÖ – CH3; H2Ö:; CH3 ÖH H3O+

Step 6: An acid/base reaction. Deprotonation of the oxonium ion reveals the carbonyl in the carboxylic acid product and regenerates the acid catalyst.

Reduction of Esters

$$LiAlH_4 + R'C(=O)OR'' \longrightarrow H-C(OH)(H)-R' + R''O-H$$

Reactions usually in Et_2O or THF followed by H_3O^+ work-ups

Reaction type: Nucleophilic Acyl Substitution then **Nucleophilic Addition**

Summary

- Carboxylic esters are reduced give 2 alcohols, one from the alcohol portion of the ester and a 1° alcohol from the reduction of the carboxylate portion.
- Esters are less reactive towards **Nu** than aldehydes or ketones.
- They can only be reduced by $LiAlH_4$ and **NOT** by the less reactive $NaBH_4$
- The reaction requires that 2 hydrides (H^-) be added to the carbonyl group of the ester
- The mechanism is an example of the **reactive system type.**
- Note the aldehyde intermediate.

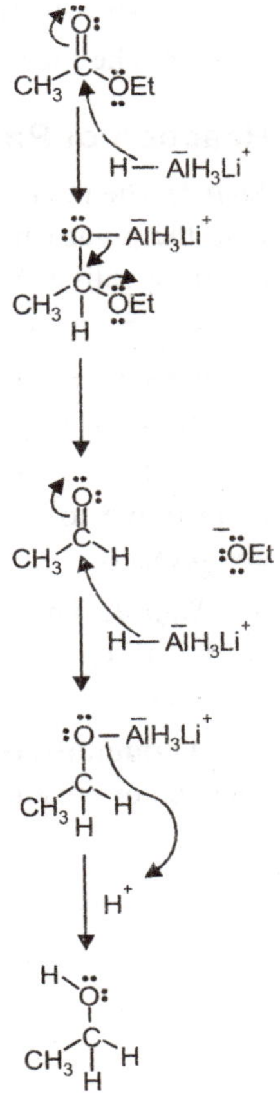

CH3–C(=O)–ÖEt + H–$\bar{A}lH_3Li^+$ → :Ö–$\bar{A}lH_3Li^+$ / CH3–C(H)–ÖEt → CH3–C(=O)–H + :ÖEt + H–$\bar{A}lH_3Li^+$ → :Ö–$\bar{A}lH_3Li^+$ / CH3–C(H)(H) → (H^+) → H–Ö: / CH3–C(H)(H)

Reaction of Lialh$_4$ with an Ester

Step 1: The nucleophilic H from the hydride reagent adds to the electrophilic C in the polar carbonyl group of the ester. Electrons from the C=O move to the electronegative O creating an intermediate metal alkoxide complex.

Step 2: The tetrahedral intermediate collapses and displaces the alcohol portion of the ester as a leaving group, this produces a ketone as an intermediate.

Step 3: Now we are reducing an aldehyde. The nucleophilic H from the hydride reagent adds to the electrophilic C in the polar carbonyl group of the aldehyde. Electrons from the C=O move to the electronegative O creating an intermediate metal alkoxide complex.

Step 4: This is the work-up step, a simple acid/base reaction. Protonation of the alkoxide oxygen creates the primary alcohol product from the intermediate complex.

Reactions of RLI and RMgX with Esters

2RLi or 2RMgX + R'C(=O)OR'' ⟶ R–C(OH)(R)–R'

Reaction usually in Et_2O followed by H_3O^+ work-up

Reaction type: Nucleophilic Acyl Substitution then **Nucleophilic Addition**

Summary

- Carboxylic esters, **R'CO$_2$R"**, react with 2 equivalents of organolithium or Grignard reagents to give tertiary alcohols.
- The tertiary alcohol that results contains 2 identical alkyl groups (from **R** in the scheme)
- The reaction proceeds via a ketone intermediate which then reacts with the second equivalent of the organometallic reagent
- Since the ketone is more reactive than the ester, the reaction *cannot* be used as a preparation of ketones.
- The mechanism is an example of the reactive system type.

Reaction of Rm$_g$x with an Ester

Step 1: The nucleophilic C in the organometallic reagent adds to the electrophilic C in the polar carbonyl group of the ester. Electrons from the C=O move to the electronegative O creating an intermediate metal alkoxide complex.

Step 2: The tetrahedral intermediate collapses and displaces the alcohol portion of the ester as a leaving group, this produces a ketone as an intermediate.

Step 3: The nucleophilic C in the organometallic reagent adds to the electrophilic C in the polar carbonyl group of the ketone. Electrons from the C=O move to the electronegative O creating an intermediate metal alkoxide complex.

Step 4: This is the work-up step, a simple acid/base reaction. Protonation of the alkoxide oxygen creates the alcohol product from the intermediate complex.

CH$_3$C(=O)OEt + CH$_3$–MgBr ⟶ $^-$O(MgBr$^+$)–C(CH$_3$)(CH$_3$)OEt ⟶ CH$_3$C(=O)CH$_3$ + BrMg ŌEt; + CH$_3$–MgBr ⟶ $^-$O(MgBr$^+$)–C(CH$_3$)$_3$; H$^+$ ⟶ H–O–C(CH$_3$)$_3$

Ammonolysis of Ester: Esters react with ammonia, in either concentrated aqueous or ethanoic solution to form acid amides.

$$R-\overset{O}{\overset{\|}{C}}-O-R + NH_3 \longrightarrow H_3C-\overset{O}{\overset{\|}{C}}-NH_2 + R_1-OH$$

ester; acetamide

$$H_3C-\overset{O}{\overset{\|}{C}}-O-CH_2-CH_3 + NH_3 \longrightarrow H_3C-\overset{O}{\overset{\|}{C}}-NH_2 + H_3C-CH_2-OH$$

ethyl acetate; acetamide; ethanol

However, the reaction is much slower than with acid chloride or anhydrides and amides are therefore more easily prepared from ester. The mechanism of Ammonolysis of ester is essentially the same as that given for the acid chlorides and anhydrides.

Reaction with Hydroxylamine: Esters react with hydroxylamine in presence of strong bases e.g ethanoic KOH to give the corresponding hydroxamic acids.

$$\underset{\text{ester}}{R-\overset{O}{\overset{\|}{C}}-O-R} + \underset{\text{hydroxylamine}}{H_2N-OH} \longrightarrow \underset{\text{hydroxamic acid}}{R-\overset{O}{\overset{\|}{C}}-NH-OH} + \underset{\text{alcohol}}{R_1-OH}$$

These compounds produce deep purple complexes with ferric ion. This reaction, therefore, forms the basic of detection of ester groups in organic substances.

Pyrolysis of esters: Esters which have a hydrogen atom on β-carbon of the alkoxy group. When heated to 500°C loose a molecule of an acid to form an alkene. Ester of primary alcohols are converted to 1 – alkenes and acid.

$$\underset{\text{ester}}{R-\overset{O}{\overset{\|}{C}}-O-CH_2-CH_2-R_1} \xrightarrow{500°C} \underset{\text{acid}}{R-\overset{O}{\overset{\|}{C}}-OH} + \underset{\text{1-alkene}}{H_2C=CH-R}$$

e.g

$$\underset{\text{propyl acetate}}{H_3C-\overset{O}{\overset{\|}{C}}-O-CH_2-CH_2-CH_3} \xrightarrow{500°C} \underset{\text{acid}}{H_3C-\overset{O}{\overset{\|}{C}}-OH} + \underset{\text{but-1-ene}}{H_2C=CH-CH_2-CH_3}$$

In this reaction the R – CO – O group and hydrogen on adjacent carbon are eliminated as acid. It may be noted that with primary alcohol esters, pyrolysis affords a single alkene, with secondary and tertiary alcohol esters where elimination of acid may take place in more than one direction, a mixture of alkene is produced. Acetates are commonly employed for the reaction.

Mechanism

The pyrolysis of esters occurs through a cyclic transistion state

$$\underset{\text{ester}}{R(H)(H)C-CH_2-O-\overset{O}{\overset{\|}{C}}-R} \longrightarrow \underset{\text{cyclic transition}}{[R_1CH\cdots H\cdots O=C(R)-O\cdots CH_2]} \longrightarrow \underset{\text{alkene}}{R_1-CH=CH_2} + \underset{\text{acid}}{R-\overset{O}{\overset{\|}{C}}-OH}$$

. **Acyloin formation**: esters react with sodium metal in a suitable solvent such as benzene or toluene to give bimolecular reduction products called Acyloin.

$$2\,H_3C{-}C(=O){-}O{-}CH_2CH_3 + 4Na \longrightarrow H_3C{-}C(ONa){=}C(ONa){-}CH_3 \longrightarrow H_3C{-}C(=O){-}CH(OH){-}CH_3$$

ethyl acetate → disodium but-2-ene-2,3-diolate → 3-hydroxybutan-2-one (acetoin, an acyloin)

It may be noted that acyloins are compounds in which the corresponding – CHOH – CO – is flanked by alkyl groups. The reaction occurs on the surface of the sodium metal and is rather similar to the bimolecular reduction ketones to pinacols

Claisen Condensation

Esters, like aldehydes and ketones, enolize in base to produce a small equilibrium concentration of the corresponding enolate anion. As in the aldol condensation, this enolate anion can function as a nucleophile towards the carbonyl carbon of another mole of ester, and the result is the formation of a bond between the a-carbon of one mole of ester and the carbonyl of a second, to give a tetrahedral intermediate, as shown below for ethyl acetate. Expulsion of ethoxide from this intermediate re-forms the carbonyl and generates the condensation product, ethyl acetoacetate. This condensation reaction between two moles of an ester is called the **Claisen condensation**.

2 $CH_3C(=O)OCH_2CH_3$ ----→ $CH_3C(=O)CH_2C(=O)OCH_2CH_3$ + $CH_3CH_2O^-$

[$CH_3CH_2O{-}C(O^-)(CH_3){-}CH_2C(=O)OCH_2CH_3$]

$CH_3CH_2O{-}C(=O){-}CH_3$ + $^-CH_2{-}C(=O){-}OCH_2CH_3$

It is also possible to utilize two *different* esters in a Claisen-type condensation reaction (a **mixed Claisen condensation**). In order to minimize *self*-condensations, generally one reactant is chosen which has **no a-hydrogens**, and that reactant is maintained in large excess over the second reactant (which will have a-hydrogens and can form the enolate anion). An example of this is the synthesis shown below. This b-keto ester can be prepared by elimination of ethoxide from the tetrahedral intermediate formed by the addition of ethoxide anion to the ketone carbonyl. Splitting this tetrahedral intermediate between the a-carbon of the ester portion of the molecule and the tetrahedral center, it is evident that it can be prepared by the condensation of the enol of ethyl acetate and ethyl benzoate, as shown below.

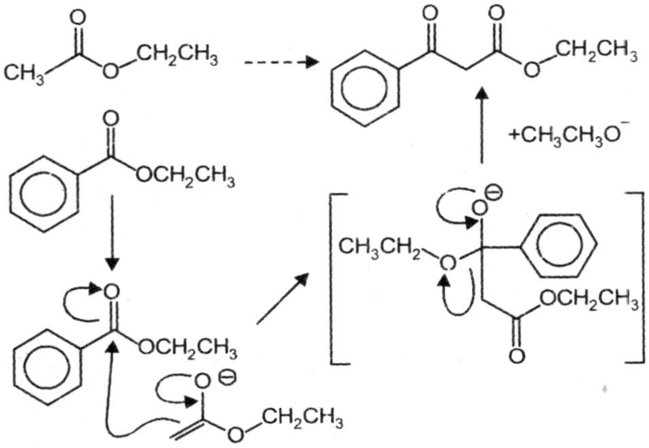

The b-keto esters which are formed in a Claisen condensation are stable in base, but readily decarboxylate in acid solution to give **simple ketones.**

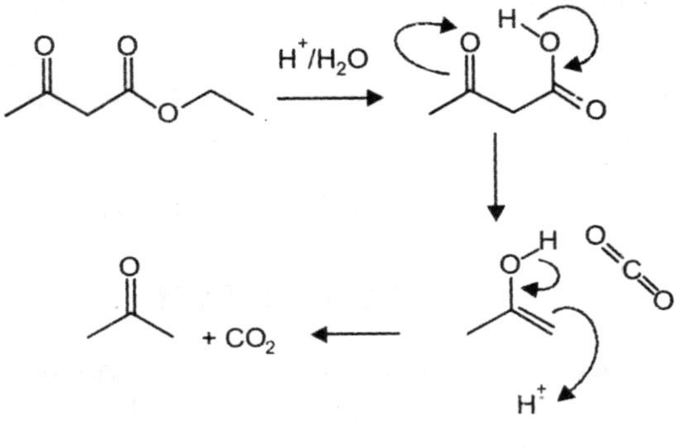

The decarboxylation reaction involves the hydrolysis of the ester to generate a b-keto carboxylic acid. These can undergo acid-catalyzed decarboxylation by the mechanism shown above to give the intermediate enol, which rapidly converts to the corresponding ketone. If the Claisen condensation involved the reaction of two moles of the **same** ester, the product formed will be a **symmetrical ketone** and **the Claisen condensation is an excellent method for the preparation of symmetrical ketones**. If a mixed Claisen condensation was utilized, more complex, unsymmetrical ketones can be prepared.

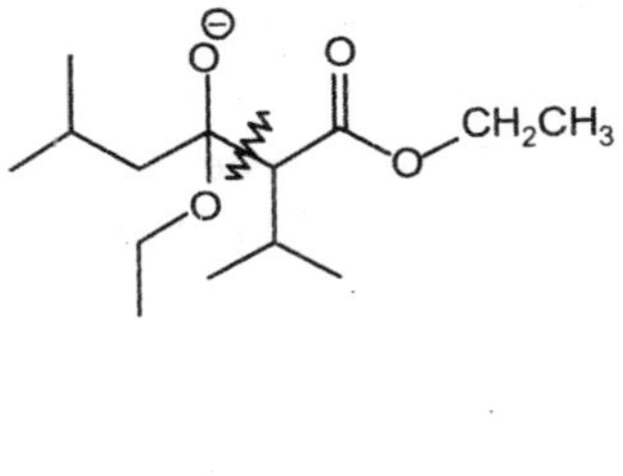

As an example of the use of a Claisen condensation in synthesis, consider the preparation of 2,6-dimethyl-4-heptanone. This molecule is a symmetrical ketone. We can make these by the decarboxylation of b-keto acid in aqueous acid. The required b-keto acid, in turn can be prepared by the Claisen condensation of two moles of ethyl 3-methylbutanoate in the presence of ethoxide anion. Remember, to determine the structure of the required ester, simply add ethoxide to the ketone carbonyl to form the tetrahedral intermediate, shown on the right, and split the compound at the bond between the tetrahedral center and the a-carbon of the ester portion of the molecule. The complete synthesis is shown below.

1. $CH_3CH_2O^-$

2. H^+/H_2O

loss of CO_2

Uses of Esters

- Esters are used as artificial perfumes or scents as they emit a sweet smell.
- Esters are used in making artificial food flavours that are added in many edible items like ice creams, soft drinks, sweets, etc.
- Esters are used as industrial solvents for making cellulose, fats, paints and varnishes.
- Esters are used as solvents in pharmaceutical industries.
- Esters are used as softeners in plastic industries and molding industries

ETHYL ACETATE

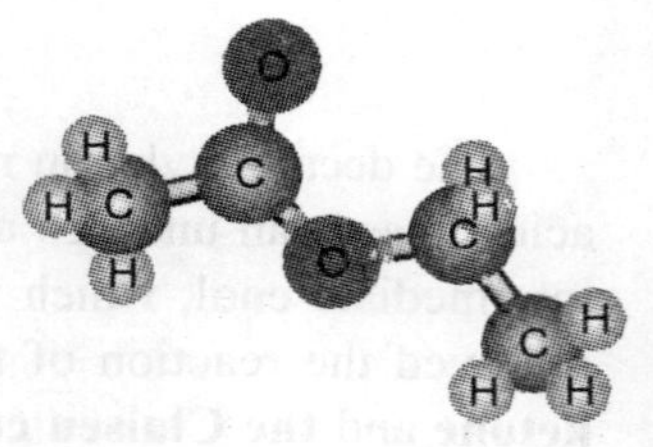

Ethyl acetate, also known as **acetic acid ethyl ester, ethyl ethanoate, acetic ether** or **acetic ester**, is a clear, flammable liquid with a characteristic, not unpleasant smell like certain glues or nail polish removers. From the chemical point of view it is an ester with formula: $CH_3COOCH_2CH_3$. It may be formed (along with acetic acid) as a contaminant in wine that has been exposed to air. Ethyl acetate is used as a solvent in glues and nail polish removers, in chemical reactions, and for extractions. Ethyl acetate is a non-polar (lipophilic) to weakly polar (hydrophilic) aprotic solvent with an 8% solubility in water. Ethyl acetate is not stable in the presence of strong aqueous bases and acids.

Ethyl acetate is an ester that is synthesized from acetic acid and ethanol in the presence of strong acids like sulfuric acid in an esterification reaction. The two reactants and the H_2SO_4 catalyst are heated under reflux for approx. 40 min.

$$CH_3CH_2OH + CH_3COOH \rightarrow CH_3COOCH_2CH_3 + H_2O$$

Because the reaction is reversible and produces an equilibrium, yield is low unless driven to the right by removal of water. The yield can also be increased by using an acid chloride, acetyl chloride (CH_3COCl), instead of the carboxylic acid. This is usually performed in the presence of a base such as pyridine (to remove HCl) and since it does not result in an equilibrium more of the ester is produced. It is unstable in the presence of strong bases like sodium hydroxide or strong acids like hydrochloric acid, and it is hydrolyzed back into ethanol and acetic acid, especially at elevated temperature.

Preparation

- Mix $50cm^3$ of ethanol and $50cm^3$ of *glacial* ethanoic acid thoroughly in a $250cm^3$ round-bottomed flask. -**Explain the meaning of the term *glacial* as applied to ethanoic acid.**
- Glacial ethanoic acid is pure ethanoic acid CH_3COOH. It is called glacial because on cooling it forms glassy crystals.
- Add *slowly with cooling* and shaking $10cm^3$ of *concentrated sulphuric acid*. -**Why must the concentrated sulphuric acid be added slowly and with cooling?**
- The concentrated sulphuric acid on dilution gives out a lot of heat; the slow addition with cooling is necessary to avoid splashing if the mixture gets hot.
- **Why is concentrated sulphuric acid used in this recipe?**
- Concentrated sulphuric acid is a ***catalyst*** for the esterification reaction

$$CH_3CH_2OH + CH_3COOH \rightleftharpoons CH_3COOCH_2CH_3 + H_2O$$

- Ensure that the mixture is *homogeneous*, then fit the flask with a *reflux water-condenser* and boil the mixture gently *for 10 minutes*. Rearrange the position of the condenser for distillation, and distil off about two-thirds of the mixture. -**Why must the mixture be homogeneous?**
- Concentrated sulphuric acid is much denser than any of the other reagents. If not well mixed initially, the solution is liable to get too hot and boil uncontrollably when mixing occurs later in the reaction.
- **What is a reflux water-condenser?**
- A reflux water condenser is a Liebig condenser arranged vertically above the reaction flask; vapors are condensed and returned to the flask, the contents of which can therefore be boiled for long periods without any loss of material.
- **Why is the reaction comparatively slow?**
- Most organic reactions are slow since they involve the breaking of strong covalent bonds. The proportion of molecular collisions that have the necessary activation energy is usually fairly low.
- Transfer the distillate to a separating funnel and add about 25 cm^3 of 30% *sodium carbonate solution*. Stopper the funnel, invert it, and shake, *opening the tap from time to time.*
- Allow the two layers to separate and carefully run off and reject the lower layer, ensuring that the sodium carbonate is removed as completely as possible.

- **-What is the function of the sodium carbonate solution?**
- The distillate contains traces of ethanoic acid and perhaps some sulphuric acid. Sodium carbonate solution removes this.
- **Why do you need to open the tap from time to time?**
- The neutralisation with sodium carbonate produces carbon dioxide gas; opening the tap releases this and avoids a build-up of pressure that might blow the stopper out of the funnel.
- Prepare a solution of 25g of calcium chloride in 25 cm^3 of water, *add it to the crude ethyl ethanoate* in the funnel, and shake vigorously. Allow the mixture to separate, and run off the lower aqueous layer as completely as possible. **-What is the function of the calcium chloride at this stage of the preparation?**
- The crude ethyl ethanoate contains traces of ethanol; the calcium chloride solution removes this since it complexes with the ethanol
- Run the ethyl ethanoate into a small conical flask, add a few lumps of granular *anhydrous calcium chloride*, and shake occasionally until the *liquid is clear*. **-What is the function of the calcium chloride at this stage of the preparation?**
- The solid calcium chloride is a dehydrating agent; it removes any remaining traces of water from the product.
- Then decant the liquid into a *clean, dry* 100 cm^3 round-bottom flask, add some *anti-bumping granules*, and arrange for distillation including a 0-100°C thermometer in the apparatus. The distilling flask should be placed in a cold water bath, which is gradually heated. *The ether that is always formed* in this reaction will distil off at 35-40°C, and may be discarded. Continue to heat, and collect the fraction that boils between 74° and 79°C. **-Why should the distillation flask be clean and dry?**
- You have just dried the mixture; you don't want it wet again.
- **What is the nature and purpose of anti-bumping granules?**
- Anti-bumping granules are small pieces of silica; broken unglazed pottery works as well. This provides a nucleus on which gas bubbles grow, therefore avoiding the sudden production of large gas bubbles which can lead to 'bumping'. This, properly called succussion, causes liquid to splash over into the condenser and therefore produce an impure product. Severe bumping can even lead to loss of material through vents or can blow a distillation apparatus apart.
- **Why is an ether formed in this reaction?**
- An ether is formed because sulphuric acid can cause the elimination of a molecule of water between two alcohol molecules. The water causes the ionisation of the sulphuric acid:

$$CH_3CH_2OH + HOCH_2CH_3 + H_2SO_4 \rightleftharpoons CH_3CH_2OCH_2CH_3 + H_3O^+ + HSO_4^-$$

Physical Properties

General	
Systematic name	Ethyl acetate
Other names	acetic acid ethyl esterethyl ethanoateacetic ester
Molecular formula	$C_4H_8O_2$
SMILES	CC(=O)OCC
Molar mass	88.11 g/mol
Appearance	Clear, colourless liquid
CAS number	[141-78-6]
Properties	
Density and phase	0.897 g/cm^3, liquid
Solubility in water	8.3 g/100 ml (20 °C)
Solubility in ethanol,acetone, diethyl ether,benzene	Miscible
Melting point	"83.6 °C (189.55 K)
Boiling point	77.1 °C (350.25 K)
Critical temperature	250.11 °C (523.26 K)
Viscosity	0.426 cP at 25 °C
Structure	
Dipole moment	1.78 D

Chemical Properties

(*i*) Alkyl alkanoates can be hydrolyzed into its component acid and alkanol. The reaction is catalysed by an acid or alkali

$$CH_3COOC_2H_{5(l)} + H_2O_{(l)} \xLeftrightarrow{\text{alkali/acid}} CH_3COOH_{(aq)} + C_2H_5OH_{(aq)}$$

Note that if an alkali is used as the catalyst, it will immediately react with the alkanoic acid formed to produce the salt of the acid.

$$CH_3COOC_2H_{5(l)} + KOH_{(aq)} \Leftrightarrow CH_3COOK_{(aq)} + C_2H_5OH_{(aq)}$$

(*ii*) When ammonia is bubbled through an ester, like ethyl ethanoate, ethanol and ethanamide, CH_3CONH_2 are produced.

$$CH_3COOC_2H_{5(l)} + NH_{3(g)} \Leftrightarrow CH_3CONH_{2(s)} + C_2H_5OH_{(aq)}$$

(*iii*) Ethyl ethanoate is reduced to ethanol by hydrogen from a strong reducing agent like lithium tetrahydridoaluminate(III), $LiAlH_4$.

$$CH_3COOC_2H_{5(aq)} + 4[H] \xrightarrow{\text{from } LiAlH_4} 2C_2H_5OH_{(aq)}$$

Uses

Ethyl acetate is a very effective poison for use in insect collecting and study (entomology). In a jar charged with ethyl acetate, the vapors will kill the collected (usually adult) insect quickly without destroying it. It also keeps the insect soft enough to allow proper mounting suitable for a collection.

Other Esters

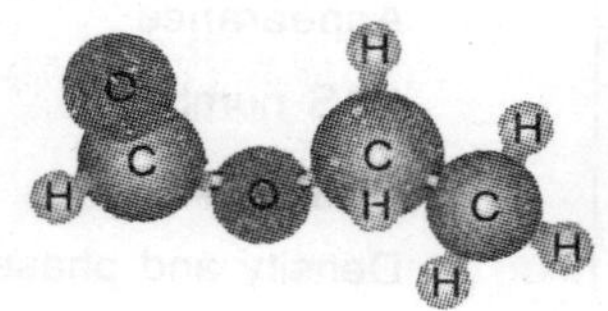

Ethyl formate, also known as **Ethyl methanoate** is an ester formed when ethanol reacts with formic acid.

Physical Properties

Boiling point: 52-54°C	Relative density of the vapour/air-mixture at 20°C (air = 1): 1.4
Melting point: -80°C	Flash point: -20°C c.c.
Relative density (water = 1): 0.92	Auto-ignition temperature: 440°C
Solubility in water, g/100 ml at 20°C: 10.5	Explosive limits, vol% in air: 2.7-16.5
Vapour pressure, kPa at 20°C: 25.6	Octanol/water partition coefficient as log Pow: 0.23
Relative vapour density (air = 1): 2.55	

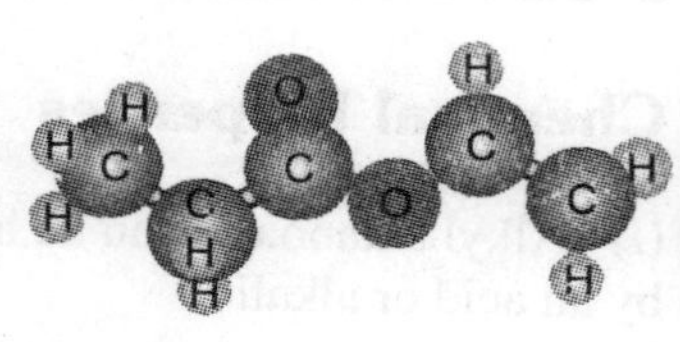

Ethyl propionate, is a clear, colourless liquid with a slightly sweetish odor. It is soluble in water and alcohol; melts at -21 C, boils at 141 C. It is used as a preservative in feed and food industry directly or in the forms of ammonium propionate, calcium and sodium propionates. It is used to manufacture various propionates which used in the reduction of pharmaceuticals, anti-fungal agents, agrochemicals, plastics, plasticizers, rubber chemicals, dyes, artificial flavours and perfumery synthetics. It is used also as a solvent and in nickel-electroplating solutions. Ethyl Propionate is used in perfumery and fragrance. Odor description: fruity rum juicy fruit.

Physical Properties

PHYSICAL STATE	clear liquid
MELTING POINT	-75 C
BOILING POINT	98 - 100 C

SPECIFIC GRAVITY	0.886 - 0.889
SOLUBILITY IN WATER	Slightly soluble
VAPOR DENSITY	
REFRACTIVE INDEX	1.383 - 1.385
NFPA RATINGS	Health: 1; Flammability: 2; Reactivity: 0
FLASH POINT	26 C
STABILITY	Stable under ordinary conditions

Ethyl Butyrate, is used as a solvent and plasticizer for cellulose. It is used flavor extracts and an ingredient of fragrance. Odor description: fruity juicy pineapple cognac.

Physical and Chemical Properties

PHYSICAL STATE	clear liquid
MELTING POINT	-100 C
BOILING POINT	120 C
SPECIFIC GRAVITY	0.875 - 0.885
SOLUBILITY IN WATER	Slightly soluble
REFRACTIVE INDEX	1.390 - 1.395
NFPA RATINGS	Health: 1; Flammability: 2; Reactivity: 0
FLASH POINT	26 C
STABILITY	Stable under ordinary conditions

Ethyl Valerate, is a member of the series of fatty acids of colourless liquid with a penetrating aroma; slightly soluble in water, soluble in alcohol, and ether; boils at 185 C . It is used in flavourings, perfumes, plasticizers and pharmaceuticals. This compound reacts with bases, oxidizing agents and reducing agents. Valeric acid and its derivatives are used as components and intermediates in perfume manufacturing; flavours, ester-type lubricants; plasticizers, pharmaceuticals, vinyl stabilizers. Ethyl **Valerate is used in fruit flavourings.**

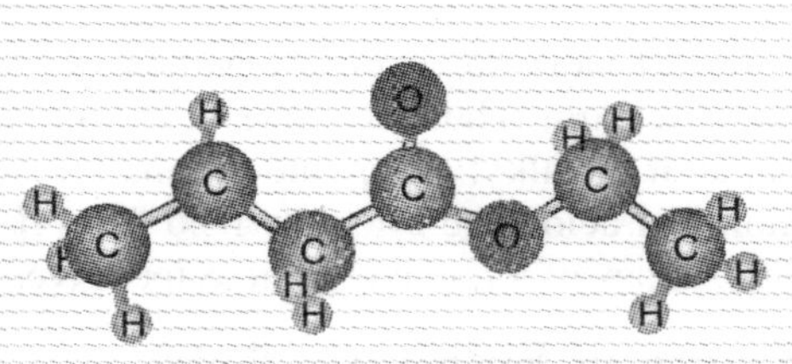

Physical and Chemical Properties

PHYSICAL STATE	clear liquid
MELTING POINT	-91 C
BOILING POINT	145 C
SPECIFIC GRAVITY	0.87 - 0.88
REFRACTIVE INDEX	1.397 ± 0.02
FLASH POINT	35 C
STABILITY	Stable under ordinary conditions

Esters of Inorganic Acids

The term ester is applied to compounds obtained by the reaction of an alcohol with an acid. This even, through the reacting with alcohol is inorganic, the product is an ester. Hence, the ester, derived form inorganic acids such as sulphuric acid, sulphurous acid, nitric acid, nitrous acid, carbonic acid, phosphoric acid, boric acid etc. are referred to as ***Inorganic Esters***.Theoretically these esters are to be regarded as the derivation of intoganci acids in which one or more inonizable H atoms, which are replaced by as many as alkyl groups. For e.g.

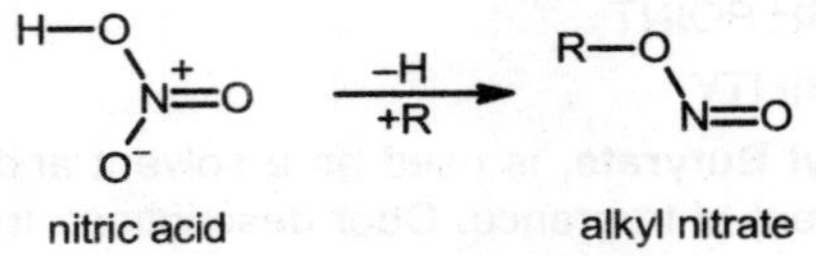

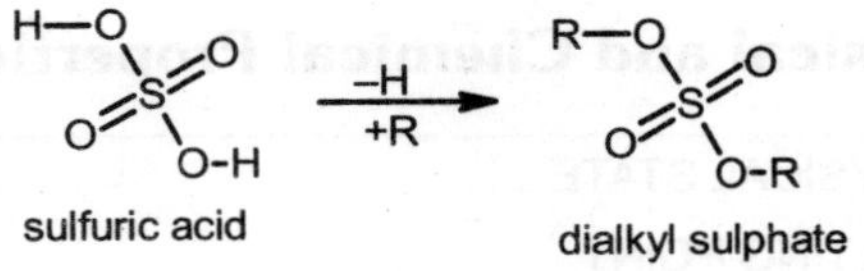

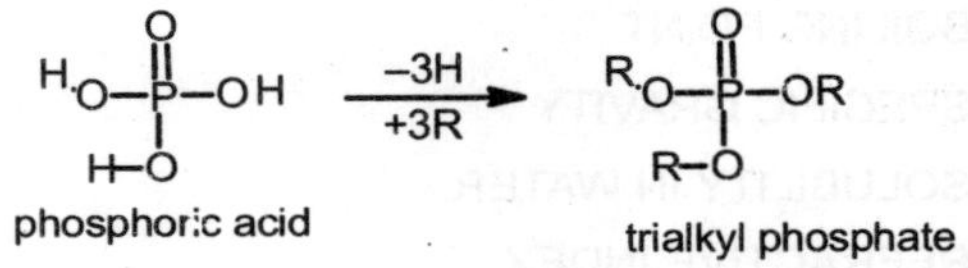

nitric acid —(−H, +R)→ alkyl nitrate

sulfuric acid —(−H, +R)→ dialkyl sulphate

phosphoric acid —(−3H, +3R)→ trialkyl phosphate

Truly speaking, the alkyl halides are also inorganic esters of hydrogen halides HBr, HI). But they differ from the esters of other inorganic acids in the sense that they do not contain oxygen atom. Further more, alkyl halides are versatile synthetic reagents and almost all classes of organic compounds can be prepared from them. Because of these reasons, alkyl halides merit a separate treatment. Here we will discuss esters of inorganic acids other than those of hydrogen halides.

In general, inorganic esters have properties similar to those of organic esters. They can be prepared by :

(a) The reaction of an alcohol with an inorganic acid chloride;†

(b) The reaction of an alcohol with inorganic acid;††

(c) The reaction of an alcohol with an inorganic anhydride.

Nomenclature

Esters of inorganic acids are named like salts, the name of alkyl group being substituted: for that of the metallic atom. Thus the name of an individual ester is written as two separate words, the first being the name of the alkyl group and second the name of the acid residue obtained by removal of replaceable H atoms of acid. The esters derived from *di.* and *tri-basic* acids are named as their *di.* and *tri-alkyl* salts. The names of a few inorganic esters given in the Table will illustrate their nomenclature.

Esters of Sulphuric Acid

Sulphuric acid forms two series of esters numerically acid and normal esters accordingly as one or both the replaceable H atoms have been substituted by alkyl groups.

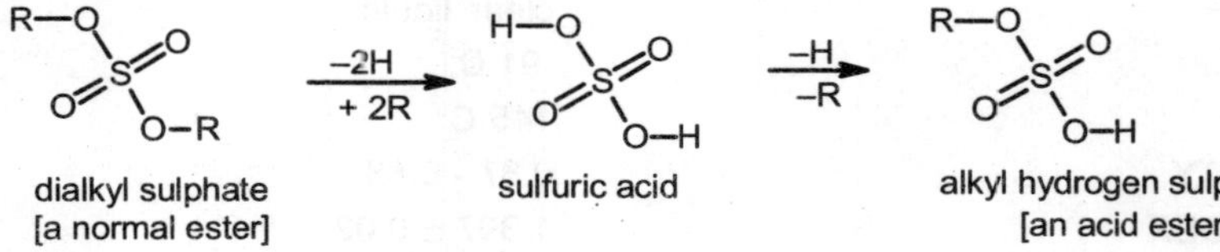

dialkyl sulphate [a normal ester] —(−2H, + 2R)— sulfuric acid —(−H, −R)→ alkyl hydrogen sulphate [an acid ester]

† The first method is a general one, while the other two methods fail when the inorganic acid is too weak.

†† The first method is a general one, while the other two methods fail when the inorganic acid is too weak.

Table 14.1. Names of some Inorganic Esters

Formula	Acid Group	Name of Acid Group	Name of Ester
$CH_3-O-S(=O)_2-O-H$	$-O-S(=O)_2-O-H$	hydrogen sulphate	methyl hydrogen sulphate
$CH_3-O-S(=O)_2-O-CH_3$	$-O-S(=O)_2-O-$	sulphate	dimethyl sulphate
$C_2H_6-O-\overset{+}{N}(-O)=O$	$-O-\overset{+}{N}(-\bar{O})=O$	nitrate	ethyl nitrate
$C_4H_9-O-N=O$	$-O-N=O$	nitrate	n-butyl nitrite
$C_2H_5-O-S(=O)-O-C_2H_5$	$-O-S(=O)-O-$	sulphite	diethyl sulphite
$C_2H_5-O-C(=O)-O-C_2H_5$	$-O-C(=O)-O-$	carbonate	diethyl carbonate
$C_2H_5-O-P(=O)(-O-C_2H_5)-O-C_2H_5$	$-O-P(=O)(-O)-O-$	phosphate	triethyl phosphate
$C_2H_6-O-B(=O)(-O-C_2H_6)-O-C_2H_6$	$-O-B(=O-)(-O-)-O-$	borate	triethyl borate

Among the ester of sulphuric acid, ethyl hydrogen sulphate, dimethyl sulphate and diethyl sulphate are important

ETHYL HYDROGEN SULPHATE
(Ethyl Sulphuric Acid)

Introduction

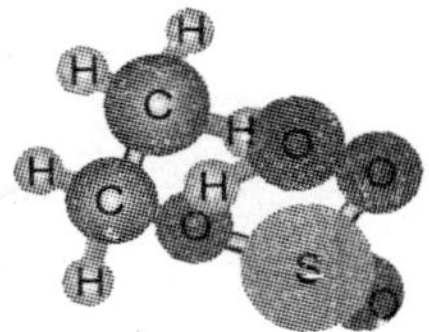

It is the acid ester of sulphuric acid, and can be prepared by the following methods: -

(*i*) By heating ethyl alcohol with success of sulphuric acid

$$\underset{\text{ethanol}}{CH_3CH_2OH} + \underset{\text{sulfuric acid}}{HO-SO_2-OH} \rightleftharpoons \underset{\text{ethyl hydrogen sulphate}}{CH_3CH_2-O-SO_2-OH} + H_2O$$

> However, in the presence of cuprous or silver salts, the reaction can be carried out at room temperature

(*ii*) By passing ethylene through conc. sulphuric acid at about 100°C (industrial)

$$H_2C{=}CH_2 + H_2SO_4 \rightleftharpoons C_2H_5OSO_3H$$

ethylene + sulfuric acid ⇌ ethyl hydrogen sulphate

Physical Properties

Molecular Formula	$C_2H_6O_4S$
Formula Weight	126.13164
Composition	C(19.04%) H(4.79%) O(50.74%) S(25.42%)
Molar Refractivity	25.20 ± 0.3 cm^3
Molar Volume	81.2 ± 3.0 cm^3
Parachor	238.7 ± 4.0 cm^3
Index of Refraction	1.532 ± 0.02
Surface Tension	74.3 ± 3.0 dyne/cm
Density	1.551 ± 0.06 g/cm^3
Polarizability	9.99 ± 0.5 $10^{-24}cm^3$
Monoisotopic Mass	125.998679 Da
Nominal Mass	126 Da
Average Mass	126.131536 Da

Chemical Properties

Ethyl hydrogen sulphate behaves as an acid as well as ester.

***(1) Acidic Nature*:** It is a genuine acid and is capable of exchanging their as yet non-replaced hydrogen for metals in the usual, manor. This ethyl hydrogen sulphate decomposes carbonates and bi-carbonates to liberate carbon dioxide.

$$C_2H_5OSO_3H + K_2CO_3 \longrightarrow C_2H_5OSO_3K + CO_2\uparrow + H_2O$$

ethyl hydrogen sulphate → ethyl potassium sulphate

$$C_2H_5OSO_3H + Na_2CO_3 \longrightarrow C_2H_5OSO_3Na + CO_2\uparrow + H_2O$$

ethyl hydrogen sulphate → ethyl sodium sulphate

Sodium and potassium derivatives of ethyl hydrogen sulphate are very reactive and are used to prepare alkyl halides thioalcohols and ethyl thioesters.

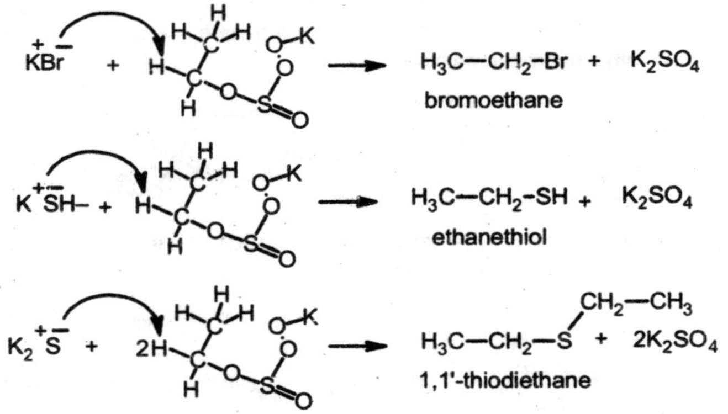

$\overset{+}{K}\overset{-}{Br}$ + $CH_3CH_2OSO_2OK$ → H_3C-CH_2-Br + K_2SO_4

bromoethane

$K^+ SH^-$ + $CH_3CH_2OSO_2OK$ → H_3C-CH_2-SH + K_2SO_4

ethanethiol

$K_2^+S^-$ + 2 $CH_3CH_2OSO_2OK$ → $H_3C-CH_2-S-CH_2-CH_3$ + $2K_2SO_4$

1,1'-thiodiethane

***(2) Hydrolysis*:** ethyl hydrogen sulphate when heated with boiling water is hydrolysed back to ethyl alcohoi and sulphuric acid.

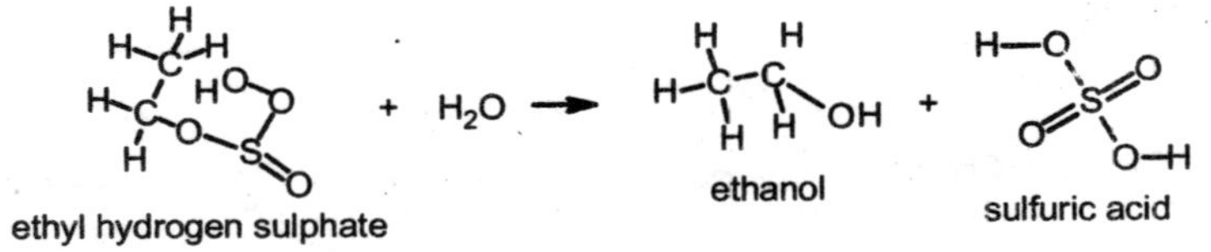

$CH_3CH_2OSO_2OH$ + H_2O → CH_3CH_2OH + H_2SO_4

ethyl hydrogen sulphate — ethanol — sulfuric acid

***(3) Action of Heat*:** When heated alone, it decomnoses to give diethyl sulphate and sulphuric acids

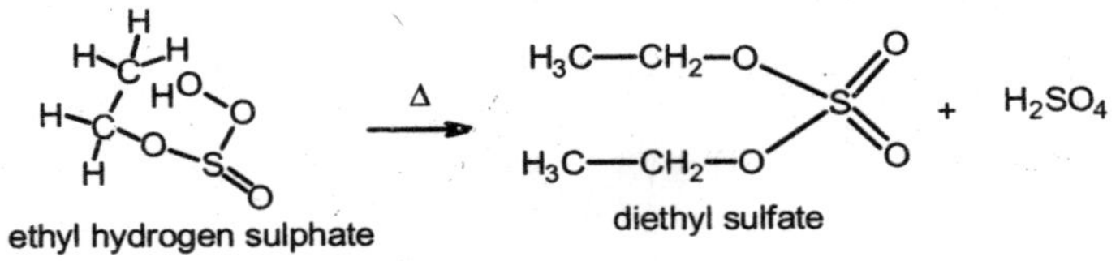

$CH_3CH_2OSO_2OH \xrightarrow{\Delta} (H_3C-CH_2-O)_2SO_2 + H_2SO_4$

ethyl hydrogen sulphate — diethyl sulfate

***(4) Action with Sulphuric acid*:** When heated in excess of sulphuric acid at 120°C it produces ethylene.

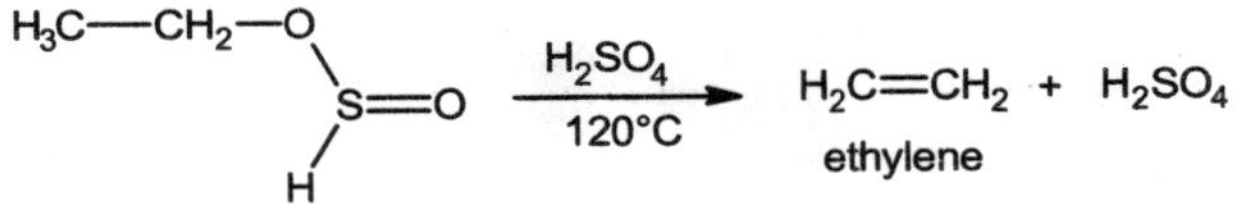

$H_3C-CH_2-O-S(=O)-H \xrightarrow[120°C]{H_2SO_4} H_2C=CH_2 + H_2SO_4$

ethylene

Mechanism

The possible mechanism of this reaction is

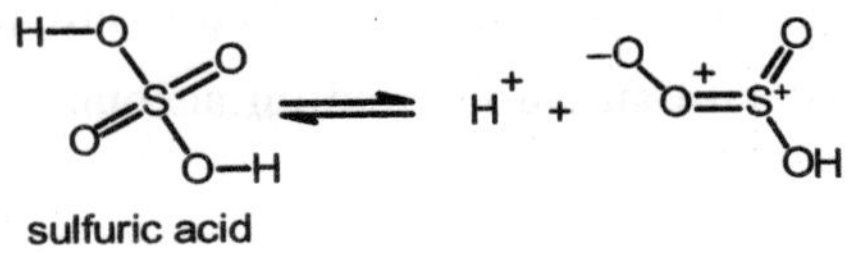

$H_2SO_4 \rightleftharpoons H^+ + {}^-O-O=\overset{+}{S}(=O)-OH$

sulfuric acid

$$O{=}S{=}O + CH_3{-}CH_2{-}O{-}S({=}O){-}OH \longrightarrow H_2C{=}CH_2 + H{-}O{-}S({=}O)({-}O^-){-}O{-}H + {}^-O{-}S^+({=}O)({=}O){-}OH$$

oxosulfane oxide; ethyl hydrogen sulfite; ethylene

$$H^+ + {}^-O{-}S^+({=}O)({=}O){-}OH \rightleftharpoons H{-}O{-}S({=}O)({=}O){-}O{-}H$$

***(5) Action with Alcohol*: When reacts with ethyl alcohol at 140°C, ethyl hydrogen sulphate yields** diethyl ether.

$$CH_3CH_2{-}O{-}S(O){-}O{-}H + CH_3CH_2{-}O{-}H \xrightarrow{140°C} H_3C{-}CH_2{-}O{-}CH_2{-}CH_3 + H_2SO_4$$

diethyl ether

Mechanism:

$$H_3C{-}CH_2{-}O{-}H + H_3C{-}CH_2{-}O{-}S({=}O){-}HO \rightleftharpoons H_3C{-}CH_2{-}\overset{+}{O}(H){-}CH_3 + {}^-O{-}S({=}O){-}OH \longrightarrow H_3C{\cdot}CH_2O(CH_2CH_3) + H_2SO_4$$

Uses: Ethyl hydrogen sulphate is frequently used to prepare ethyl halides, ethyl thio alcohols and ethyl thioether in a fairly good yield.

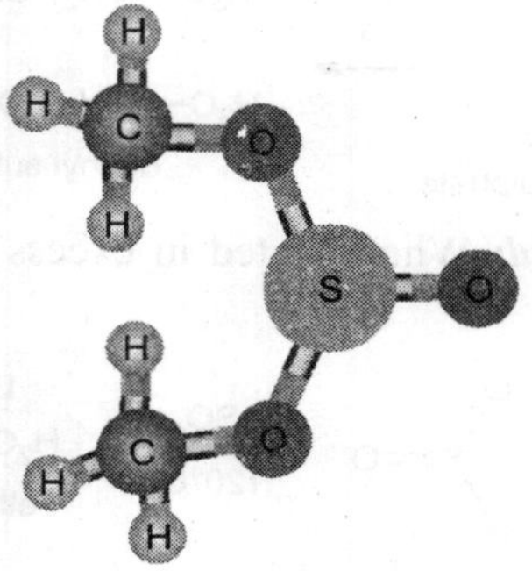

DIMETHYL SULPHATE

Methyl hydrogen sulphate is then distilled under reduced pressure to yield di methyl sulphate

It is the most important ester of sulphuric acid on account of its great value as a methylating agent.

Preparation

(*a*) This is prepared by the reaction of methyl alcohol and concentrated sulphuric acid and then heating the mixture

$$CH - OH + H_2SO_4 \xrightarrow{\Delta} CH_3 - O - SO_2OH + H_2SO_4$$

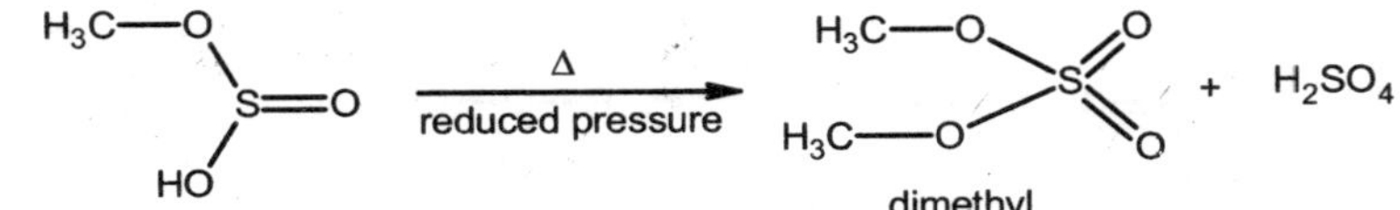

$H_3C—O—S(=O)—HO$ $\xrightarrow[\text{reduced pressure}]{\Delta}$ $(H_3C—O)_2S(=O)_2$ + H_2SO_4

methyl hydrogen sulphate; dimethyl sulphate

(*b*) It can also be obtained by the action of methyl iodide and silver sulphate

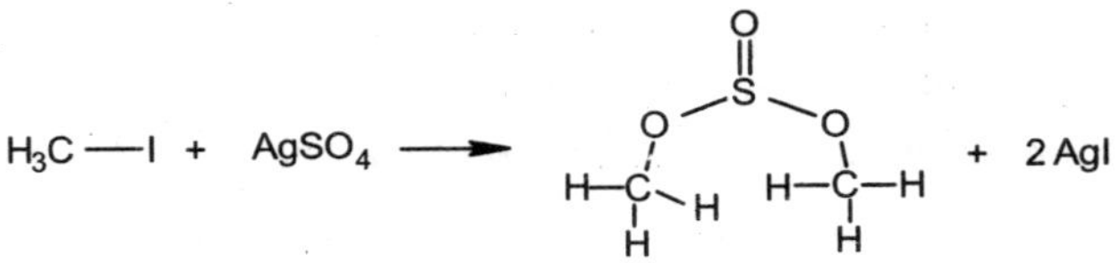

$H_3C—I$ + $AgSO_4$ ⟶ $(CH_3O)_2S=O$ + 2 AgI

dimethyl sulphate

(*c*) Technically it is prepared by passing sulphur trioxide (2 mols) into methyl alcohol (2 mols)

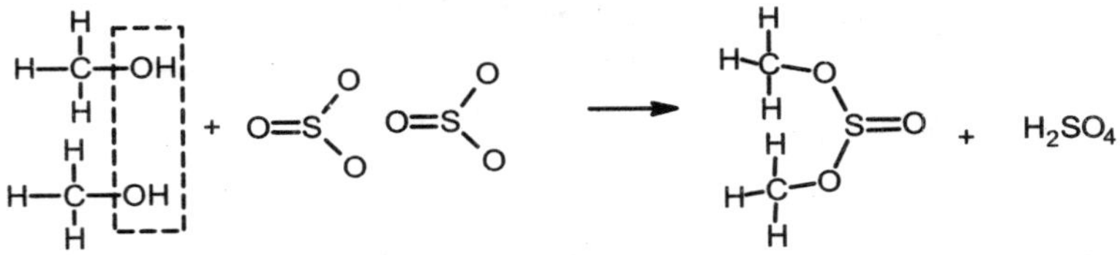

$2\,CH_3OH$ + $O=SO_2$ $O=SO_2$ ⟶ $(CH_3O)_2S=O$ + H_2SO_4

methyl alcohol; SO_3 molecule; dimethyl sulphate

Physical Properties

Molecular Formula	$C_2H_6O_3S$
Formula Weight	110.13224
Composition	C(21.81%) H(5.49%) O(43.58%) S(29.11%)
Molar Refractivity	$23.66 \pm 0.3\ cm^3$
Molar Volume	$83.7 \pm 3.0\ cm^3$
Parachor	$221.9 \pm 4.0\ cm^3$
Index of Refraction	1.476 ± 0.02
Surface Tension	49.2 ± 3.0 dyne/cm
Density	$1.314 \pm 0.06\ g/cm^3$
Polarizability	$9.38 \pm 0.5\ 10^{-24}cm^3$
Monoisotopic Mass	110.003764 Da
Nominal Mass	110 Da
Average Mass	**110.132231 Da**

Chemical Properties

1. A Methylating Agent: In the presence of sodium and potassium hydroxide, di methyl sulphate react with hydroxyl and amino compounds, whereby the replaceable H atoms in then is situated by methyl group. This reaction which is reflected to Methylating may be repeated by the following equation.

$R{-}OH$ + dimethyl sulphate + $Na{-}OH \longrightarrow R{-}O{-}CH_3$ + sodium methyl sulphate + H_2O

Although only of the methyl groups takes part in the reaction, dimethyl sulphate is a valuable methylating agent. In this respect it is superior to methyl iodide since it reacts with greater rapidity and gives yields Dimethyl sulphate is particularly effective for the methylation of phenols and carbohydrates, through amines, through and carbolic acid may be methylated with it

Examples: (a) Methylation of Phenols to give methyl phenyl ether

phenol + dimethyl sulphate + $NaOH \longrightarrow$ methoxybenzene + sodium methyl sulphate + H_2O

Mechanism

Methylation of phenol is nucleophilic substitution reaction and follows the following mechanism

phenol + $NaOH \longrightarrow$ $C_6H_5{-}O{-}Na$ + H_2O

$C_6H_5{-}O{-}Na$ + dimethyl sulphate $\longrightarrow$ methoxybenzene + sodium methyl sulphate

(*b*) Methylation of aniline to give *N*-methylamine

aniline + dimethyl sulphate + $NaOH \longrightarrow C_6H_5{-}NH{-}CH_3$ + sodium methyl sulphate + H_2O

Mechanism

$$C_6H_5NH_2 + (CH_3O)_2SO_2 \longrightarrow [C_6H_5N^+H_2CH_3]\; O{=}S^{+}{-}O{-}CH_3$$

$$[C_6H_5N^+H_2CH_3]\; O{=}S^{+}{-}O{-}CH_3 + NaOH \longrightarrow C_6H_5N^+H_2CH_3 + H_3C{-}O{-}S({=}O){-}O{-}O{-}Na + H_2O$$

Similarly N-methylamine in presence of excess of dimethyl sulphate in presence of sodium hydroxide give methyl esters.

$$R{-}\overset{O}{\overset{\|}{C}}{-}OH + (CH_3O)_2SO + Na{-}OH \longrightarrow R{-}\overset{O}{\overset{\|}{C}}{-}O{-}CH_3 + Na{-}O{-}S({=}O){-}O{-}CH_3 + H_2O$$

(*d*) Thiols may also be methylated with dimethyl sulphate in presence of sodium hydroxide

$$\underset{\text{ethanethiol}}{H_3C{-}CH_2{-}SH} + (CH_3O)_2SO + NaOH \longrightarrow H_3C{-}CH_2{-}S{-}CH_3 + Na{-}O{-}S({=}O){-}O{-}CH_3 + H_2O$$

Dimethyl sulphate is a valuable methylating agent and is particularly useful for determining the size of the ring of the carbohydrates

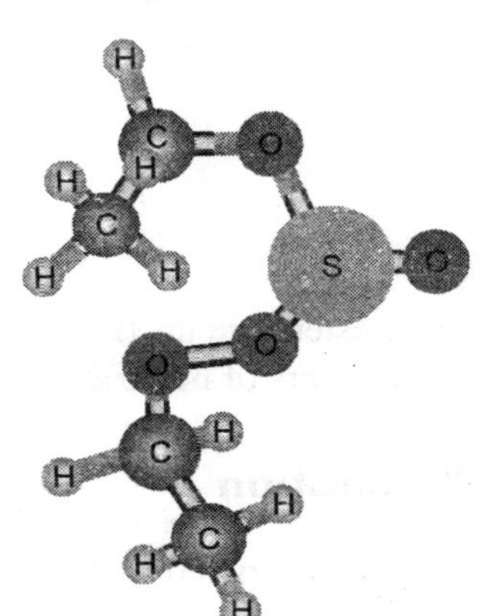

DIETHYL SULPHATE OR ETHYL SULPHATE

Preparation

It can be prepared by the same method as an applicable to dimethyl sulphate. However, industrially diethyl sulphate is produced by passing ethylene in cold conc. sulphuric acid.

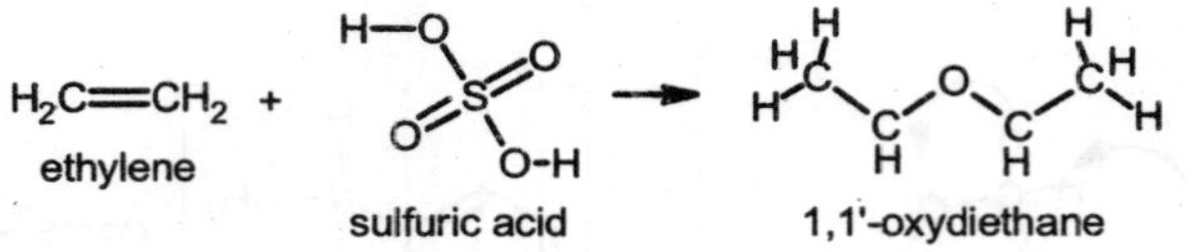

$H_2C{=}CH_2$ + sulfuric acid → 1,1'-oxydiethane

ethylene; sulfuric acid; 1,1'-oxydiethane

Physical Properties

Molecular Formula	$C_4H_{10}O$
Formula Weight	74.1216
Composition	C(64.82%) H(13.60%) O(21.59%)
Molar Refractivity	22.32 ± 0.3 cm^3
Molar Volume	100.9 ± 3.0 cm^3
Parachor	210.9 ± 4.0 cm^3
Index of Refraction	1.360 ± 0.02
Surface Tension	19.0 ± 3.0 dyne/cm
Density	0.734 ± 0.06 g/cm^3
Polarizability	8.85 ± 0.5 $10^{-24}cm^3$
Monoisotopic Mass	74.073165 Da
Nominal Mass	74 Da
Average Mass	74.123212 Da

Uses

Diethyl Sulphate is used as an ethylation agent in organic synthesis

ESTERS OF NITRIC ACID

Since nitric acid has only one replaceable hydrogen atom, it forms only one type of esters commonly known alkyl nitrates.

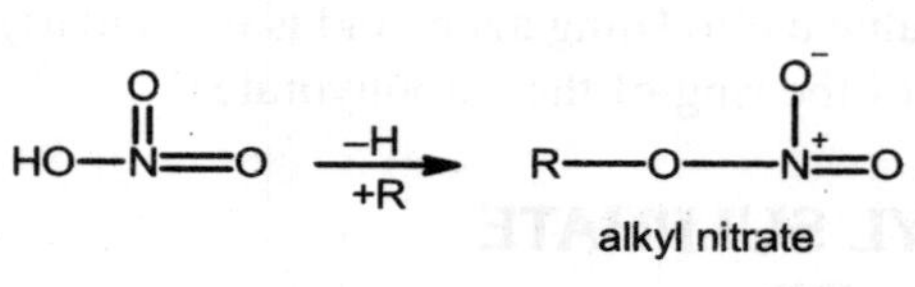

$$HO{-}N(=O){=}O \xrightarrow[+R]{-H} R{-}O{-}\overset{+}{N}(O^-){=}O$$

alkyl nitrate

Or

These esters are unstable and sometimes especially in contact with nitric acid decomposes violently. Of the esters of nitric acid only ethyl nitrate is important and will be discussed here.

Preparation

It may be prepared by the following methods

(1) By the action of concentrated nitric acid in presence of urea at low temperature.

ethanol + nitric acid —urea→ ethyl nitrate + H_2O

Urea is added to destroy any nitrous acid formed as a result of the reduction of nitric acid by ethyl alcohol.

(2) By heating ethyl alcohol iodide with ethanoic silver nitrite solution

$$H_3C—CH_2—I + AgNO_3 \xrightarrow{\text{ethanol}} C_2H_5ONO_2 + AgI$$

Physical Properties

Molecular Formula	*$C_2H_5NO_3$*
Formula Weight	91.066
Composition	C(26.38%) H(5.53%) N(15.38%) O(52.71%)
Molar Refractivity	19.08 ± 0.3 cm^3
Molar Volume	80.6 ± 3.0 cm^3
Parachor	189.9 ± 4.0 cm^3
Index of Refraction	1.388 ± 0.02
Surface Tension	30.7 ± 3.0 dyne/cm
Density	1.128 ± 0.06 g/cm^3
Polarizability	7.56 ± 0.5 $10^{-24}cm^3$
Monoisotopic Mass	91.026943 Da
Nominal Mass	91 Da
Average Mass	91.066591 Da

Chemical Properties

(1) Hydrolysis: - On boiling with water, ethyl nitrate is hydrolysed to form ethyl alcohol and nitric acid

ethyl nitrate + H_2O → ethanol + nitric acid

(2) Reduction: When reduced with tin and hydrochloric acid, it is converted to ethyl alcohol and hydroxylamine.

$$\underset{\text{ethyl nitrate}}{CH_3CH_2-O-\overset{+}{N}(=O)-O^-} + 6[H] \xrightarrow{S_N/HCl} CH_3CH_2-OH + \underset{\text{hydroxylamine}}{H_2N-OH} + H_2O$$

This action shows closely that in ethyl nitrate molecule, the ethyl group is linked nitrogen through an oxygen atom

ESTERS OF NITROUS ACID

ETHYL NITRITE

Preparation

(*i*) By the distillation of a mixture of ethyl alcohol, sodium nitrite and con. Hydrochloric acid or sulphuric acid.

$$NaNO_2 + HCl \longrightarrow NaCl + HNO_2$$

$$H_3C-CH_2-OH + HNO_2 \longrightarrow \underset{\text{nitroethane}}{H_3C-CH_2-\overset{+}{N}(=O)-O^-} + H_2O$$

(*ii*) By the action of ethyl nitrogen trioxide (nitrous anhydrous in ethyl alcohol)

(*iii*) By the reaction of ethyl iodide with potassium nitrite

$$H_3C-CH_2-I + KNO_2 \longrightarrow \underset{\text{nitroethane}}{H_3C-CH_2-\overset{+}{N}(=O)-O^-} + KI$$

Physical Properties

Molecular Formula	$C_2H_5NO_2$
Formula Weight	75.0666
Composition	C(32.00%) H(6.71%) N(18.66%) O(42.63%)
Molar Refractivity	17.33 ± 0.3 cm^3
Molar Volume	74.3 ± 3.0 cm^3

Parachor	170.2 ± 4.0 cm^3
Index of Refraction	1.382 ± 0.02
Surface Tension	27.5 ± 3.0 dyne/cm
Density	1.009 ± 0.06 g/cm^3
Polarizability	6.87 ± 0.5 $10^{-24}cm^3$
Monoisotopic Mass	75.032028 Da
Nominal Mass	75 Da
Average Mass	75.067286 Da

Chemical Properties

It is reaction are similar to those of ethyl nitrate.

(I) Hydrolysis: Upon hydrolysis with alkalies (NaOH or KOH) ethyl nitrite forms ethyl alcohols and nitrous acid.

$$CH_3CH_2NO_2 \text{ (nitroethane)} + H_2O \longrightarrow H_3C-CH_2-OH \text{ (ethanol)} + HNO_2$$

$$HNO_2 + NaOH \longrightarrow Na-O-N=O \text{ (sodium nitrite) or } NaNO_2$$

(II) Reduction: When reduced with tin and hydrochloric acid, ethyl nitrite yields nitrite ethyl alcohol and ammonia

$$CH_3CH_2NO_2 + 6[H] \xrightarrow{S_N/HCl} CH_3CH_2OH + NH_3 + H_2O$$

Uses

1.Ethyl nitrite is easily broken and reacts with organic compounds in the same as nitrous acid and is used with advantage in 'organic synthesis as the reaction can carried out in the absence of water.

> The above reaction furnishes evidence that in ethyl nitrite molecule, the ethyl group is not directly linked to the N atom but is joined to it through O atom

2. It accelerates the pulse rate and lowers the blood lure and is, therefore, used as medicine for the treatment of asthma and heart diseases.

ISOAMYL NITRIATE

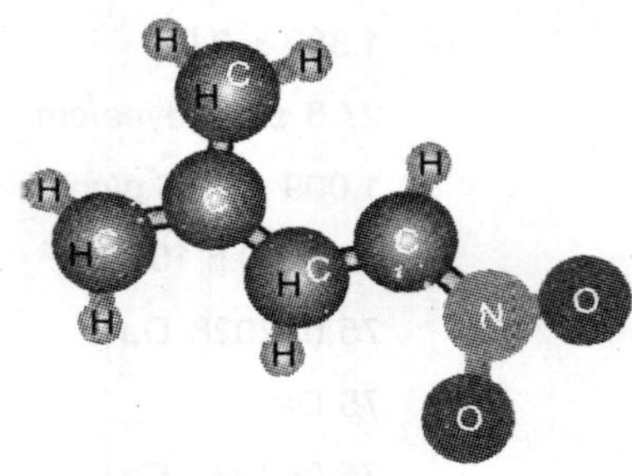

Preparation

(i) By heating a mixture of isoamyl alcohol and sodium nitrite and conc. sulphuric acid

$$NaNO_2 + H_2SO_4 \rightarrow NaHSO_4 + HNO_2$$

$$(H_3C)_2CH-CH_2-CH_2-OH + H-N^{+}(=O)O^{-} \longrightarrow \text{isoamyl nitrite}$$

isoamyl alcohol; nitrous acid; isoamyl nitrite

$$(H_3C)_2CH-CH_2-CH_2-OH + O=N-O-N=O \longrightarrow (H_3C)_2CH-CH_2-CH_2-CH_2-O-N=O$$

isoamyl alcohol; isomethyl nitrite

Physical Properties

Boiling Point	97-99°C
Molecular Formula	$C_5H_{11}NO_2$
Formula Weight	117.14634
Composition	C(51.26%) H(9.46%) N(11.96%) O(27.32%)
Molar Refractivity	31.19 ± 0.3 cm^3
Molar Volume	124.2 ± 3.0 cm^3
Parachor	287.0 ± 4.0 cm^3
Index of Refraction	1.416 ± 0.02
Surface Tension	28.4 ± 3.0 dyne/cm
Density	0.943 ± 0.06 g/cm^3
Polarizability	12.36 ± 0.5 $10^{-24}cm^3$
Monoisotopic Mass	117.078979 Da
Nominal Mass	117 Da
Average Mass	117.148253 Da

Esters of Carbonic Acid

These are called alkyl carbonates as they are derivatives of carbonic acid.

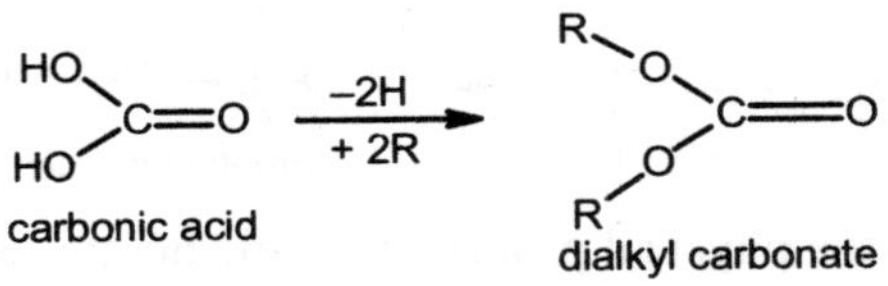

$$(HO)_2C{=}O \xrightarrow[+\,2R]{-2H} (RO)_2C{=}O$$

carbonic acid → dialkyl carbonate

Alkyl carbonates can be prepared by the action of phosgene or alcohols or alcohols or on sodium alkoxide. Thus,

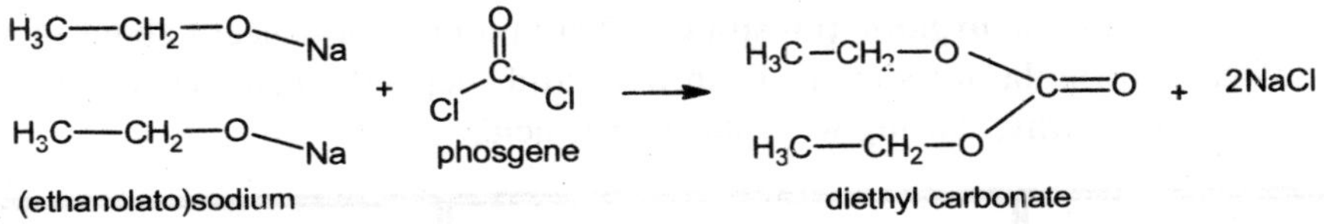

$$2\,H_3C{-}CH_2{-}O{-}Na + Cl{-}CO{-}Cl \longrightarrow (H_3C{-}CH_2{-}O)_2C{=}O + 2NaCl$$

(ethanolato)sodium + phosgene → diethyl carbonate

Di alkyl carbonates are colourless liquids

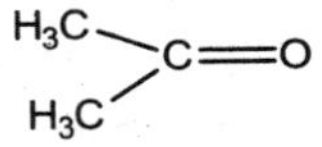

$(H_3C)_2C{=}O$ dimethyl carbonate b.p 90 - 91°C

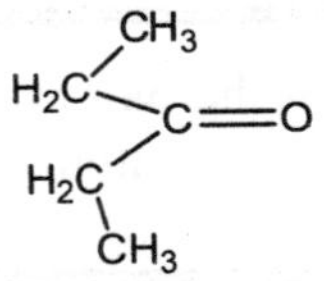

$(CH_3{-}H_2C)_2C{=}O$ diethyl carbonate b.p 126°C

They react with Grignard's reagent to give tertiary alcohols;

Thus

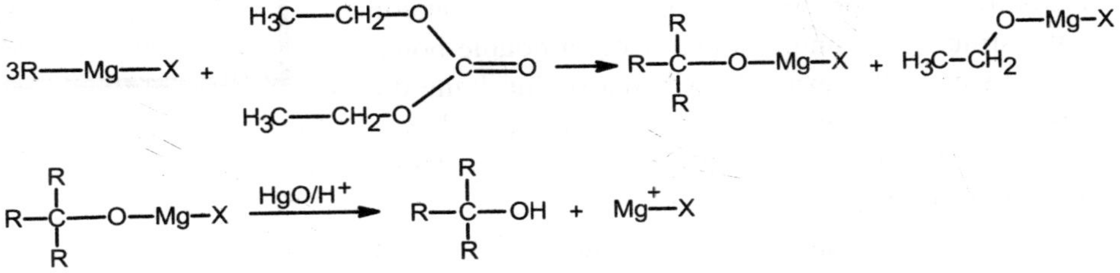

$$3R{-}Mg{-}X + (H_3C{-}CH_2{-}O)_2C{=}O \longrightarrow R_3C{-}O{-}Mg{-}X + H_3C{-}CH_2{-}O{-}Mg{-}X$$

$$R_3C{-}O{-}Mg{-}X \xrightarrow{HgO/H^+} R_3C{-}OH + \overset{+}{Mg}{-}X$$

Esters of Boric Acid

These are called alkyl borates.

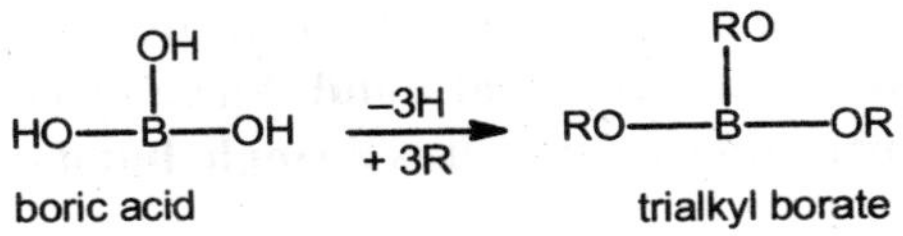

$$HO{-}B(OH){-}OH \xrightarrow[+\,3R]{-3H} RO{-}B(OR){-}OR$$

boric acid → trialkyl borate

Alkyl borates are prepared by reaction of alcohols with boric acid, sometimes sulphuric acid is added as a catalyst.

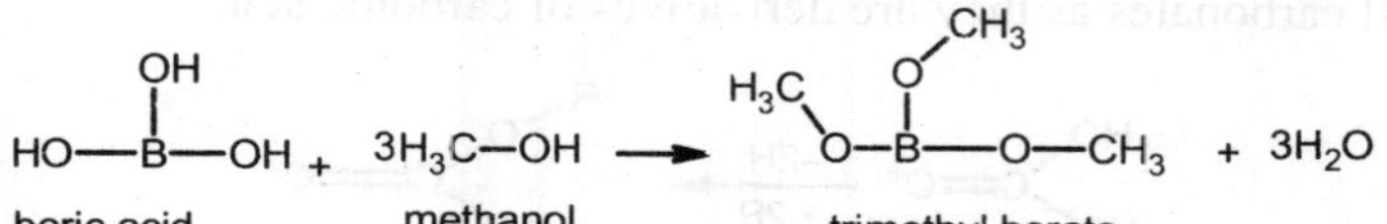

$$B(OH)_3 + 3H_3C{-}OH \longrightarrow B(OCH_3)_3 + 3H_2O$$

boric acid + methanol → trimethyl borate

They are colourless liquids, the physical properties of the borates are shown below in the table.

Esters of Phosphoric Acid

Compounds of this type are the **organic phosphates** that you may have heard about in recent years. Two versions of each formula are shown. Both are valid, depending on what you are trying to emphasize. Let me explain why, starting with phosphoric acid.

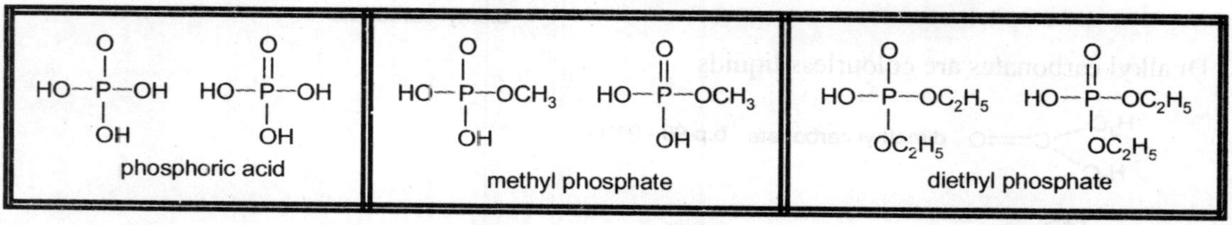

phosphoric acid	methyl phosphate	diethyl phosphate
$O{-}P(OH)_3$ and $O{=}P(OH)_3$	$O{-}P(OH)_2(OCH_3)$ and $O{=}P(OH)_2(OCH_3)$	$O{-}P(OH)(OC_2H_5)_2$ and $O{=}P(OH)(OC_2H_5)_2$

The first structural formula is the structural formula of phosphoric acid. In the past we simply used the molecular formula H_3PO_4. The structural formula shows how the oxygen atoms and the hydrogen atoms are arranged.

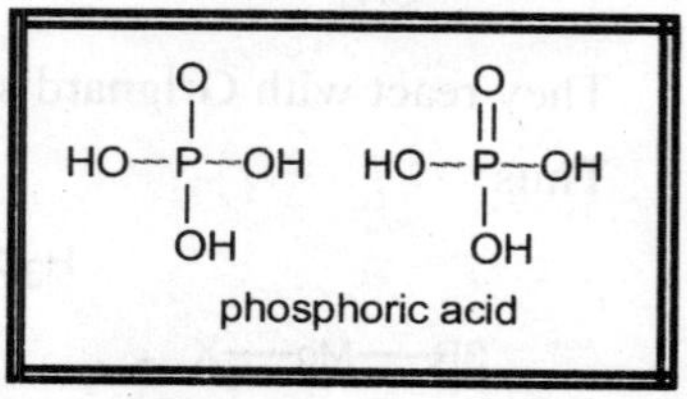

If you look at the formula on the left with a critical eye, you may notice that the oxygen on the top has only one bond to the phosphorus below it. That doesn't seem right for oxygen. That might create a little confusion for you, which I hope to explain. Sometimes you will see the structural formula drawn using a double bond, as shown in the right side of this example. The reason for doing this is that it gives oxygen the two bonds that it is "supposed" to have. Unfortunately, that puts ten valence electrons around phosphorus. However, it is not really a double bond. Instead, it is what is called **coordinate covalent bond**.

phosphoric acid: $O{-}P(OH)_3$ and $O{=}P(OH)_3$

This diagram (which is also shown in Example 19) will help clarify that. We say that oxygen can form two bonds because it needs to gain two electrons. In this particular case, the top oxygen atom gains both of those electrons from the phosphorus atom (red dots here, xs in workbook). The oxygen atom is not contributing any electrons to that bond. So it is simply two electrons from the phosphorus being shared with the oxygen. Since both electrons come from the phosphorus, it is called a **coordinate covalent bond**. Since there are only two electrons in the bond (one pair of electrons), it is a **single bond**.

H:O:P:O:H (electron-dot diagram of phosphoric acid, with O above P and O–H below)

The formation of a phosphate ester from phosphoric acid also involves the splitting out of water in an intermolecular dehydration reaction. This results in the compounds such as those shown below (and in Example 18). In these examples, alkyl groups are attached to one or two of the oxygens which are bonded to phosphorus. We will be dealing with such examples later on when we deal with nucleic acids such as RNA and DNA because the phosphate esters are a very important part of those compounds.

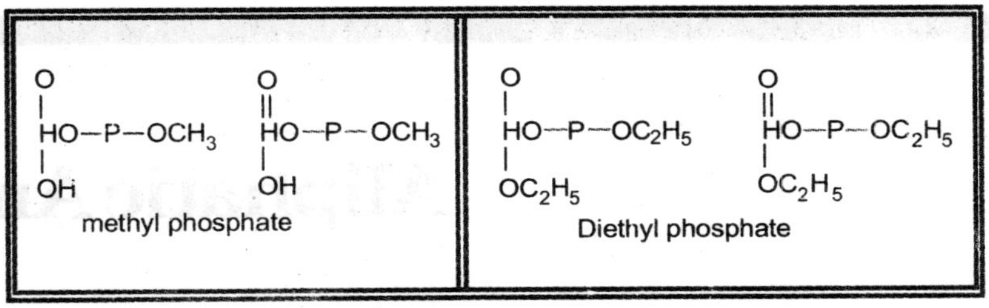

O | HO—P—OCH_3 | OH O ‖ HO—P—OCH_3 | OH methyl phosphate	O | HO—P—OC_2H_5 | OC_2H_5 O ‖ HO—P—OC_2H_5 | OC_2H_5 Diethyl phosphate

Two examples of phosphate esters are shown above. These are named for you, and I am sure you can figure out how we came up with those names just by looking at the names and the structural formulas. Although you should be able to describe and recognize the structural formulas of phosphate esters, you will not be expected to name them.

CHAPTER

15

Aliphatic Amines

Amine are typical organic bases structurally' related to the inorganic base ammonia.

They contain alkyl groups attached directly to nitrogen atom. They could be thought of as derivatives of ammonia (NH.) in which one, two or all three hydrogen have been replaced by alkyl groups. Thus,

NH_3	CH_3—NH_2	CH_3—NH—CH_3	$(CH_3)_3N$
ammonia	methanamine	*N*-methylmethanamine	*N*·*N*-dimethylmethanamine

Primary, Secondary and Tertiary Amines

Amines are designated as *primary* (1°), *secondary* (2°), or *tertiary* according as one, two or three hydrogen atoms in ammonia molecule have been substituted by alkyl groups (R). The general formula and the characteristic functional groups of the three classes of amines are stated below.

Class	*General Formula*			*Functional Group*			
Primary (1°)	R—N(H)(H)	or	RNH_2	—N(H)(H)	or	—Nh_2	amino
Secondary (2°)	R—N(R)(H)	or	R_3NH	—N(—)(H)	or	>NH	imino
Tertiary (3°)	R—N(R)(R)	or	R_3N	—N(—)(—)	or	>N—	tertiary nitrogen atom

In secondary and tertiary amines, the alkyl groups bonded to the nitrogen atom may be same or different. When all the alkyl groups are same the amine is referred to as a Simple Amide or Symmetrical Amine; and when two or all the three alkyl groups are different, the amine is often called, **a Mixed Amine or asymmetrical Amine.**

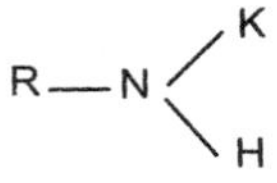

R—N(K)(H) simple 2° amino

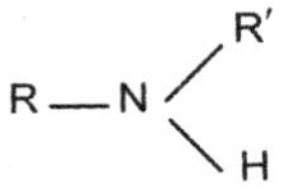

R—N(R′)(H) mixed 2° amine

R—N(R′)(R″) R′—N(R″)(R‴) mixed 3° amines

Tetra alkylammonium Salts or Quaternary Salts. The ammonium ion is a tetrahedrally symmetrical ion having a structure, quite similar to that of methane. It is possible to replace one or more of the H atoms of the ammonium ion by alkyl groups. When all the four hydrogen of ammonium ion have been replaced by alkyl groups, we have *telraalkylammonium ion or quoternary ammonium ion.*

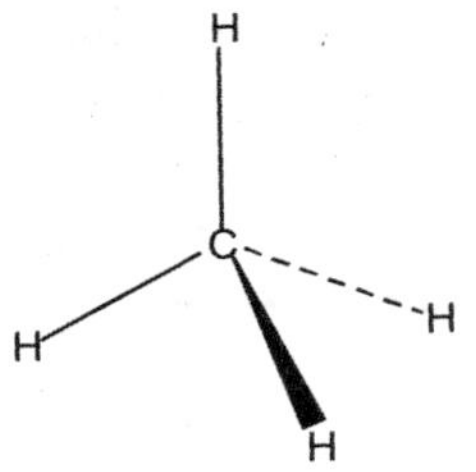

methane 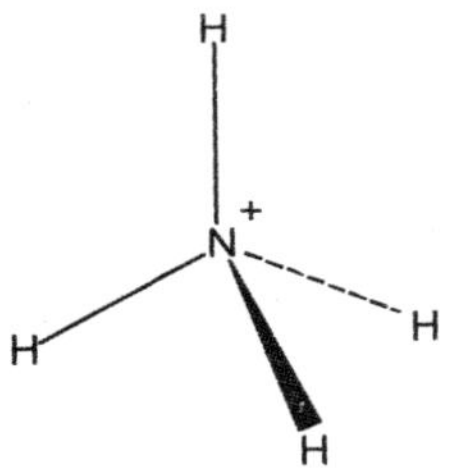 ammonium ion 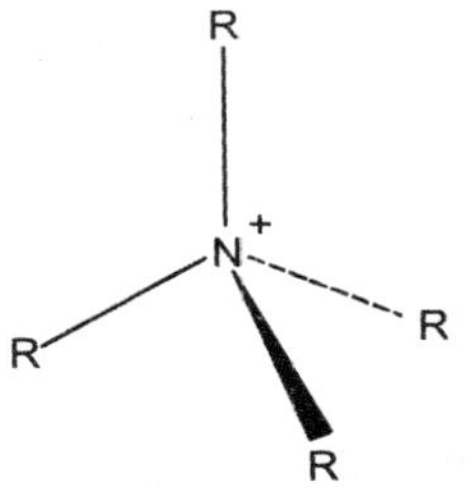 tetraalkylammonium ion

The tetraalkylammonium ion has "built in " ionic charge and interacts with anions salts analogous to ammonium salts.

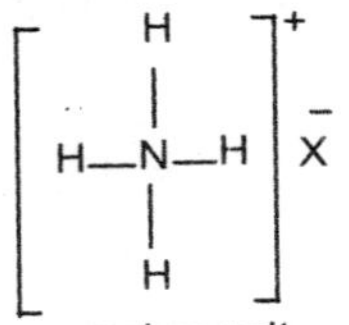

$[NH_4]^+ X^-$ ammonium salt

$[NR_4]^+ X^-$ tetraalkylammonium salt

$[(CH_3)_4N]^+ Cl^-$ tetramethylammonium chloride

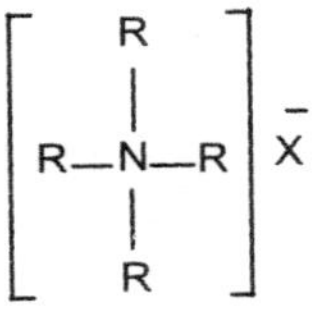

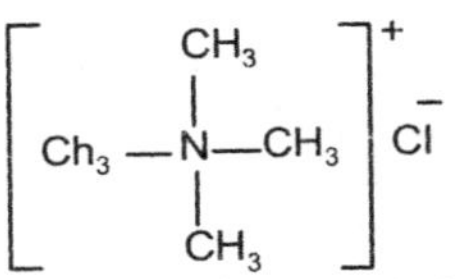

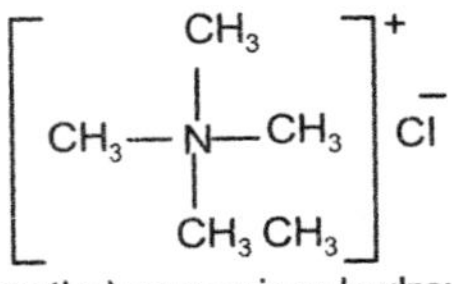

$[(CH_3)_3N(CH_2CH_3)]^+ Cl^-$ ethyltrimethylammonium hydroxid

Here X may be Cl, I, HSO_4, or OH. For example

Diamines and Triamines. It is possible that a single organic molecule may contain two or more amino groups in its structure. If two are present, the compound is named as a *diamine*; if tnere are three -NHz groups present it is referred to as a *triamine,* and so on

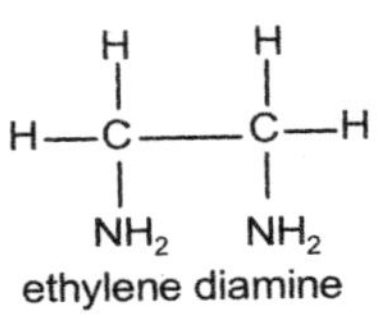

H—CH(NH_2)—CH(NH_2)—H ethylene diamine

H—CH(NH_3)—CH(NH_3)—CH(NH_2)—H propylene triamine

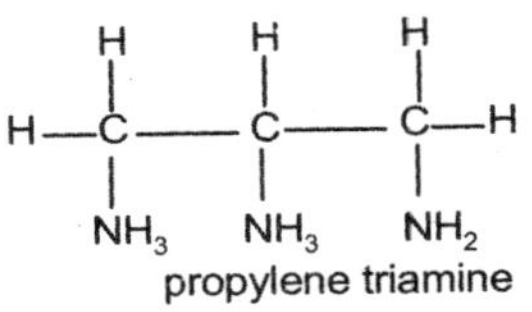

The term 'amine" is widely Used in organic chemistry for the carbon attached directly to N has single bonds to other groups. Thus,

$$R{-}C(=O){-}N\langle^{H}_{H} \quad \text{or} \quad R{-}CO{-}NH_2 \qquad \textit{IS NOT AN AMINE}$$

$$R{-}C(=O){-}CH_2{-}N\langle^{H}_{H} \quad \text{or} \quad R{-}CO{-}CH_2{-}NH_2 \qquad \textit{IS AN AMINE}$$

Any compound, regardless of structure, which has a 1°, 2°, or 3° function in its molecule is termed an amine. This large class of organic compounds includes many substances that are biologically important. Both as free materials and in chemical combination with other molecules, amines are widely distributed in nature. They as particularly present in the nervous systems of vertebrates. Adrenaline which is rapidly released in the body during times of stress is a secondary-amine. Benzedrine is a primary amine which produces frenetic "highs" in the nervous system. Benadryl, a tertiary amine, is useful in inducing sleep as also in treating the symptoms of allergies. Some of the most valuable drugs (connine, nicotine, and amphetamine) obtained from plant extracts contain one or more basic nitrogens in their structures and have, therefore, been assigned the general name alkaloids

Nomenclature

Amines are named both by Common and [IUPAC] systems:

(1) Common System: In this system the name of. an individual amine is derived by..' listing the alkyl group(s) attached to N atom in order of increasing size and adding the word *amine*. The name of

Common names of amines

Structure	Name	Structure	Name
CH_3-NH_2	metylamine	$CH_3-NH-CH_3$	dimethylamine
$CH_3-CH_2-NH_3$	ethylamine	$CH_3-N(CH_3)-CH_3$	trimethylamine
$CH_3-CH_2-CH_2-NH_2$	*n*-propylamine	$CH_3-CH_2-NH-CH_3$	methylethylamine
$CH_3-CH(CH_3)-NH_3$	isopropylamine	$CH_3-CH_2-N(CH_3)-CH_3$	dimethylethylamine
$CH_3-C(CH_3)_2-NH_3$	*tert*-butylamine	$CH_3-CH_2-N(CH(CH_3)_2)-CH_3$	methylethyl-isopropylamine

the amine emerges as a single continuous one word. If an alkyl'group occurs twice or thrice on theN-atom, the prefix *di-* or *tri-* is placed before the name of the alkyl group.

The prefix sec- or tert- is used just to indicate the type of amine and has no structural bearing as in the case of aicohols.

(2) IUPAC System: This system adopts the common names of lower members. However, in practice the alkyl groups on nitrogen are written in alphabetical order and not in order of size or complexity. For example

$H_3C-CH_2-NH-CH_3$

ethyl methyl amine

$H_3C-N(CH_3)-CH(CH_3)_2$

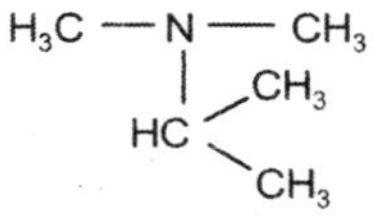

ethyl isopropyl methyl amine

When two of the alkyl groups attached to N atom are identical they are written as di alyl and alphabetized. Thus

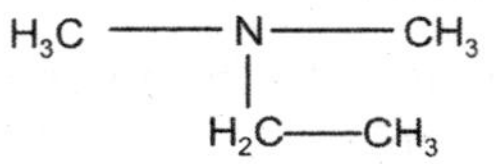

$H_3C-N(CH_3)-CH_2-CH_3$

dimethylethanamine

$H_3C-CH_2-N(CH(CH_3)_2)-CH(CH_3)_2$

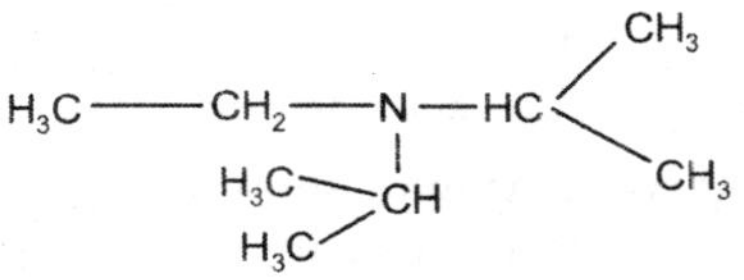

diisopropylethylamine

The IUPAC System is only used to name long chain substituted amines. These names; are derived by regarding the amino group (–NHI) as a substituent and naming the given amine:j aa aminoalkane. As usual tbe longest continuous hydrocarbon chain containing the amint. group is the parent structure and the position on this chain is indicated by arabic numerals.

The counting of the carbons is done from the end that gives lowest number to tbe amino grou"l For example,

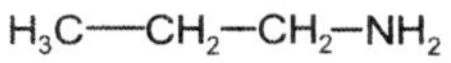

$H_3C-CH_2-CH_2-NH_2$

propan-1-amine

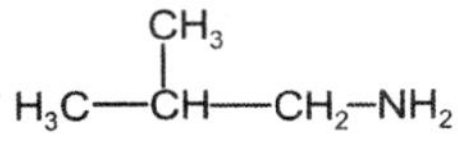

$H_3C-CH(CH_3)-CH_2-NH_2$

2-methylpropan-1-amine

$H_3C-CH(NH_2)-CH(CH_3)-CH_2-CH_3$

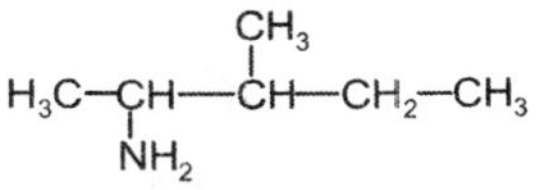

3-methylpentan-2-amine

Further, if the amino group contains alkyl groups (R) as substituent, the word N-alkyl or N, N-dialkyl is prefixed to the word ‘amino’. Thus,

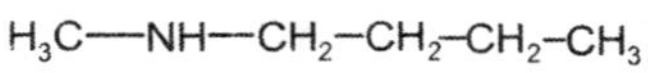

$H_3C-NH-CH_2-CH_2-CH_2-CH_3$

N-methylbutan-1-amine

$H_3C-N(CH_3)-CH(CH_3)-CH_2-CH_2-CH_2-CH_2-CH_3$

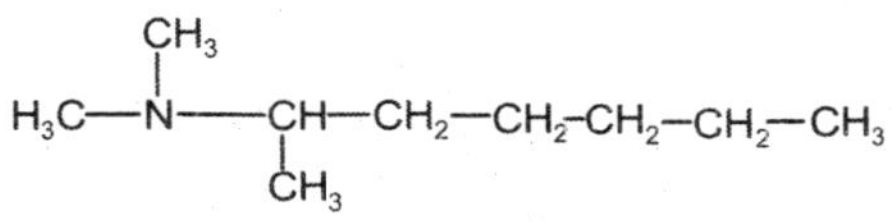

N,N-dimethylheptan-2-amine

If the amino group (-NH.) or substituted amino group (-NHR or -NR.) is part of a molecule .at contains another functional group, the second functional group (say -COOH -OH) determines the parent name of the compound.

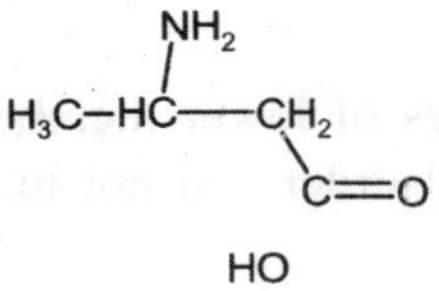

$H_3C-CH(NH_2)-CH_2-C(=O)-OH$
3-aminobutanoic acid

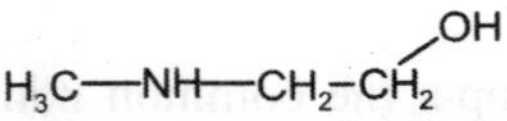

$H_3C-NH-CH_2-CH_2-OH$
2-(methylamino)ethanol

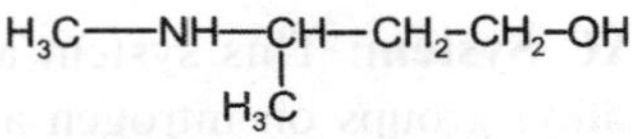

$H_3C-NH-CH(CH_3)-CH_2-CH_2-OH$
3-(methylamino)butan-1-ol

Isomerism

Amines exhibit the following types of isomerism.

1. Chain Isomerism. Like other classes of organic compounds, amines show chain isomerism arising from the different structure of the alkyl groups bonded to the amine function Thus,

$H_3C-CH_2-CH_2-CH_2-NH_2$
butan-1-amine

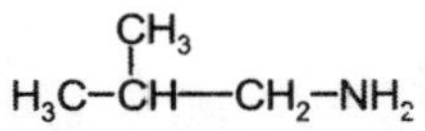

$H_3C-CH(CH_3)-CH_2-NH_2$
2-methylpropan-1-amine

2. Position Isomerism. Amines show position isomerism due to different ,location the amine function -NH., -NHR or -NR, on the carbon chain. For example,

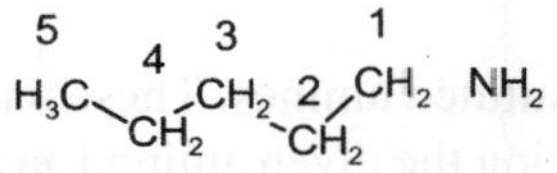

$\overset{5}{H_3C}-\overset{4}{CH_2}-\overset{3}{CH_2}-\overset{2}{CH_2}-\overset{1}{CH_2}-NH_2$
pentan-1-amine

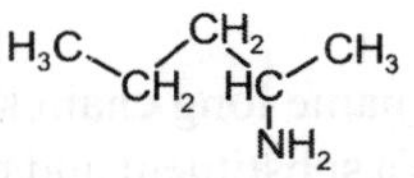

$H_3C-CH_2-CH_2-CH(NH_2)-CH_3$
pentan-2-amine

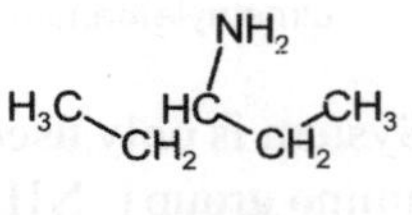

$H_3C-CH_2-CH(NH_2)-CH_2-CH_3$
pentan-3-amine

3. Metamerism in 'amines is exhibited because of, the different alkyl groups bonded the amine function, molecular formula. of the amine remaining the same.

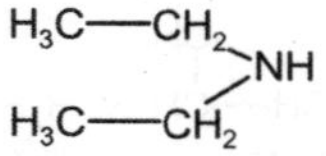

$H_3C-CH_2-NH-CH_2-CH_3$
N-ethylethanamine
diethyl amine

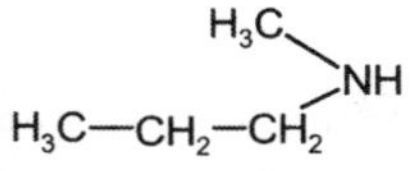

$H_3C-CH_2-CH_2-NH-CH_3$
N-methylpropan-1-amine

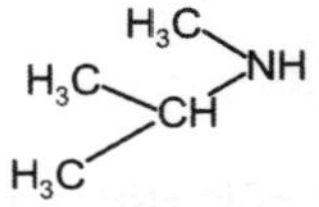

$(H_3C)_2CH-NH-CH_3$
N-methylpropan-2-amine
methyl isopropyl amino

4. Same Molecular Formula represents 1°,1°. 3° amine. Amines show another type of isomerism in which the same molecular formula may stand for a primary, a secondary and a tertiary amine. Thus C_2H_9N represents the following amines

$CH_3-CH_3-CH_3-NH_2$
propylamine
1°

$CH_3-NH-CH_3-CH_3$
ethylmethylamine
2°

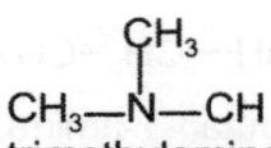

$CH_3-N(CH_3)-CH$
trimethylamine
3°

Structure

The electronic configuration of N atom can be written as Is^1, $2s^2$, $2p^1{}_v$, $2p_y1$, $2p_s1$, The 2s orbital and three 2p orbitals hybridize to give four *Sp3* orbitals. Thus as in case of carbon atom, the four s1" orbitals of nitrogen are directed toward ~ the four corners of a regular tetrahedron.

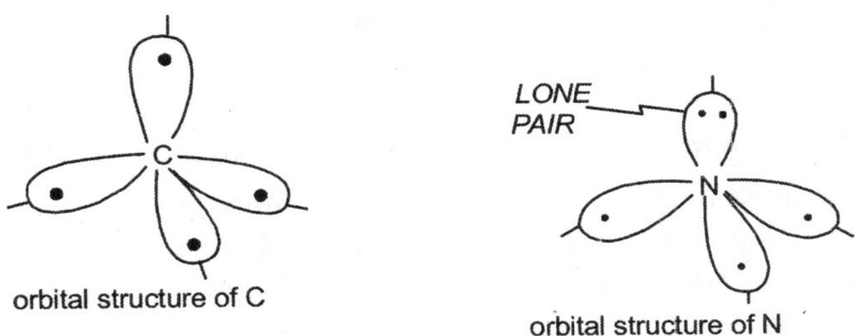

In contrast to carbon, three of the *sp3* hybrid orbitals in nitrogen are occupied by single electrons while the fourth has a pair of electrons. The three half-filled orbitals are. capable of forming *a* bonds by overlap with s orbital of hydrogen atoms or sp' orbitals of carbons of alkyl groups (R). Thus the orbital structure of 1°,2° and 3° amines can be drawn as

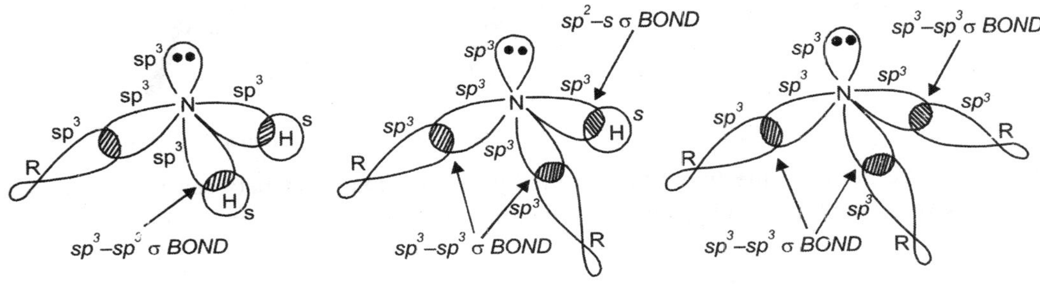

Fig. 15.1. Showing the Orbital structure of 1°, 2° and 3°

Simplifying, the structure of the three types of aplines RNH. (1°), RaNH (2°) and RaN(3°) may be shown as in Fig. 15.1.

It may be noted that in contrast to tetrahedral carbon in methane. The fourth vertex of tetrahedral nitrogen' in amines is occupied by the unshared electron pair. Thus amines have the pyramidal molecules.

The N-atom occupies the apex of the pyramid, while the rest of the H atoms or R groups are at the corners of triangular bases.

The geometry of amines is supported by .the electron diffraction studies of trimethylamines which have shown that its molecule is pyramidal having C-N-C bond angle of 108°.

This is very nearly the same as the H - N - H bond angle in ammonia and H - C - H and in tetrahedral carbon in methane.

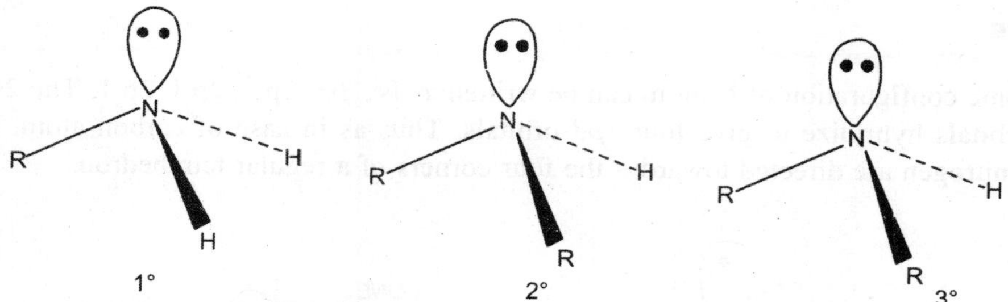

Fig. 15.2 : Showing the Structures representation of three types of amines

The value of bond angle of 107° in ammonia is accounted by the ;act that lone pair-bond pair interactions are greater than bond pair-bond pair interactions, thereby contracting the angle from normal tetrahedral value of 109-5° to 107°C.(in ammonia) The bond angle m try methylamine is 108°C which is slightly more than the bond angle of ammonia. This can be attributed to greater.

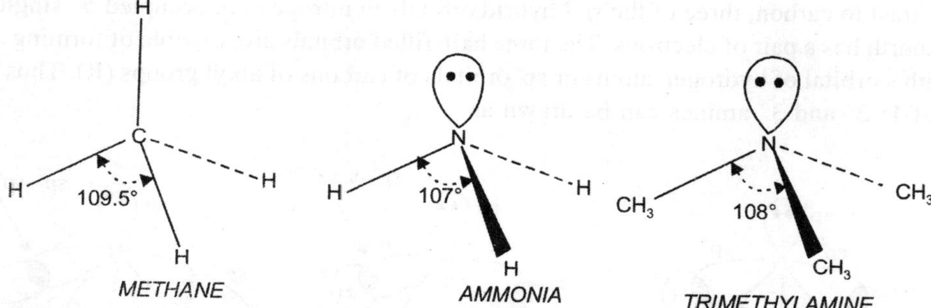

steric hindrance between the two adjacent methyl groups than that between two hydrogen atoms, thereby widening the bond angle from 107° (in ammonia) to 108° (in trimethylamine).

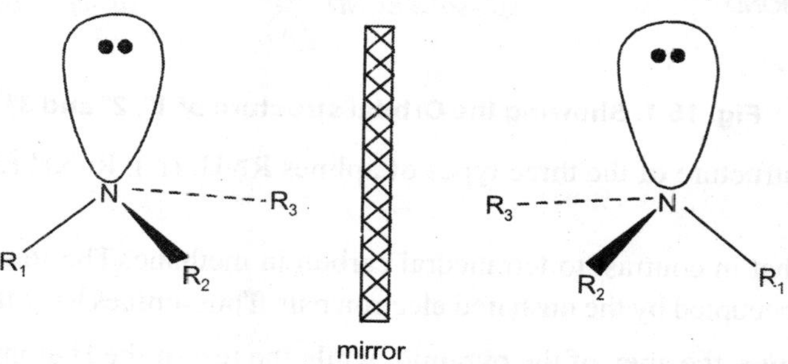

Fig. 15.3 : Mirror image forms of a tertiary amine.

Why Asymmetric Amines cannot be Resolved into Enantioners? Let us examine the molecule of an amine containing three different alkyl groups attached to nitrogen in light of the orbital structure discussed above. It will be revealed (Fig. 15.3) that this is not super imposable on its mirror image. Therefore, such an amine is asymmetric or dissymmetric and should exist in two enantiomeric forms

(1 and II) similar to compounds with an asymmetric carbon atom. But all attempts to resolve this type of amine *e.g.,* ethylmethyl-*n*-propylamine, have failed. Spectroscopic evidence, however, suggests that an asymmetric amine undergoes a rapid inversion of FORM I to its mirror image FORM II. . This inversion of I to II could be visualized to take place through. a planar transition slate like the umbrella in a windstorm.

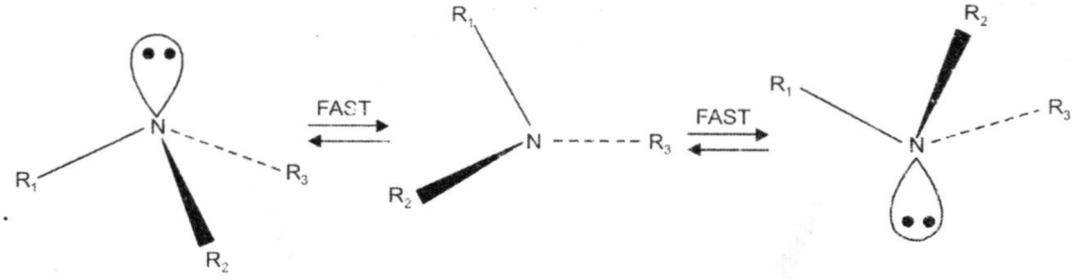

I (ORIGINAL FORM) III (PLANAR TRANSITION FORM) II (INVERTED FORM)

The inversion from form I to II requires only a small amount of energy (ΔE_{inv} = 5 K/*cal.mole*)). This energy barrier of 5 kcal/mole is easily met at room temperature by the thermal collisions of amine molecules. Therefore, a fast transformation between the forms I and II takes place. For [hat reason. no resolution of the enantiomers of asymmetric amines is possible.

However, the inversion of enantiomers in asymmetric amines can be suppressed by ammonium cation formation. The coordination of the nonbonding electron pair with a proton fixes them in a bond and thus prevents the inversion.

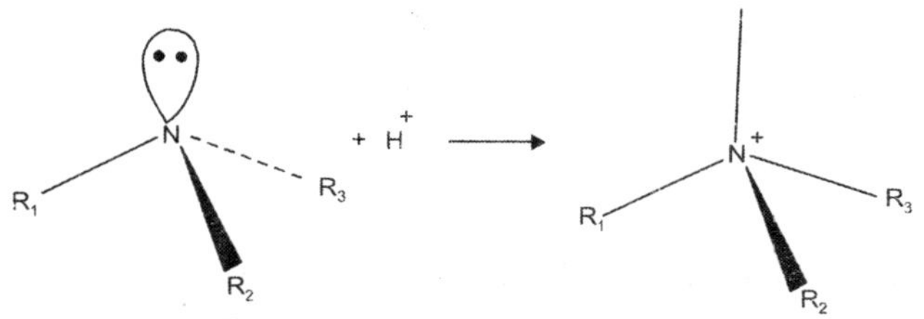

asymmetric 3° amine can undergo inversion

substituted ammonium cation does not undergo inversion

Substututed ammonium cations therefore are capable of exhibiting enantiomerism

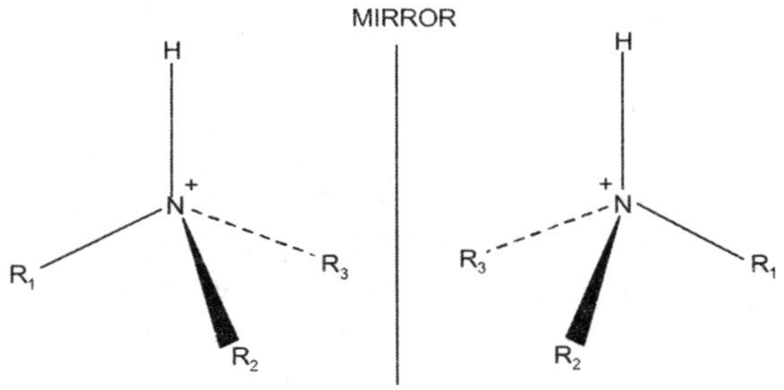

MIRROR

Enantiomers of substituted ammonium cation

This has been confirmed experimentally. Alkylbenzylmethylphenylammonium iodide in which four different groups are attached to nitrogen atom of the substituted ammonium ion has been resolved into two enantiomeric forms each of which is optically active.

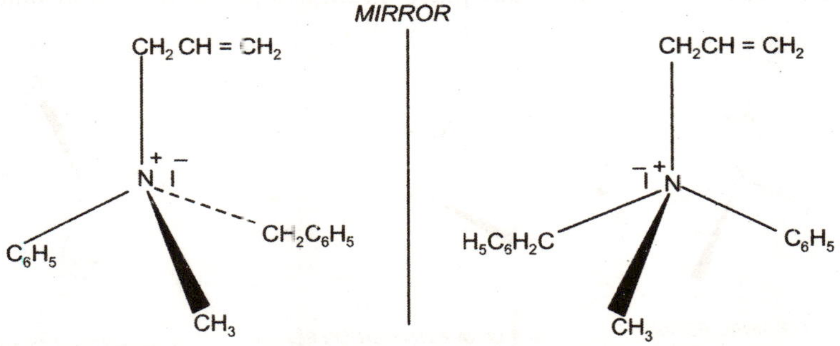

MIRROR

$CH_2CH=CH_2$ ... $\overset{+}{N}\,\overset{-}{I}$ ... C_6H_5, $CH_2C_6H_5$, CH_3 | $CH_2CH=CH_2$ ... $\overset{-}{I}\,\overset{+}{N}$ ... $H_5C_6H_2C$, C_6H_5, CH_3

Enantionmetric forms of alkylbenzylmethylphenylammonium iodide

Methods of Preparation of Primary Amines

The synthetic methods for the preparation of primary amines (R-NH.) are listed below.

1. Reaction of Alkyl halides with Ammonia. Primary amines can be obtained by reaction, of alkyl halides with excess of ammonia.

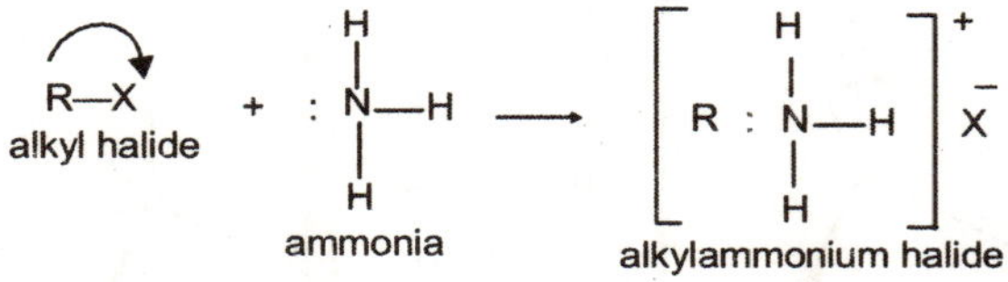

$$\underset{\text{alkyl halide}}{R{-}X} + \underset{\text{ammonia}}{:NH_3} \longrightarrow \underset{\text{alkylammonium halide}}{[R:NH_3]^+\bar{X}}$$

The resulting alkylammonium salt (RNH.X) on subsequent treatment with (NaOH) liberates the free primary amine.

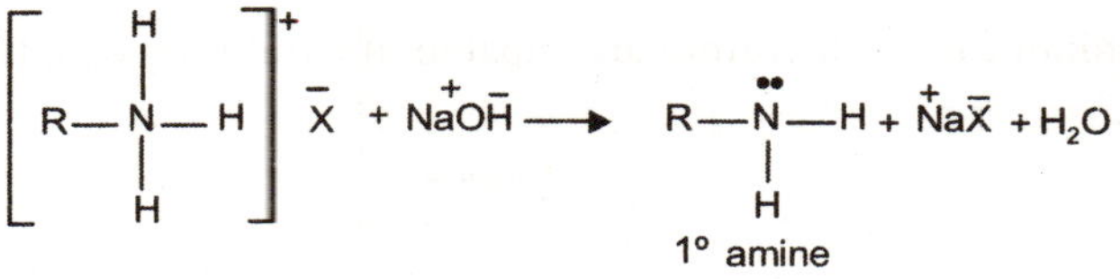

$$[R{-}NH_3]^+\bar{X} + \overset{+}{Na}\overset{-}{OH} \longrightarrow \underset{1^\circ\text{ amine}}{R{-}\ddot{N}H_2} + \overset{+}{Na}\overset{-}{X} + H_2O$$

Although this method appears so simple it is not suitable for laboratory synthesis replacement of hydrogens on N atom in ammonia docs not stop at the first stage when a primary amine is produced as shown above. The process of alkylation of hydrogen on N atom continues to form a secondary and a tertiary amine. The tertiary amine reacts with a fourth molecule of alkyl halide to produce a tetraalkylammonium salt. The following series of reaction illustrate the progressive alkylation of ammonia to form secondary amines, tertiary amines with tetraalkylammonium salts.

(*i*) $\underset{\text{1° amine}}{RNH_2} + RX \longrightarrow \underset{\text{dialkylammonium halide}}{R_3\overset{+}{N}H_2\overset{-}{X}}$

$R_3\overset{+}{N}H_2\overset{-}{X} + NaOH \longrightarrow \underset{\text{2° amine}}{R_3NH} + Na\overset{+}{}\overset{-}{X} + H_2O$

(*ii*) $\underset{\text{2° amine}}{R_2NH} + RX \longrightarrow \underset{\text{trialkylammonium halide}}{R_3\overset{+}{N}H\overset{-}{X}}$

$R_3\overset{+}{N}H\overset{-}{X} + NaOH \longrightarrow \underset{\text{3° amine}}{R_3N} + NaX + H_2O$

(*iii*) $\underset{\text{3° amine}}{R_3N} + RX \longrightarrow \underset{\text{tetraalkylammonium salt}}{R_4\overset{+}{N}\overset{-}{X}}$

$R_4\overset{+}{N}\overset{-}{X} + NaOH \longrightarrow$ *No Reaction*

The resultant mixture of a primary amine, a secondary amine, a tertiary amine and a quaternary salt is not easily separated. Hence this method of preparing a primary amine by reaction of alkyl halide on ammonia is of limited application. It is generally used on industrial scale where special processes arc employed for the isolation of the primary amine from the rest of the components.

2. Reaction of Alcohols with Ammonia. When a mixture of excess of ammonia and alcohol vapour is passed over a dehydrating catalyst such as alumina or thoria at 300-400° a primary amine results.

$$\underset{\text{methanol}}{CH_3-OH} + \underset{\text{ammonia}}{H-NH_2} \xrightarrow[300-400°]{-ThO_2} \underset{\substack{\text{methylamine}\\ \text{(1° amine)}}}{CH_3-NH_2} + H_2O$$

Unfortunately the reaction does not stop here. The methylamine produced in the first step reacts with a second molecule of methanol to give a secondary amine. The secondary amine further reacts with a third molecule of methanol to yield a tertiary amine.

$$CH_3-NH_2 + \underset{\text{methanol}}{CH_3-OH} \xrightarrow[300°-400°]{ThO_2} \underset{\substack{\text{dimethylamine}\\ \text{(2° amine)}}}{(CH_3)_2NH} + H_2O$$

$$(CH_3)_2NH + \underset{\text{methanol}}{CH_3-OH} \xrightarrow{\text{similar conditions}} \underset{\substack{\text{trimethylamine}\\ \text{(3° amine)}}}{(CH_3)_3N} + H_2O$$

Thus the resulting mixture contains primary, secondary and tertiary amines. However when a large excess of ammonia vapours are used in the initial mixture, primary amine is the chief product.

3. Reduction of Nitroalkanes. Primary amines can be obtained by reduction of nitroalkanes which can best be affected with lithium aluminium hydride.

$$R-CH_3-NO_2 + 6H \xrightarrow[\text{ether}]{LiAlH_4} R-CH_3-NH_3 + 2H_3O$$

nitroalkane — alkylamine

This reduction can also be carried by using hydrogen in the presence of a metal catalyst nickel or platinum), or with metal and acid (Sn/HCl or Fe/HCl). For example,

$$CH_3-CH_3-CH_3-NO_3 \xrightarrow{Fe/HCl} CH_3-CH_3-CH_3-NH_3$$

1- nitropropane — *n*-propylamine

$$CH_3-CH_3-\underset{NO_2}{CH}-CH_3 \xrightarrow[\text{ether}]{LiAlH_4} CH_3-CH_2-\underset{NH_3}{CH}-CH_3$$

2- nitrobutane — 2- aminobutane

However, nitroalkanes are not easily prepared. Only the lower members nitro methane, nitroethane and the nitropropanes are commonly available from vapour-phase nitration of alkanes. Therefore, this method although most useful for the synthesis of aromatic amine has limited utility for the preparation-of alkylamines.

4. Reduction of Amides. Primary amines are formed by the reduction of amides with lithium aluminium. hydride,

$$R-\overset{O}{\overset{\|}{C}}-NH_2 \xrightarrow[\text{ether}]{LiAlH_4} R-CH_2-NH_2$$

amide — a primary amine

$$CH_3-CH_2-\overset{O}{\overset{\|}{C}}-NH_2 \xrightarrow{LiAlH_4} CH_3-CH_3-CH_2-NH_3$$

propionamide — *n*-propylamine

The amide required in this method can be readily prepared by the reaction of the: corresponding acyl halide with ammonia. For example, the following sequence shows below this reaction may be used to good advantage for obtaining ethylamine from ethyl alcohol.

$$CH_3CH_2OH \xrightarrow{H_2Cr_2O_7} CH_3COOH \xrightarrow{SOCl_2} CH_3-\overset{O}{\overset{\|}{C}}-Cl \xrightarrow{NH_3} CH_3-\overset{O}{\overset{\|}{C}}-NH_3 \xrightarrow{LiAlH_4} CH_3CH_3-NH_2$$

ethyl alcohol — accetic acid — accetyl chloride — acctamide — ethylamine

5. Reduction of Oximes of Aldehydes and Ketones: Good yields of primary amines are obtained by reducing of oximes of aldehydes and ketones. The common reducing agents are sodium and absolute ethanol or hydrogen and nickel catalyst.

$$\begin{matrix}R\\R\end{matrix}\!\!>C=N-OH \xrightarrow{2[H]} \begin{matrix}R\\R\end{matrix}\!\!>CH-NHOH \xrightarrow{2[H]} \begin{matrix}R\\R\end{matrix}\!\!>CH-NH_2 + H_2O$$

an oxime — a primary amine

e.g.,

$$CH_3(CH_2)_5-\underset{}{\overset{CH_3}{C}}=N-OH \xrightarrow{2[H]} CH_3(CH_3)_5-\overset{CH_3}{CH}-NHOH \xrightarrow{2[H]} CH_3(CH_3)_5-\overset{CH_3}{CH}-NH_3 + H_2O$$

an oxime — 2-aminooctane

6. Reduction of Alkyl Cyanides or Nitriles. Alkyl nitriles are reduced to primary amines by catalytic hydrogenation (nickel catalyst), sodium metal in ethanol (Mendius Reaction) or lithium aluminium hydride in ether.

$$\underset{\text{alkyl nitrile}}{R{-}C{\equiv}N} + 4H \longrightarrow \underset{1^{\circ}\text{ amine}}{R{-}CH_2{-}NH_2}$$

This method is particularly useful for the preparation of alkylamines because alkyl; nitriles are easily made by the reaction of alkyl halides with cyanide ion. For example,. n-butylamine can be synthesised from n-propyl bromide as follows.

$$\underset{n\text{-propylbromide}}{CH_3CH_2CH_2{-}Br} \xrightarrow{\text{NaCN}} \underset{n\text{-butyronitrile}}{CH_3CH_2CH_2{-}C{\equiv}N} \xrightarrow[\text{ether}]{\text{LiAlH}_4} \underset{n\text{-butylamine}}{CH_3CH_2CH_2CH_2{-}NH_3}$$

OR

$$CH_3CH_2CH_2{-}C{\equiv}N \xrightarrow[140^{\circ}.130\text{ atm}]{H_3/Ni} CH_3CH_2CH_2CH_2{-}NH_2$$

The above sequence converts an alkyl halide to a 10 amine with extension of the carbon chain.

7. Reductive Amination of Aldehydes and Ketones. Primary amines may also be prepared by passing a mixture of an aldehydes or a ketone and a large excess of ammonia and hydrogen over nickel at 150°C

$$\underset{\text{an aldehyde}}{R{-}\overset{\overset{\displaystyle O}{\|}}{C}{-}H} + NH_3 + H_2 \xrightarrow[300\text{ atm}]{\text{Ni, }150^{\circ}} \underset{\text{primary amine}}{R{-}CH_2{-}NH_3} + H_3O$$

$$\underset{\text{a ketone}}{R{-}\overset{\overset{\displaystyle O}{\|}}{C}{-}CH_3} + NH_3 + H_3 \xrightarrow{\text{similar conditions}} \underset{\text{an alkylmethylamine}}{R{-}\overset{\overset{\displaystyle CH_3}{|}}{CH}{-}NH_3} + H_3O$$

This is reaction probably takes place by formation of Schiff's base first, which on catalysis hydrogenation yields a primary amines.

$$>C{=}O + H_3N \xrightarrow{-H_2O} \underset{\text{Schiff's base}}{>C{=}NH} \xrightarrow{+H_3\text{ (cat)}} \underset{\text{primary amine}}{>CH{-}NH_3}$$

This process of introducing an amino group into the molecule of an aldehydes or ketone is termed reductive Amination. It may be illustration by the following example.

$$\underset{\text{pentan-2-one}}{H_3C{-}\overset{\overset{\displaystyle O}{\|}}{C}{-}CH_2\text{-}CH_2\text{-}CH_3} + NH_3 + H_2 \xrightarrow[300\text{ atm}]{\text{Ni, }150\text{ °C}} \underset{\text{pentan-2-amine}}{H_3C{-}\overset{\overset{\displaystyle H_2N}{|}}{CH}{-}CH_2\text{-}CH_2\text{-}CH_3} + H_2O$$

8. Hydrolysis of Isocyanates and Isocyanides. Primary amines are produced by the hydrolysis of alkyl isocyanides with an aqueous mineral acid.

$$\underset{\text{alkyl iscyanide}}{R-\overset{+}{N}\equiv\overset{-}{C}} + 2H_2O \xrightarrow{HCl} \underset{1^\circ\text{ amine}}{R-NH_3} + \underset{\text{formic acid}}{HCOOH}$$

$$\underset{\text{propyl iscyanide}}{CH_3CH_3CH_2-\overset{+}{N}\equiv\overset{-}{C}} + 2H_2O \longrightarrow \underset{\text{propylamine}}{CH_3CH_3CH_3-NH_3} + HCOOH$$

Alkyl isocyanates when hydrolysed with alkali also yield primary amines.

$$\underset{\text{alkyl isocyanate}}{R-N=C=O} + 2KOH \longrightarrow \underset{1^\circ\text{ amine}}{R-NH_3} + K_3CO_3$$

$$\underset{\text{ethyl isocyanate}}{CH_3CH_2-N=C=O} + 2KOH \longrightarrow \underset{\text{ethylamine}}{CH_2CH_2-NH_2} + K_2CO_3$$

The isocyanate required for the is obtained from alkyl halides by interaction with silver isocyanate.

$$CH_3CH_2-I + Ag-N=C=O \rightarrow CH_3CH_2-N=C=O + AgI$$

9. Hoffman Degradation: It is a simple laboratory method for conversion of an amide to primary amine having one less carbon atom. The amide is treated with bromine is in the presence of alkali (NaOH) and the overall may be written as.

$$\underset{\text{an amide}}{R-\overset{O}{\overset{\|}{C}}-NH_2} + Br_2 + 3NaOH \longrightarrow \underset{1^\circ\text{ amine}}{R-NH_2} + 2NaBr + NaHCO_3 + H_2O$$

For example,

(i) $$\underset{\text{propionamide}}{CH_3-CH_3-\overset{O}{\overset{\|}{C}}-NH_3} \xrightarrow{Br_2/NaOH} \underset{\text{ethylamine}}{CH_3-CH_3-NH_3}$$

(ii) $$\underset{\textit{tert}\text{-butylacetamide}}{CH_3-\underset{CH_3}{\overset{CH_3}{C}}-CH_2-\overset{O}{\overset{\|}{C}}-NH_2} \xrightarrow{Br_2/NaOH} \underset{\text{neopentylamine}}{CH_3-\underset{CH_3}{\overset{CH_3}{C}}-CH_3-NH_2}$$

This reaction, also know~ as *Hofmann Rearrangement*, gives an amine with one I carbon, in contrast to straight reduction of amide with LiAlH4, which yields an amine with I same number of carbon atoms.

10. Curtius Degradation. Primary amines can be prepared in good yield by treating acyl azides ($RCON_2$) in acidic or alkaline medium. The over all reaction which proceeds the elimination of nitrogen from acyl azide yields a 1° amine containing one carbon less, is called *Curtius Degradation*

$$\underset{\text{acyl azide}}{R-\overset{O}{\overset{\|}{C}}-NH_2} \xrightarrow{-N_2} \underset{\text{alkyl isocyanate}}{R-N=C=O} \xrightarrow{H_2O/H^+} \underset{1^\circ\text{ amine}}{R-NH_2}$$

Alkyl isocyanate is obtained as an intermediate which is hydrolysed to give the respective primary amine.

The acid azide required in this method is prepared from the corresponding ac chloride by reaction with sodium azide ($Na+N_3$-). Thus the sequence leading to the production of a primary amine from a carboxylic acid would be:

$$\underset{\text{carboxylic acid}}{R-\overset{O}{\overset{\|}{C}}-OH} \longrightarrow \underset{\text{acyl chloride}}{R-\overset{O}{\overset{\|}{C}}-Cl} \xrightarrow{\overset{+}{Na}\overset{-}{N_3}} \underset{\text{acyl azide}}{R-\overset{O}{\overset{\|}{C}}-N_3} \longrightarrow \underset{\text{isocyanate}}{R-N=C=O} \longrightarrow \underset{1^\circ\text{ amine}}{R-NH_3}$$

The conversion RCOOH RNH. can also be carried with the help of Schmidt Reaction which is a modification of Curtius Degradation. In this reaction, a carboxylic acid ill warmed with sodium azide ($Na^+N_3^-$) and concentrated sulphuric acid. The carboxylic acid is directly converted to the corresponding primary amine without the necessity of isolating alkyl azide.

$$R-\overset{O}{\overset{\|}{C}}-OH \xrightarrow[\text{Conc. } H_2SO_4 \text{ warm}]{\overset{+}{Na}\overset{-}{N_3}} R-NH_3 + N_2 + CO_2$$

Schmidt reaction has the advantage of experimental simplicity as compared to the Hofmann and *Curtius Degradation*

11. Gabriel Synthesis: In this method advantage is taken of the fact that phthalimide readily forms potassium phthalimide when treated with potassium hydroxide. The potassium phthalimide when reacted with an alkyl halide is converted to N-alkylphthalimide. This upon hydrolysis with hydrochloric acids yields a primary amine.

$$\underset{\text{phthalimide}}{C_6H_4(CO)_2NH} + KOH \longrightarrow \underset{\text{potassium phthalimide}}{C_6H_4(CO)_2\overset{-}{N}\overset{+}{K}} + H_2O$$

$$\underset{\text{pot phthalimide}}{C_6H_4(CO)_2\overset{-}{N}\overset{+}{K}} + R-X \xrightarrow{-KX} \underset{N\text{ - alkylphthalimide}}{C_6H_4(CO)_2N-R} \xrightarrow{H_2O/H^+} \underset{\text{phthalic acid}}{C_6H_4(COOH)_2} + \underset{\text{primary amines}}{R-NH_3}$$

N-alkylpbthalimide is often resistant to hydrolysis but it readily reacts with (H_2N-NH_2) to give phthalolhydrazide and. a primary amine.

$$C_6H_4(CO)_2N-R + H_2N-NH_2 \xrightarrow[(60-80\ ^\circ C))]{\text{warm}} \underset{\text{pthaloylhydrazide}}{C_6H_4(CO)_2(NH)_2} + R-NH_2$$

Gabriel Syntheis provides an indirect method of carrying on the transformation of RX ? RNH_2 without formation of secondary or tertiary amine as by products.

12. Forms Chrloroamine and Gridnard reagent: Primary amines can be obtained by coupling chloromaine with Grignard reagents.

$$\underset{\text{Grignard reagent}}{R\text{—}MgCl} + \underset{\text{chloramine}}{Cl\text{—}H_2} \xrightarrow{\text{ether}} \underset{1° \text{ amine}}{R\text{—}NH_2} + MgCl_2$$

Chloramine required for the process is derived from ammonia and sodium hypochlorite.

$$NH_3 + \underset{\text{sodium hypochlorite}}{\overset{-\ +}{ClONa}} \longrightarrow \underset{\text{chloramine}}{NH_2\text{—}Cl} + Na\overset{+}{\bar{O}}H$$

13. Ritter Reaction: It is an important method for preparing primary amines having a tertiary group. Thus tetra-butyl alcohol is reacted with hydrogen cyanide in the presence of conc. Sulphuric to give tert-butylamine.

$$\underset{\textit{tert- butyl alcohol}}{(CH_3)_3C\text{—}OH} + H_2SO_4 + HCN \longrightarrow \underset{\textit{tert- butylamine } (1° \text{ amine})}{(CH_3)_3C\text{—}NH_2}$$

If the carbonium ion derived from alcohol be represented as R^+, the course of the above reaction would be as follows:

$$ROH \xrightarrow{H^+} \overset{+}{R} + HC{=}N: \longrightarrow HC{=}\overset{+}{N}\text{—}R \xrightarrow{HS\bar{O}_4} \overset{OSO_3H}{HC}{=}NR \xrightarrow{H_2O} \overset{O}{\overset{\|}{HC}}NHR \xrightarrow{\bar{O}H} HC\bar{O} + \underset{1° \text{ amine}}{RHN_2}$$

Methods of Preparation of Secondary Amines

The general methods of secondary amines (R - NH - R) are discussed below.

$$\underset{1° \text{ amine}}{R\text{—}\ddot{N}H_2} + \underset{\text{alkyl halide}}{R\text{—}X} \longrightarrow \underset{\text{dialky!ammonium salt}}{[R\text{—}NH_2\text{—}R]^+\bar{X}}$$

$$[R\text{—}NH_2\text{—}R]^+\bar{X} + \overset{+\ -}{NaOH} \longrightarrow \underset{2° \text{ amine}}{R\text{—}\ddot{N}H\text{—}R} + \overset{+\ -}{NaX} + H_2O$$

1. Alkylation of Primary amine in ethanolic solution is heated with one mole of alkyl halide, a tetra alkyl ammonium salt is produced. This when treated with NaOH solution gives the free secondary amine

For example

$$\underset{\text{ethylamine}}{C_2H_5{-}NH_3} + \underset{\text{ethyl iodide}}{C_2H_5I} \longrightarrow \underset{\text{diethylammonium iodide}}{[C_2H_5{-}NH_3{-}C_2H_5]^+I^-}$$

$$[C_6H_5{-}NH_3{-}C_2H_5]^+I^- + NaOH \longrightarrow \underset{\text{diethylamine}}{C_2H_5{-}\dot{N}H{-}C_2H_5} + \overset{+}{Na}\bar{I} + H_3O$$

This is not a method of choice for the preparation of secondary amines. The produce invariably contains tertiary amine and tetra alkylammonium salt.

2. Reduction of N - substituted amides: N- substituted amides when reduced with LiAlH4 yield secondary amines.

$$\underset{\textit{N-}\text{ alkylamide}}{R{-}\overset{O}{\overset{\|}{C}}{-}NH{-}R'} + 4H \xrightarrow{LiAlH_4} \underset{2^\circ\text{ amine}}{R{-}CH_3{-}NH{-}R'} + H_2O$$

$$e.g.,\quad \underset{\textit{N-}\text{methylpoprionamide}}{CH_3{-}CH_2{-}\overset{O}{\overset{\|}{C}}{-}NH{-}CH_3} + 4H \xrightarrow{LiAlH_4} \underset{\text{methylpropylamine}}{CH_3{-}CH_2{-}CH_2{-}NH{-}CH_3} + H_2O$$

3. Reduction of Alkyl Isocyanides or Iso nitriles: When alkyl isocyanides or isonitriles are reduced with nascent hydrogen alkyl methylamines are obtained.

$$\underset{\text{alkyl isocyanide}}{R{-}\overset{+}{N}{\equiv}C} + 4H \longrightarrow \underset{\substack{\text{alkylmethylamine}\\(2^\circ\text{ amine})}}{R{-}NH{-}CH_3}$$

4. Reduction of Schiff's Bases: Primary amines react with aldehydes and ketones to give Schiff's bases. The Schiff's bases so obtained upon reduction with hydrogen and platinum yield secondary amines.

$$\underset{\text{aldehyde}}{R{-}\overset{H}{\overset{|}{C}}{=}O} + H_2N{-}R' \longrightarrow \underset{\text{Schiff's base}}{R{-}CH{=}N{-}R'} + H_2O$$

$$R{-}CH{=}N{-}R' + 3/2\ H_2 \xrightarrow{Pt} \underset{2^\circ\text{ amine}}{R{-}CH_2{-}NH{-}R'}$$

5. Reductive Alkyllation of Aldehydes and Ketones: When a primary amine is reacted with an aldehydes or ketone, imines are formed. The imines are readily reduced with hydrogen and Raney nickel, or sodium borofluoride to give a secondary amine.

$$\underset{1^\circ\text{ amine}}{R{-}NH_3} + \underset{\text{ketone}}{O{=}C\langle^{R}_{R}} \longrightarrow \underset{\text{imine}}{R{-}N{=}C\langle^{R}_{R}} + H_2O$$

$$R{-}N{=}C\langle^{R}_{R} + H_3 \xrightarrow{\text{Raney Ni}} \underset{2^\circ\text{ amine}}{R{-}NH{-}CH\langle^{R}_{R}}$$

$$e.g.,\ \underset{\text{methylamine}}{CH_3NH_2} + \underset{\text{acetone}}{O{=}C(CH_3)_2} \xrightarrow{=H_2} \underset{\text{imine}}{CH_3N{=}C(CH_3)_2} \xrightarrow[Ni]{+H_2} \underset{\text{isopropyl methylamine}}{CH_3NH{-}CH(CH_3)_2}$$

The advantage of this method over alkylation with alkyl halide is that over alkylatic with the formation of tertiary amine and tetraalkylammonium salt cannot take place.

6. Hydrolysis of p-Nitroso-N. N-dialkylanilines: In this method aniline is first heated with an alkyl halide to form N, N dialkylaniline. This is then nitrosated by reaction with nitrous acid when p-nitroso-N. N-dialkylaniline is obtained which upon boiling with sodium hydroxide yields a pure secondary amine.

This is not a method of choice for the preparation of secondary amines. The product invariably contains tertiary amine and tetraalkylammonium salt.

(2) Reduction of N-substituted amldes. N.substituted amides when reduced with $LiAlH_4$ yield secondary amines.

$$\underset{N\text{- alkylamide}}{R{-}\overset{O}{\overset{\|}{C}}{-}NH{-}R'} + 4H \xrightarrow{LiAlH_4} \underset{2^\circ\text{ amine}}{R{-}CH_2{-}NH{-}R'} + H_2O$$

$$e.g.,\ \underset{N\text{- methylpoprionamide}}{CH_3{-}CH_2{-}\overset{O}{\overset{\|}{C}}{-}NH{-}CH_3} + 4H \xrightarrow{LiAlH_4} \underset{\text{methylpropylamine}}{CH_3{-}CH_2{-}CH_2\ NH{-}CH_3} + H_2O$$

(3) Reduction or Alkyl Isocyanides or lsonitriles. When alkyl isocyanides or iwni are reduced with nascent hydrogen, alkylmethylamines are obtained.

$$\underset{\text{alkyl isocyanide}}{R{-}\overset{+}{N}{\equiv}\overset{-}{C}} + 4H \longrightarrow \underset{\substack{\text{alkylmethylamine}\\(2^\circ\text{ amine})}}{R{-}NH{-}CH_3}$$

This method is of academic interest only.

(4) Reduction of Schiff's Bases. Primary amines react with aldehydes and keton~s to! Schiff's bases. The Schiff's bases so obtained upon reduction with hydrogen and platinum yield;secondary amines.

$$\underset{\text{aldehyde}}{R{-}\overset{H}{\overset{|}{C}}{=}O} + H_2N{-}R' \longrightarrow \underset{\text{Schiff's base}}{R{-}CH{=}N{-}R'} + H_2O$$

$$R{-}CN{=}N{-}R' + 3/2\ H_2 \xrightarrow{Pt} \underset{2^\circ\text{ amine}}{R{-}CH_2{-}NH{-}R'}$$

(5) Reductive Alkylation of Aldehydes and Ketones. When a primary amine is rea~ with an aldehyde or a ketone, imines are formed. The imines are readily reduced with hydrq and Raney nickel, or sodium borofluoride to give a secondary amine.

$$R{-}NH_2 + O{=}C\begin{matrix}R\\R\end{matrix} \longrightarrow R{-}N{=}C\begin{matrix}R\\R\end{matrix} + H_2O$$

1° amine, ketone, imine

$$R{-}N{=}C\begin{matrix}R\\R\end{matrix} + H_2 \xrightarrow{\text{Raney Ni}} R{-}NH{-}CH\begin{matrix}R\\R\end{matrix}$$

2° amine

e.g.,
$$\underset{\text{methylamine}}{CH_3NH_2} + \underset{\text{acetone}}{O{=}C\begin{matrix}CH_3\\CH_3\end{matrix}} \xrightarrow{-H_2O} \underset{\text{imine}}{CH_3N{=}C\begin{matrix}CH_3\\CH_3\end{matrix}} \xrightarrow[Ni]{+H_2} \underset{\text{isopropyl methylamine}}{CH_3NH{-}CH\begin{matrix}CH_3\\CH_3\end{matrix}}$$

The advantage of this method over alkylation with alkyl halide is that over alkylation with the formation of tertiary amine and tetraalkylammonium salt cannot take place.

(6) Hydrolysis of p-Nitroso-N. N-dialkylanilines. In this method aniline is first heat with an alkyl halide to form *N*. N-dialkylaniline. This is then nitrosated by reaction with nitrous acid when p-nitroS9-N, N-dialkylaniline is obtained which upon boiling with sodium hydro yields a pure secondary amine.

(*i*)
$$\underset{\text{aniline}}{C_6H_5NH_2} \underset{-HX}{\xrightarrow{R-X}} \underset{\textit{N, N-dialkylaniline}}{C_6H_5NR_2} \underset{-H_2O}{\xrightarrow{HO-NO}} \underset{\textit{p-nitroso-N, N-dialkylaniline}}{ON{-}C_6H_4{-}NR_2}$$

(*ii*)
$$ON{-}C_6H_4{-}NR_2 + HO{-}H \xrightarrow{NaOH} \underset{\textit{p-nitrosophenol}}{ON{-}C_6H_4{-}OH} + \underset{\text{2° amine}}{R_2NH}$$

(7) Hydrolysis of Dialkylcyaoamides. Pure secondary amines may be obtained by synthesizing dialkylcyanamides by the action of alkyl halide on sodium cyanamide.The dialkylcyanamide is then hydrolysed by aqueous acids or bases to produce a secondary amine.

(*i*)
$$\underset{\text{sodium cyanamide}}{\overset{1+}{Na_2} : \overset{3-}{\ddot{N}}{-}C{\equiv}N} \longrightarrow 2\overset{1+}{Na} + \underset{\text{cyanamide ion}}{:\overset{2-}{\ddot{N}}{-}C{\equiv}N}$$

$$\underset{\text{cyanamide ion}}{N{\equiv}C{-}\overset{2-}{\ddot{N}}:} + R{-}X \xrightarrow{-\bar{X}} N{\equiv}C{-}\overset{-}{\ddot{N}}{-}R \xrightarrow{R-X} \underset{\text{dialkyl cyanamide}}{N{\equiv}C{-}N\begin{matrix}R\\R\end{matrix}} + \bar{X}$$

(*ii*)
$$\begin{matrix}R\\R\end{matrix}N{-}C{\equiv}N \xrightarrow[H^+]{+2H_2O} NH_3 + \underset{\substack{\textit{N,N-dialkylcarbonic acid}\\\textit{(unstable)}}}{\begin{matrix}R\\R\end{matrix}N{-}\overset{O}{\overset{\|}{C}}{-}OH} \longrightarrow \underset{\text{2° amine}}{\begin{matrix}R\\R\end{matrix}NH} + CO_2$$

(8) From Chloroamines and Grignard Reagents. Secondary amines can also be prepared by reaction between the corresponding chloroamines (RNHCl) and Grignard reagents. The chloroamine needed for the reaction is obtained by the reaction of a primary amine with sodium hypochlorite (Na+ClO-).

$$R{-}NH_2 + Cl\bar{O}\overset{+}{N}a \longrightarrow R{-}NH{-}Cl + \overset{+}{N}a\bar{O}H$$
1° amine — chloroamine

$$R{-}NH{-}Cl + R'{-}MgCl \xrightarrow{\text{ether}} R{-}NH{-}R' + MgCl_2$$
2° amine

This method is particularly suited for the preparation of secondary amines having ~o tertiary alkyl groups, such as di-tert-butyl amine, (CHa).C-NH-C(CH.)a. .

Methods of Preparation of Tertiary Amlnes

The methods listed below are available for the preparation of tertiary amine (R3N).

1. Alkylation of a Secondary Amine. When a secondary amine in ethanoic solution is heated with a calculated amount of the appropriate alkylhalide, the corresponding trialkylammonium salt is obtained. This on treatment with NaOH liberates the tertiary amine.

$$R_2\ddot{N}H \ (\text{2° amine}) + R{-}X \ (\text{alkyl halide}) \xrightarrow{\Delta} [R_3\ddot{N}{-}H]^+ \bar{X} \ (\text{trialkylammonium salt})$$

$$[R_3N{-}H]^+ \bar{X} + \overset{+}{N}a\bar{O}H \longrightarrow R_3N: \ (\text{3° amine}) + \overset{+}{N}a\bar{X} + H_2O$$

This is a fairly practical method for the preparation of tertiary amines. The tetraalkyl ammonium salt is also produced which is stable to alkali. Thus upon distillation of the react mixture after treatment with alkali yields the tertiary amine.

(2) Reductino of *N*, N-Disubstituted Amides. Tertiary amines may be prepared by reduction of *N*, N-disubstituted amides with $LiAIH_4$

$$R{-}\overset{O}{\overset{\|}{C}}{-}NR_2 \ (\textit{N, N}\text{-disubstituted amide}) + 4[H] \xrightarrow{LiAlH_4} R{-}CH_2{-}NR_2 \ (\text{3° amine}) + H_2O$$

e.g.,
$$CH_3CH_2{-}\overset{O}{\overset{\|}{C}}{-}N(CH_3)_2 \ (\textit{N, N}\text{-dimethylpropionamide}) + 4[H] \xrightarrow{LiAlH_4} CH_3CH_2{-}CH_2{-}N(CH_3)_2 \ (\text{dimethylpropylamine}) + H_2O$$

(3) Reductive Alkylation. Tertiary amines can be obtained by methylation of a primary or secondary amine by a mixture of formic acid and formaldehyde. This is referred to Eschweiler-Clarks reaction.

$$R{-}NH_2 \ (\text{1° amine}) + 2\,CH_2O \ (\text{formaldehyde}) + 2\,HCOOH \ (\text{formic acid}) \longrightarrow R{-}N(CH_3)_2 \ (\text{3° amine}) \mid 2\,H_2O + 2\,CO_2$$

$$\textit{e.g.,}\ (CH_3)_3C{-}NH_2 \ (\textit{tert}\text{-butylamide}) + 2CH_2O + 2HCOOH \longrightarrow (CH_3)_3C{-}N(CH_3)_3 \ (\text{dimethyl-}\textit{tert}\text{-butylamine}) + 2H_2O + 2CO_2$$

(4) Decomposition or Tetraalkylammonium Hydroxides. Tertiary amines are best prepare by decomposing the tetraalkylammonium hydroxide upon heating strongly. The required tetra alkyl ammonium hydroxide is obtained by treating corresponding tetraalkylammonium salt with moist silver oxide (AgOH). For example,

$$\underset{\text{tetramethylammonium iodide}}{(CH_3)_4\overset{+}{N}\,\overset{-}{I}} + \overset{+}{Ag}\overset{-}{O}H \longrightarrow \underset{\text{tetramethylammonium hydroxide}}{(CH_3)_4\overset{+}{N}\,\overset{-}{O}H} + AgI\downarrow$$

$$(CH_3)_4\overset{+}{N}\,\overset{-}{O}H \xrightarrow{\Delta} \underset{\text{trimethylamide}}{(CH_3)_3N:} + CH_3{-}OH$$

With a higher tetraalkytammonium hydroxide, however, the decomposition follows different route

$$\underset{\text{Tetraethylammonium Hydroxide}}{H{-}CH_2{-}CH_2{-}\overset{+}{N}(C_2H_5)_3\,\overset{-}{O}H} \xrightarrow{\Delta} \underset{\text{Triethylamine}}{(C_2H_5)_3N} + \underset{\text{Ethylene}}{CH_2{=}CH_2} + H_2O$$

(5) From Chloroamines and Grignard Reagents. Tertiary amines can also be prepared by coupling the corresponding chloramine (R_2NCl) with Grignard Reagents in ether. The required chloroamine is derived from a secondary amine by reaction with sodium hypochlorite (Na+ClO)

$$\underset{2^\circ\text{ amine}}{R_2N{-}H} + Cl\overset{-}{O}\overset{+}{Na} \longrightarrow \underset{\text{chloroamine}}{R_2N{-}Cl} + \overset{+}{Na}\overset{-}{O}H$$

$$R_2N{-}Cl + \underset{\text{Grignard reagent}}{R'{-}MgCl} \xrightarrow{\text{ether}} \underset{3^\circ\text{ amines}}{R_2N{-}R'} + MgCl_2$$

This is the only satisfactory method for the preparaticn of tertiary amines having two or more tertiary alkyl groups *e.g.*, tri-tert-butylamine, $[(CH_3){-}C]_3N$.

How to Separate Mixtures of Amines

When the mixture contains salts of primary, secondary and tertiary amines along with the tetraalkylammonium salt, it is first distilled with potassium hydroxide solution. The three amines are salt free and distil over. The tetraalkylammonium hydroxide, however, remains unaffected by alkali and being non-volatile is left behind. Thus,

$$\underset{1^\circ\text{ amine salt}}{R{-}\overset{+}{N}H_3\,\overset{-}{Cl}} + \overset{+}{K}\overset{-}{O}H \longrightarrow \underset{\substack{1^\circ\text{ amine}\\ \text{(volatile)}}}{R{-}NH_2} + \overset{+}{K}\overset{-}{Cl} + H_2O$$

$$\underset{2^\circ\text{ amine salt}}{R_2\overset{+}{N}H_2\,\overset{-}{Cl}} + \overset{+}{K}\overset{-}{O}H \longrightarrow \underset{\substack{2^\circ\text{ amine}\\ \text{(volatile)}}}{R{-}NH{-}R} + \overset{+}{K}\overset{-}{Cl} + H_2O$$

$$R_3\overset{+}{N}H\,\overset{-}{Cl} + \overset{+}{K}\overset{-}{O}H \longrightarrow R_3N + \overset{+}{K}\overset{-}{Cl} + H_2O$$

3° amine salt → 3° amine (volatile)

$$R_4\overset{+}{N}\,\overset{-}{Cl} + \overset{+}{K}\overset{-}{O}H \longrightarrow \textit{No Rection}$$

tetralkylammonium salt (nonvolatile)

The distillate contains the free amines i.e. primary secondary and tertiary which may then be separate by following methods.

(1) Fractional Distillation. A mixture of primary, secondary and tertiary amines can be separated into the individual components by fractional distillation since their boiling points are fairly apart. For example,

$$\underset{1^\circ \text{ amine (bp } 17^\circ)}{CH_3CH_2NH_3} + \underset{2^\circ \text{ amine (bp } 56^\circ)}{(CH_3CH_2)_2NH} + \underset{3^\circ \text{ amine (bp } 98^\circ)}{(CH_3CH_3)N} \xrightarrow{F.D.} \begin{matrix}(1)\ CH_2CH_2NH_2 \\ (2)\ (CH_3CH_2)_2NH \\ (3)\ (CH_3CH_2)_3N\end{matrix}$$

This is the most satisfactory method and is used industrially. The success of this method is primarily due to the high efficiency of the modern industrial fractionation ensures almost complete separation of the amines.

(2) By reaction with Diethyl oxalate; Hofmann Method. The given mixture of 1°,2° 3° amines is heated with diethyl oxalate, when

(i) the 1° amine forms a dialkyl oxamide which is crystalline solid.

$$\underset{\text{diethyl oxalate}}{\begin{matrix}CO-OC_2H_5 \\ | \\ CO-OC_2H_5\end{matrix}} + \underset{1^\circ \text{ amine}}{\begin{matrix}H-HN-R \\ H-HN-R\end{matrix}} \xrightarrow{\Delta} \underset{\substack{\text{dialkyloxamide} \\ \textit{(crystalline solid)}}}{\begin{matrix}CO-NH-R \\ | \\ CO-NH-R\end{matrix}} + 2C_2H_5OH$$

(ii) the 2° amine forms dialkyloxamic ester which is an oily liquid.

$$\underset{\text{diethyl oxalate}}{\begin{matrix}CO-O_2CH_5 \\ | \\ CO-OC_2H_5\end{matrix}} + H-N\begin{matrix}R \\ R\end{matrix} \xrightarrow{\Delta} \underset{\substack{\text{dialkyloxamie ester} \\ \textit{(olly liquid)}}}{\begin{matrix}CO-N\begin{matrix}R \\ R\end{matrix} \\ | \\ CO-OC_2H_5\end{matrix}} + C_2H_5OH$$

The second ester group does not react presumably because of steric hindrance.

(iii) the 3° amine does not react as it does not contain a replaceable hydrogen atom on nitrogen. The reaction mixture is now fractionally distilled. The 3° amine distils over and forms the first fraction. This is followed by dialkyl oxamic ester which forms the second fraction.

The dialkyloxamide remains behind in the distillation flask.

The dialkyloxamide and the dialkyloxamic ester separated as above are hydrolysed wit. KOH to generate the free amines which are distilled off.

$$\underset{\text{diethyl oxamide}}{\begin{matrix}CO{-}NH{-}R\\|\\CO{-}NH{-}R\end{matrix}} + 2\,KOH \longrightarrow \underset{\text{pot oxalate}}{\begin{matrix}CO{-}\overset{-}{O}\overset{+}{K}\\|\\CO{-}\overset{-}{O}\overset{+}{K}\end{matrix}} + \underset{1^\circ\text{ amine}}{2\,R{-}NH_2}$$

$$\underset{\text{diethyl oxamic ester}}{\begin{matrix}CO{-}NR_2\\|\\CO{-}OC_2H_5\end{matrix}} + 2\,KOH \longrightarrow \begin{matrix}CO{-}\overset{-}{O}\overset{+}{K}\\|\\CO{-}\overset{-}{O}\overset{+}{K}\end{matrix} + \underset{2^\circ\text{ amine}}{R_2NH} + C_2H_5OH$$

(2) Reaction with Benzensulphonyl chloride: Hinsberg method. This method makes use of the different reaction of 1° , 2° and tertiary amines benzenesulphonyl chloride (hinsberg reagent) i.e.

$$C_6H_5{-}\overset{O}{\underset{O}{\overset{\|}{\underset{\|}{S}}}}{-}Cl \quad \text{or} \quad C_6H_5{-}\overset{O}{\underset{O}{\overset{\|}{\underset{\|}{S}}}}{-}Cl \quad \text{or} \quad C_6H_5{-}SO_2{-}Cl$$

The given mixture of three amines is treated with benzenesulphonyl chloride and then with aqueous sodium hydroxide.

(*i*) 1° amine forms *N*-alkyl-benzenesulphonamide which is soluble in base.

$$\underset{\substack{\text{benzenesulphonyl}\\ \text{chloride}}}{C_6H_5{-}SO_2{-}Cl_2} + \underset{1^\circ\text{ amine}}{H_2N{-}R} \longrightarrow \underset{\text{N- alkyl-benezenesulphonamide}}{C_6H_5{-}SO_2{-}NH{-}R} + HCl$$

$$C_6H_5{-}SO_2{-}NH{-}R + NaOH \longrightarrow \underset{\substack{\text{N-alkyl-sodium benzene}\\ \text{sulphonamide (soluble)}}}{[C_6H_5{-}SO_2{-}\bar{N}{-}R]\overset{+}{Na}} + H_2O\cdot$$

(*ii*) 2° amine forms *N*, N-dialkylbenzenesulphonamide which is insoluble in NaOH solution. This is so because the resulting sulphonamide has no replaceable hydrogen atom to yield the sodium salt.

$$\underset{\substack{\text{benzenesulphonyl}\\ \text{chloride}}}{C_6H_5{-}SO_2{-}Cl} + \underset{2^\circ\text{ amine}}{H{-}NR_2} \longrightarrow \underset{\substack{\text{N, N-dialky-benzene-sulphonamide}\\ \text{(insoluble)}}}{C_6H_5{-}SO_2{-}NR_2} + HCl$$

(*iii*) 3° amine does not react with benzenesulphonyl chloride at all.

The reaction mixture obtained after treatment with benzenesulphonyl chloride in the presence of NaOH is now distilled. The tertiary amine passes over first. The remaining mixture is filtered and the filtrate on acidification gives the sulphonamide of the 1° amine, while the solid residue is the sulphonamide of the 2° amine. The two sulphonamides thus isolated are hydrolysed with hydrochloric acid to regenerate the individual 1° and 2° amines which are fractionally distilled.

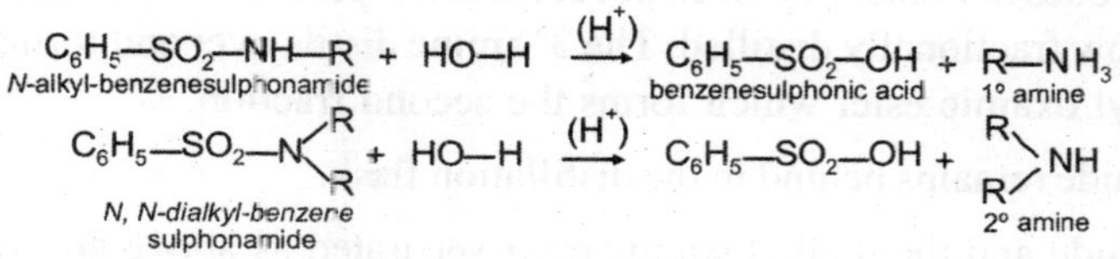

$$C_6H_5-SO_2-NH-R + HO-H \xrightarrow{(H^+)} C_6H_5-SO_2-OH + R-NH_2$$

N-alkyl-benzenesulphonamide — benzenesulphonic acid — 1° amine

$$C_6H_5-SO_2-N\begin{matrix}R\\R\end{matrix} + HO-H \xrightarrow{(H^+)} C_6H_5-SO_2-OH + \begin{matrix}R\\R\end{matrix}\!>NH$$

N, N-dialkyl-benzene sulphonamide — 2° amine

Nowadays benzenesulphonic chloride has been replaced by p-toluenesulphonyl chloride, CH.-C.H,-SO,Cl, since the substituted sulphuramides thus formed are stable solids which can be easily crystallised.

Physical Properties

(1) The loweramines are colourless gases at room temperature. Those with four or more carbons are volatile liquids. The higher. molecular-weight amines are colourless solids.

(2) Odour low.molecular-weight amines, such as methylamine and dimethylamine, have strong odours like the smell of ammonia. The odour resembles that of a decaying fish as the size of the alkyl group attached to nitrogen increases. In general, aliphatic amines are said to possess 'fishy' or fishlike odours.

(3) They are all lighter than water.

(4) Boiling points. Simple amines are polar compounds. Primary amines have two polar N-H bonds, secondary amines have one 'polar N -H bond, while tertiary amines have none.

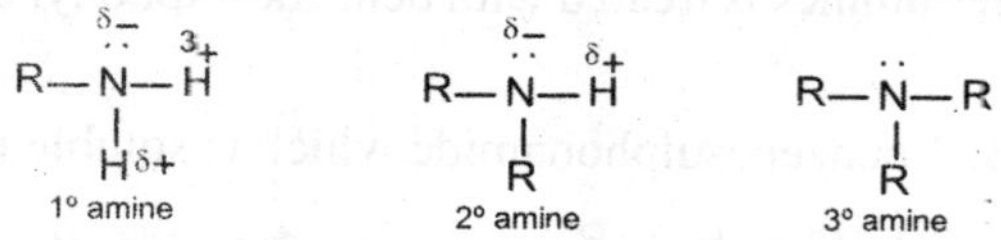

$$R-\overset{\delta-}{\ddot{N}}(-H^{\delta+})-H^{\delta+} \qquad R-\overset{\delta-}{\ddot{N}}(-R)-H^{\delta+} \qquad R-\ddot{N}(-R)-R$$

1° amine — 2° amine — 3° amine

Therefore, primary and secondary amines are polar and capable of forming intermolecular hydrogen bonds by virtue of the presence of unshared electron pair on nitrogen atom. Tertiary amines are also polar but they arc not capable of intermolecular hydrogen bonding.

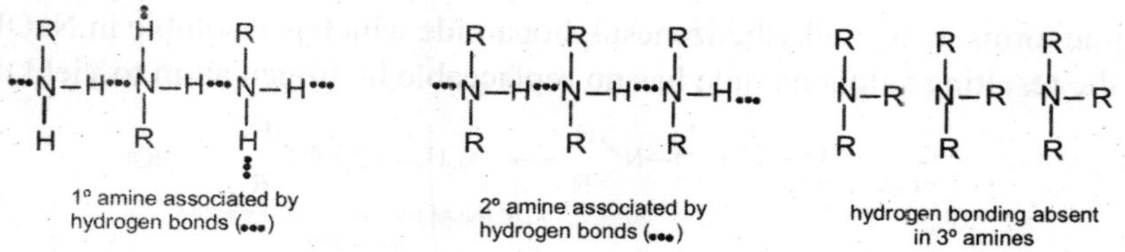

1° amine associated by hydrogen bonds (•••) — 2° amine associated by hydrogen bonds (•••) — hydrogen bonding absent in 3° amines

The hydrogen bonds formed by amines (N-H...N) are not as strong as the hydrogen bonds for alcohols (O-H...O) of comparable molecular weight. This is so because nitrogen is less electronegative than oxygen.

On account of their less degree of intermolecular association; the boiling points of primary and secondary amines are lower than those of alcohols of comparable molecular weight.

The boiling points of tertiary amines which form no hydrogen bonds are near those of alkanes of comparable molecular weight.

Table 15.1 : Boiling points of Amines, Alcohols and Alkanes of comparable molecular weight

Structure	Name	Molecular Weight	Boiling Point °C
CH_3-OH	methyl alcohol	32	65
CH_3-NH_3	methylamine	31	− 6
CH_3-CH_3	ethane	30	− 89
CH_3-CH_3-OH	ethyl alcohol	46	78
$CH_3-CH_3-NH_3$	ethylamine	45	17
$CH_3-NH-CH_3$	dimethylamine	45	7
$CH_3-CH_3-CH_3$	propane	44	−42
$CH_3-CH_3-CH_3-OH$	*n*-propyl alcohol	60	97
$CH_3-CH_2-CH_3-NH_3$	*n*-propylamine	59	48
$CH_3-CH_2-NH-CH_3$	ethylmethylamine	59	35
$CH_3-CH_3-CH_2-CH_3$	*n*-butane	58	−0·5
$CH_3-\underset{\underset{CH_3}{\vert}}{CH}-OH$	isopropyl alcohol	60	82
$CH_3-\underset{\underset{CH_3}{\vert}}{CH}-NH_3$	isopropylamine	59	33
$CH_3-\underset{\underset{CH_3}{\vert}}{N}-CH_3$	trimethylamine	59	4
$CH_3-\underset{\underset{CH_3}{\vert}}{CH}-CH_3$	isobutane	58	−12

(5) Solubility: All amines including tertiary amines are capable of hydrogen bond associates with water molecules

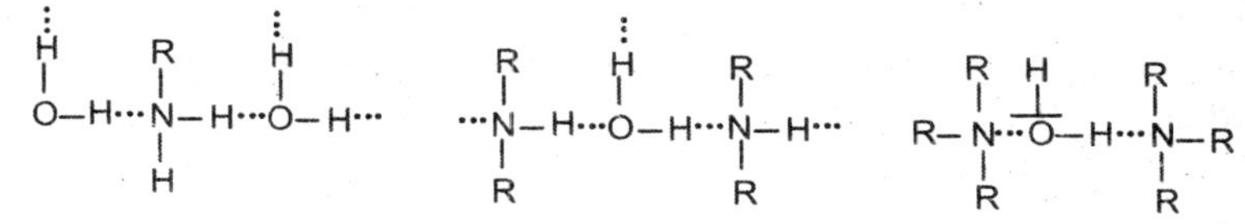

$$\cdots H-O-H\cdots \underset{H}{\overset{R}{N}}-H\cdots \overset{H}{O}-H\cdots$$

1° amine associated with water

$$\cdots\underset{R}{\overset{R}{N}}-H\cdots\overset{H}{O}-H\cdots\underset{R}{\overset{R}{N}}-H\cdots$$

2° amine associated with water

$$R-\underset{R}{\overset{R}{N}}\cdots\overset{H}{O}-H\cdots\underset{R}{\overset{R}{N}}-R$$

3° amine associated with water

Therefore low - molecular weight amines (less than six carbon atoms) are very soluble with water. The water solubility of amines deceases with increasing size of the alkyl group.

Amine	CH_3NH_2	$(CH_3)_2NH$	$(CH_3)_3NH$	$CH_2C_2NH_2$	$(CH_3CH_2)_2NH$	$(CH_3CH)_{33}N$
Solubility in 100 g H_2O	very soluble	very soluble	91g	very soluble	very soluble	14 g

Large aliphatic amines are soluble in organic solvents such as benzene, ethanol, ether and chloroform.

(6) Many amines are capable of producing serious physiological effects. The low-molecular weight volatile amines are irritants and cardiac stimulants, like ammonia.

Basic of Amines

The unshared electron pair on the nitrogen atom of ammonia and amines is responsible for their basic character. Ammonia when dissolved in water abstracts a proton from water molecule to form ammonium ion and hydroxide ion

$$H_3N: + H-O-H \rightleftharpoons [H-\overset{H}{\underset{H}{N}}:H]^+ + :\ddot{O}-H$$

Ammonia, Water, Ammonium Ion, Hydroxide Ion

Similarly, all the three type of amines (1°, 2° and 3°) react reversibly with water to form the alkyl-substituted ammonium ions and hydroxide ions.

$$R-\ddot{N}H_2 + H-O-H \rightleftharpoons R-\overset{+}{N}H_3 + \overline{O}H$$

1° amine → alkylammonium ion

$$R_2\ddot{N}H + H-O-H \rightleftharpoons R_2\overset{+}{N}H_2 + \overline{O}H$$

2° amine → dialkylammonium ion

$$R_3N: + H-O-H \rightleftharpoons R_3\overset{+}{N}H + \overline{O}H$$

3° amine → trialkylammonium ion

Evidently. aqueous solutions of ammonia and amines are alkaline in nature because of the formation of hydroxide ions as a result of the above equilibria. The value of K_b. (also called dissociation constant of the base) gives a measure of the base strength. Therefore equation (I) for a 1° amine the value of κ. may be expressed as

$$K_b = \frac{[R\overset{+}{N}H_3]\,[\overline{O}H]}{[RNH_3]\,[H_2O]}$$

Since water is present in large excess its concentrated remains practically constant and the above expression is therefore written as

$$K_b = \frac{[R\overset{+}{N}H_3][\overline{O}H]}{[RNH_2]}$$

It is clear from this expression that larger the value of K_b. the stronger would be the base. The value of K" for ammonia and a few amines are given below.

A look at tile values of Kb listed in the table shows that all amines are stronger bases than ammonia. It is also revealed that 2° amines are more basic than 1° amines but 3° amine the value of K. for methylamine is 4.2 × 10^{-}, that

Structure	Name	K_b	Structure	Name	K_b
NH_3	ammonia	$0{\cdot}2 \times 10^{-4}$			
$CH_3—NH_3$	methylamine	$4{\cdot}2 \times 10^{-4}$	$CH_3CH_2—NH_2$	ethylamine	$4{\cdot}7 \times 10^{-4}$
$CH_3—NH—CH_3$	dimethylamine	$6{\cdot}0 \times 10^{-4}$	$CH_3CH_2—NH—CH_2CH_3$	diethylamine	$3{\cdot}5 \times 10^{-4}$
$CH_3—N(CH_3)—CH_3$	trimethylamine	$0{\cdot}6 \times 10^{-4}$	$CH_3CH_2—N(CH_2CH_3)—CH_2CH_3$	triethylamine	$5{\cdot}5 \times 10^{-4}$

dimethylamine is slightly higher (6'0 x 10- 4) and for trimethylamine the value of *Kb* again!: declines to as low as 0.6 X 1Q--1.

The sequence of Basic strengths in general is

$$\underset{1^\circ \text{ amine}}{R—NH_2} < \underset{2^\circ \text{ amine}}{R—NH—R} > \underset{3^\circ \text{ amine}}{R—N(R)—R}$$

Interpretation of Relative Strengths of Amines. The differences in the base strengths of amines are accounted for by the degree of availability of the unshared electron pair on the nitrogen of an amine. Alkyl groups are electron releasing compared to hydrogen atom. Thus when in a primary amine hydrogen of ammonia is replaced by an alkyl group, electrons will shift toward the nitrogen relative to the situation in ammonia.

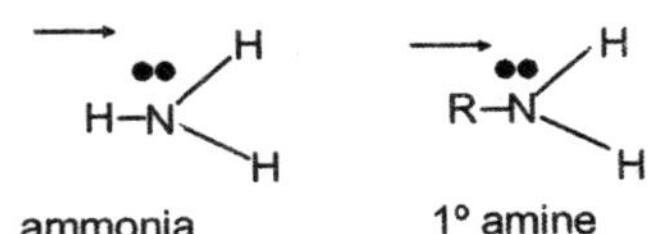

Fig 70 showing the shift to electrons to nitrogen indicated by arrow enhances the availability of unshared electron pair of hydrogen

Hence there will be more electrons close to nitrogen in RNH. than in ammonia, a, the unshared electron pair on nitrogen is more available to react with proton from water molecule. That is why methylamine is a stronger base than ammonia following the same line of argument dimethylamine having two electron. releasing methyl group! attached to the nitrogen atom, is a stronger base than methylamine. If so, trimethylamine having three methyl groups attached to nitrogen should be expected to be more basic than dimethylamine. Bit actually trimethylamine is considerably less basic than dimethylamine... Why so? In methylamine and dimethylamine the 'electronic effect' described above increases the base strength the amine. However, in trimethylamine the overcrowding of the three methyl groups attached to nitrogen causes the 'steric effect' to dominate over the 'electronic effect'. This 'steric effect' retards the protonation of nitrogen which results in an appreciably lower base strength of trimethylamine.

Chemical Properties

The chemical reactions of amines are due to (i) their basic character; and (ii) reactive hydrogen atoms attached to nitrogen. Some of these reactions are listed below.

(1) Formation of Salts with Mineral Acids. Owing to the presence of an unshared electron pair on nitrogen, amines act as Lewis bases. Thus they react with aqueous mineral acids to form substituted ammonium salts. For illustration with aqueous hydrochloric acid,

(*i*) 1° amines from alkylammonium solid:

$$\underset{\text{1° amine}}{R{-}\ddot{N}H_2} + \overset{+\;-}{HCl} \longrightarrow \underset{\text{an alkylammonium chloride}}{\left[R{-}NH_2{:}H\right]^+ \bar{Cl}}$$

or $$R{-}NH_3 + \overset{+\;-}{HCl} \longrightarrow R{-}\overset{+}{N}H_3\bar{Cl}$$

e.g., $$CH_3{-}NH_2 + \overset{+\;-}{HCl} \longrightarrow \underset{\text{methylammonium chloride}}{CH_3{-}\overset{+}{N}H_3\bar{Cl}}$$

(*ii*) 2° amines form dialkylammonium salts:

$$\underset{\text{2° amine}}{R_2\ddot{N}H} + \overset{+\;-}{HCl} \longrightarrow \underset{\text{a dialkylammonium chloride}}{\left[R{-}NHR{:}H\right]^+ \bar{Cl}}$$

or $$R{-}NH(R) + \overset{+\;-}{HCl} \longrightarrow R{-}\overset{+}{N}H_2\bar{Cl}$$

e.g., $$\underset{\text{dimethylamine}}{CH_3{-}NH(CH_3)} + \overset{+\;-}{HCl} \longrightarrow \underset{\text{dimethylammonium chloride}}{CH_3{-}\overset{+}{N}H_2(CH_3)\bar{Cl}}$$

(*iii*) 3° amines from trialkylammonium salts:

$$\underset{\text{3° amine}}{R_3\ddot{N}} + \overset{+\;-}{HCl} \longrightarrow \underset{\text{trialkylammonium chloride}}{\left[R_3N{:}H\right]^+ \bar{Cl}}$$

or $$R_3N + \overset{+\;-}{HCl} \longrightarrow R_3N\overset{+\;-}{HCl}$$

e.g., $$\underset{\text{trimethylamine}}{(CH_3)_3N} + \overset{+\;-}{HCl} \longrightarrow \underset{\text{trimethylammine chloride}}{(CH_3)_3N\overset{+\;-}{HCl}}$$

The amine salts are general. Well defined crystalline substances with -sharp melting points and are often useful for the identification of amines. Further, these amine salts, unlike the amines, are water soluble and regenerate the amines on treatment with a strong alkali such as NaOH e.g

$$\underset{\text{alkylammonium sal}}{R{-}\overset{+}{N}H_3\bar{Cl}} + NaOH \longrightarrow \underset{\text{1° amine}}{R{-}NH_2} + H_2O + \overset{+\;-}{NaCl}$$

This property is made of in the purifying amines form non-basic contaminants.

(2) Alkylation Reaction. The unshared electron pair on the nitrogen atom makes amines behave as strong nucleophiles. Thus when a primary amine is treated with an alkyl halide in a basic medium such as ammonia, the halogen atom of the alkyl halide is displace yielding dialkylammonium salt in the presence of ammonia at once gives a secondary amine.

$$R-\ddot{N}H-H + R-X \longrightarrow R_2\overset{+}{N}H_2 + \bar{X}$$

1° amine → Dialkylammonium cation

$$R_2\overset{+}{N}H-H + :NH_3 \longrightarrow R_2N{:}H + \overset{+}{N}H_4$$

dialkylammonium cation → 2° amine

The net alkylation reaction of a primary amine with alkyl halide may be written as

$$\underset{(1° \text{ amine})}{R-NH_2} + \underset{\text{ethyl halidide}}{R-X} \longrightarrow \underset{2° \text{ amine}}{R-NH-R} + \underset{\text{halogen acid}}{HX}$$

The secondary amine undergoes similar reaction with a second molecule of alkyl halide to form .a tertiary amine. The tertiary amine finally reacts with a third molecule of alkyl halide to produce the tetraalkylammonium salt. These multiple alkylation reactions of amines may be illustrated by the action of methyl bromide on methylamine.

$$\underset{\substack{\text{methylamine}\\(1° \text{ amine})}}{CH_3-\ddot{N}H_3} + CH_3-Br \longrightarrow \underset{\substack{\text{dimethylamine}\\(2° \text{ amine})}}{CH_3-\ddot{N}H-CH_3} + HBr$$

$$CH_3-\ddot{N}H-CH_3 + \underset{(2\text{nd molecule})}{CH_3-Br} \longrightarrow \underset{\text{trimethylamine } (3° \text{ amine})}{(CH_3)_2\ddot{N}-CH_3} + HBr$$

$$(CH_3)_2\ddot{N}-CH_3 + \underset{(3\text{rd molecule})}{CH_3-Br} \longrightarrow \underset{\text{tetramethylammonium bromide}}{[(CH_3)_2N(CH_3)CH_2]^+ \bar{Br}}$$

From the above equations it is clear that a primary amine will need three molecules of alkyl halide) per molecule of amine, secondary amine two molecules, and a tertiary amine only one molecule to form the final tetraalkylammonium salt. Thus by experimentally determining the amount of alkyl halide actually used for the conversion of a given amine to the tetralkylammonium salt, it is possible to distinguish between primary, secondary and tertiary amines. Also, the alkylation of a primary and secondary amine under regulated conditions provides an industrial method for the preparation of secondary and tertiary amine respectively.

(3) Acylation Primary and secondary amines react with acyl halides (RCOCl) and carboxylic anhydrides whereby the hydrogen atom attached to nitrogen atom is replaced by : acyl group (RCO-). By this process of *acylation* a primary amine produces *N-substituted* amide while a secondary amine yields *N*, N-disubstituted amide. A tertiary amine having no: available hydrogen on nitrogen, does not react at all either with acyl halides or carboxylic. anhydrides.

$$\underset{1^\circ \text{ amine}}{R-NH-H} + \underset{\text{acyl chloride}}{R'-\overset{O}{\overset{\|}{C}}-Cl} \longrightarrow \underset{N\text{-alkylamide}}{R-NH-\overset{O}{\overset{\|}{C}}-R'} + HCl$$

$$\underset{2^\circ \text{ amine}}{R-\overset{R}{\overset{|}{N}}-H} + R'-\overset{O}{\overset{\|}{C}}-Cl \longrightarrow \underset{N,\ N\text{-dialkylamide}}{R-\overset{R}{\overset{|}{N}}-\overset{O}{\overset{\|}{C}}-R'} + HCl$$

$$\underset{\substack{3^\circ \text{ amine} \\ \text{(no available H-atom)}}}{R-\overset{R}{\overset{|}{\underset{R}{\underset{|}{N}}}}} + R'-\overset{O}{\overset{\|}{C}}-Cl \longrightarrow \text{No Reaction}$$

For example,

$$\underset{\text{ethylamine}}{C_2H_6-NH_2} + \underset{\text{acetyl chloride}}{CH_2-CO-Cl} \longrightarrow \underset{N\text{-ethylacetamide}}{C_2H_5-NH-CO-CH_3} + HCl$$

$$\underset{\text{diethylamine}}{C_2H_6-\underset{C_2H_6}{\underset{|}{N}}-H} + CH_2-CO-Cl \longrightarrow \underset{N,\ N\text{-diethylacetamide}}{C_2H_6-\underset{C_2H_6}{\underset{|}{N}}\ CO-CH_3} + HCl$$

$$\underset{\text{triethylamine}}{C_2H_6-\underset{C_2H_6}{\underset{|}{N}}-C_2H_6} + CH_3COCl \longrightarrow \text{No reaction}$$

As started above, primary and secondary amines also react with carboxylic anhydride to give substituted amides

$$\underset{1^\circ \text{ amines}}{R-NH_2} + \begin{matrix} CH_3-CO \\ CH_3-CO \end{matrix}\!\!>\!O \longrightarrow \underset{N\text{- alkylacetamide}}{R-NH-COCH_3} + CH_3COOH$$

$$\underset{2^\circ \text{ amines}}{R-NH-R} + \begin{matrix} CH_3-CO \\ CH_3-CO \end{matrix}\!\!>\!O \longrightarrow \underset{N,\ N\text{- dialkylacetamide}}{R-\underset{COCH_3}{\underset{|}{N}}-R} + CH_3COOH$$

In all the above reactions acylation of amines is carried out effectively in the presence of tertiary amines which help the removal of the acid (HCl or CH3COOH). The tertiary amine which is most often used as"acid scavenger" is the aromatic base pyridine. Thus

$$R-NH_3 + CH_3-CO-Cl + \underset{\text{pyridine}}{C_5H_5N} \longrightarrow R-NH-CO-CH_3 + \underset{\text{pyridinium chloride}}{C_5H_5\overset{+}{N}H\,\overset{-}{Cl}}$$

In **Schotten-Baumann** procedure of acylation of amines a mixture of amine, cold aqueous base (NAOH or Na_2CO_3) is shaken together. The aqueous alkali acts a scavenger for the HX produced.

$$R{-}NH_3 + CH_3COCl + NaOH \xrightarrow{\text{cold}} R{-}NH{-}COCH_3 + NaCl + H_2O$$

Mechanism of Acylation of Amines: The unshared electron pair on nitrogen plays a key in amide formation by acylation of amines. The reaction takes place by the nucleophilic attack of the primary or secondary amine on to the carbonyl group of acyl chloride. That is accompanied by the loss of CI– and a proton.

$$R{-}\overset{H}{\underset{H}{N}}: + \underset{\text{Acetyl Chloride}}{CH_3{-}C(Cl){=}O} \xrightarrow{-\bar{C}l} R{-}\overset{H}{\underset{H}{N}}{-}\overset{CH_3}{C}{=}O \xrightarrow{-H^+} \underset{\textit{N-alkylacetamide}}{R{-}NH{-}\overset{O}{\overset{\|}{C}}{-}CH_3}$$

The above mechanism of acylation o. amines also shows why only one of the hydrogens of a primary amine is replaced by acyl group in actual practice. The N-acylamine produced is far less basic than the original amine, because of electron-withdrawal from nitrogen atom by induction and resonance. Therefore, further acylation of N-acylamine not take place and as a result only one acyl group can be introduced into a primary amines.

$$\underset{\text{a sulphonic acid}}{R{-}\overset{O}{\overset{\|}{\underset{\underset{O}{\|}}{S}}}{-}OH} \xrightarrow[\text{or } SOCl_2]{PCl_5} \underset{\text{a sulphonyl chloride}}{R{-}\overset{O}{\overset{\|}{\underset{\underset{O}{\|}}{S}}}{-}OH} + HCl$$

(4) Sulphonylation. Sulphonyl chlorides are the acid chlorides produced by the reaction of phosphorus pentachloride or tbionyl chloride ($SOCl_2$).

$$\underset{1^\circ\text{ amine}}{R{-}NH_3} + \underset{\text{benzenesulphonyl chloride}}{C_6H_5{-}SO_3{-}Cl} \longrightarrow \underset{\textit{N-alkylbenzonesulphonamide}}{R{-}NH{-}SO_3{-}C_6H_5} + HCl$$

$$\underset{2^\circ\text{ amine}}{R{-}\underset{R}{\underset{|}{N}}H} \quad C_6H_5{-}SO_3{-}Cl \longrightarrow \underset{\text{N, N-dialkylbenzenesulphoamide}}{R{-}\underset{R}{N}{-}SO_3{-}C_6H_5} + HCl$$

$$\underset{\substack{3^\circ\text{ amine}\\ \text{(no available H-atom)}}}{R{-}\underset{R}{\underset{|}{N}}{-}R} + C_6H_5{-}SO_3{-}Cl \longrightarrow \text{No Reaction}$$

Most sulphonamides are crystalline solids. Their preparation and determination of melting point is used for the identification of primary and secondary amines.

Hinsburg Test for distinguishing r, 2° aod 3° amines. The reactions of amines benzenesulphonyl chloride form the basis of the classical Hinsberg Test to determine the c: of an amine. In this test the given amine is treated with bcnzenesulphonyl chloride (Hins Reagent) in the presence of cold aqueous NaOH.

(a) A PRIMARY AMINE produces N-alkylsulphonamide which contains an acidic hydrogen and hence dissolves in NaOH solution to form the soluble sodium salt.

The acidification of the solution gives a precipitate of the nonbasic free sulphonamide.

$$\underset{1^\circ\text{ amine}}{RNH_2} + C_6H_5\text{—}SO_2\text{—}Cl \xrightarrow[H_2O]{NaOH} \underset{N\text{-alkylsulphonamide}}{R\text{—}\overset{\text{acidic hydrogen}}{NH}\text{—}SO_2\text{—}C_6H_5} \xrightarrow{NaOH} \underset{\substack{\text{sodium salt}\\ \text{(water soluble)}}}{R\text{—}\overset{\overset{+}{Na}}{\bar{N}}\text{—}SO_2\text{—}C_6H_5}$$

$$R\text{—}\overset{\overset{+}{Na}}{\bar{N}}\text{—}SO_2\text{—}C_6H_5 \xrightarrow[H_2O]{\overset{+}{H}} \underset{\substack{\text{free sulphonamide}\\ \text{(insoluble in acid)}}}{R\text{—}NH\text{—}SO_2\text{—}C_6H_5}$$

The unusual acidity of N-alkylsulphonamides id due to the powerful electron drawinginductive effect of the sulphonyl group (-SO_2)

***(b)* A Secondary Amine** Yields *N, N* - dialkylsulphonamide which has no acidic hydrogen precipitates immediately as it is incapable of forming soluble sodium salt NaOH

***(c)* A Tertiary Amine** having no available hydrogen does not react and remain as insoluble material.

To apply the Hinsburg test, the amine to be tested is shaken with benzenesulphonyl chloride and cold aqueous NaOH. If a homogeneous solution results which on acidification forms a precipitate, the amine is primary. If in insoluble material is produced in the reaction mixture straightaway (without acidification), the amine is either secondary (the insoluble material being dialkylsulphonarnide) or tertiary (insoluble material being unreacted amine). To distinguish between the secondary and tertiary amines, the insoluble material is acidified. A tertiaryamine if present will form a soluble quaternary salt and dissolve, while a sulphonamide from a secondary amine being neutral remains insoluble material.

Table 15.3 : Summary of Hinsberg Test

Reactants	Reagent	Product	Reagent	Result
$R\text{—}NH_2 + C_6H_5\text{—}SO_2\text{—}Cl$ (1° amine)	$\xrightarrow[H_2O]{NaOH}$	*Clear solution*	$\xrightarrow{\text{acidify}}$	*Insoluble material*
$R\text{—}NH(R) + C_6H_5\text{—}SO_2\text{—}Cl$ (2° amine)	$\xrightarrow[H_2O]{NaOH}$	*Insoluble material*	$\xrightarrow{\text{acidify}}$	*Insoluble material*
$R\text{—}N(R)\text{—}R + C_6H_5\text{—}SO_2\text{—}Cl$ (3° amine)	$\xrightarrow[H_2O]{NaOH}$	*Insoluble material*	$\xrightarrow{\text{acidify}}$	*Clear solution*

In modern practice benzenesulphonyl chloride is also replaced by p-toluenesulphonyl chloride,

$$H_3C\text{—}C_6H_4\text{—}S(=O)\text{—}O\text{—}Cl$$

$$\overset{+}{N}a\overset{-}{NO_2} + HCl \longrightarrow \underset{\text{nitrous acid}}{HO-N=O} = H_2O$$

Primary, Secondary and Tertiry amines give characteristic reactions with nitrous acid.

(*i*) *Primary amines (RNH_3)* which have two available hydrogen atoms on nitrogen react with nitrous acid in aqueous solution to form the corresponding alcohols.

$$\underset{1^\circ\text{ amine}}{R-NH_2} + \underset{\text{nitrous acid}}{O=N-OH} \rightarrow \underset{2^\circ\text{ amin}}{R-N=N-OH} \rightarrow \underset{\text{alcohol}}{R-OH} + N_2\uparrow$$

The reaction is not that simple as stated above. In fact, primary amines react with nitrous acid to give mixtures of a variety of products which contain alcohols, alkenes, and even halides, if halide ion is present. For example, if the primary amine taken is n-propylamine, the products will include n-propyl alcohol, isopropyl alcohol, propene, propyl chloride and the acidific reaction conditions employed.

The formation of various products obtained from n- propylamine by the above mechanism may be indicated as below:

$$\underset{\text{n-propylamine}}{CH_3CH_3CH_2NH_2} \xrightarrow[(HCl)]{HONO} \left| \begin{array}{c} CH_3CH_2CH_3{-}OH \\ CH_3{-}\underset{|}{\overset{OH}{CH}}{-}CH_3 \\ CH_3{-}CH{=}CH_2 \\ CH_3CH_3CH_2{-}Cl \\ CH_2{-}\overset{Cl}{\underset{}{CH}}{-}CH_3 \end{array} \right| + N_2 + H_2O$$

It may be noted particularly that whatever may be the products of the reaction, nitrogen is always evolved.

$$CH_3{-}CH_2{-}\overset{+}{CH_3} \left| \begin{array}{l} + H_2\ddot{O} \longrightarrow CH_2{-}CH_2{-}CH_3{-}\overset{+}{\ddot{O}}H_2 \xrightarrow{-H^+} CH_3CH_2CH_2OH \\ - H^+ \longrightarrow CH_3{-}CH{=}CH_2 \\ + Cl^- \longrightarrow CH_3{-}CH_2{-}CH_2Cl \end{array} \right.$$

$$CH_3{-}\overset{+}{CH}{-}CH_3 \left| \begin{array}{l} + H_2\ddot{O} \longrightarrow CH_3{-}\overset{\overset{+}{O}H_3}{CH}{-}CH_3 \xrightarrow{-H^+} CH_3{-}\overset{OH}{CH}{-}CH_3 \\ - H^+ \longrightarrow CH_3{-}CH{=}CH_3 \\ + Cl^- \longrightarrow CH_3{-}\overset{Cl}{CH}{-}CH_3 \end{array} \right.$$

It may be noted particularly that what ever may be the products of the reactions nitrogen is always evolved.

MECHANISM. The various products obtained from primary amines nitrous acid are formed by the following steps.

(*a*) Nucleophilic addition of the amine RNH_2 to nitrogen-oxygen double bond nitrous acid to give the adduct

R—N(H)(H): + N(OH)=O ⟶ R—N(H)—N(OH)—O
1° Amine

(*b*) Protonation of I followed by elimination of the molecules of water to form alkyldiazomium ion.

R—N(H)(H)—N(O)(OH): —H^+⟶ R—N(H)(H)—N(OH)(OH): —$-2H_2O$⟶ R—N⁺≡N:
I II Alkyldiazonium ion III

(*c*) Alkyldiazomium ion III is unstable and looses nitrogen (N≡N) to generate carbonium ion IV

R—N⁺≡N: ⟶ R⁺ + :N≡N
III Carbonium ion IV Nitrogen

The carbonium ion IV may rearrange to give more stable carbonium ion V (not shown).

(*d*) The carbonium ion IV and V may react with water (H.O) to form alcohols, lose a proton to form alkanes, or add halide ion (Cl-) to yield alkyl halides.

(*ii*) *Secondary amines* (R.NH) react with nitrous acid (HO–N=0) to give neutral, yellow oily substances called *Nitrosuamines*. The process which involves the replacement of the available hydrogen on nitrogen by nitroso group (–N = O) is termed Nitrosation

R—N(R)—H + HO—N=O ⟶ R—N(R)—N—O + H_2O
2° amine Nitrous acid Nitrosoamin (*yellow oil*)

When warmed with a little phenol and a few drops of conc H2SO4 nitrosoamines produce a green solution which turns deep blue or violet on alkalificiltion with NaOH. This reaction known as Liebermann Nitrosoamine Reaction is used as a test for secondary amines.

Nitrosoamines are generally poisonous and are considered to be cancer .causing agents. . In some foods sodium nitrite is used as preservative and in the stomach it combines with hydrochloric acid to produce nitrous acid. This can react with any secondary amines present, in foods to form a dangerous nitrosoamine. Therefore, the intake of sodium nitrite in any form should be avoided.

(*ii*) *Tertiary amines* (R_3N) having no available hydrogen on nitrogen fail to give a replacement reaction with nitrous acid. They simply form soluble trialkylammonium salts of nitrous acid.

$$R_3N: + HO-N=O \longrightarrow [R_3\overset{+}{N}-H]\ \bar{O}-N=O$$

3° amine → Trialkylammonium nitrite (*a soluble salt*)

The tertiary amine can be regenerated form the salt solution by addition of alkali

The reaction of aliphatic amines with nitrous acid is used as a test for distinguishing primary, secondary and tertiary amines.

(*a*) *Primary amines* react with rapid evolution of nitrogen gas.

$$R-NH_2 \xrightarrow{HONO} \underset{\text{carbonium ion}}{R^+} + N_2\uparrow$$

(*b*) *Secondary amines* form an insoluble yellow oily layer.

$$R_2NH \xrightarrow{HONO} \underset{\text{nitrosoamine}}{R_2N-NO}$$

(*c*) *Tertiary omines* form a homogeneous solution.

$$R_3N \xrightarrow{HONO} \underset{\text{soluble salt}}{[R_3NH]^+\bar{O}-N=O}$$

(6) Reaction with Aldebydes and Ketone. *(a) Primary* amines add readily to the C=O group of aldehydes and ketones to form .x-hydroxyamines, also called *Carbiolamines*. These undergo spontaneous elimination of a water molecule yielding substituted imines or *Schiff bases*. The reaction is catalyzed by acids.

$$\underset{\text{1° amine}}{R-NH_2} + \underset{\text{Aldehyde}}{R'-CH=O} \xrightarrow{H^+} \underset{\alpha\text{-hydroxyamine}}{R-NH-CH(OH)-R'} \xrightarrow{-H_2O} \underset{\text{an aldimine}}{R-N=CH-R'}$$

$$R-NH_2 + R'_2C=O \xrightarrow{H^+} R-NH-C(OH)R'_2 \xrightarrow{-H_2O} \underset{\text{a ketimine}}{R-N=CR'_2}$$

(*b*) *Secondary amines* react with aldehydes and ketones that have o:-H atoms when transitory o-hydroxyamine at once splits out a water molecule to form, Compounds known as *enamines*. Thus

$$\underset{\text{2° amine}}{R_2N-H} + \underset{\text{Ketone having }\alpha\text{-h atom}}{R'-CH(R')-C(R')=O} \overset{H^+}{\rightleftharpoons} R-N(R)-C(R')(\ddot{O}H)-CH(R')-R' \overset{-H_2O}{\rightleftharpoons} \underset{\text{Enamine}}{R_2N-CR'=CR}$$

In enamines the basic nitrogen is attached directly to an olefinic carbon.. These compounds are useful synthetic intermediates.

(c) Tertiary amines having no hydrogen on nitrogen naturally cannot react with aldehydes and ketones.

(7) Reaction with Carbon disulphide. *(a) Primary amines* when reacted with carbon disulphide (CS_2) form dithiocarbamic acids which on subsequent treatment with $HgCl_2$ give an alkyl isocyanate or 'mustard oil'.

The isothiocyanate thus produced has a pungent 'mustard like' odour which can be easily rccognised. Hence this reaction *(Hofmann Mustard Oil Reaction)* is used as a test for primary amines.

(b) Secondary amines react with carbon disulphide to give dithiocarbamic acid which is not decomposed by $HgCl_2$ to form alkyl isothiocyanate.

The isothiocyanate thus produced has a pungent 'mustard like' odour which can be easily recognised. Hence this reaction (*Hofmann Mustard Oil Reaction*) is used as a test for primary amines.

(*c*) Tertiary amines having no available hydrogen atom on nitrogen do not react with CS_2.

$$\underset{\text{1° amine}}{R-NH-H} + S{=}C{=}S \longrightarrow \underset{\text{Dithiocarbamic acid}}{R-NH-C(=S)-SH} \xrightarrow{HgCl_2} \underset{\text{Alkylisothiocyanate}}{R-N{=}C{=}S} + HgS + 2HCl$$

(8) Halogenation. Primary and secondary amines react rapidly with a solution of hypochlorous acid in the presence of a base to produce N-chloroamines or N, N-dichloroamines.

$$\underset{\text{2° amine}}{R-NR-H} + S{=}C{=}S \longrightarrow \underset{\text{Dithiocarbamic acid}}{R-NR-C(=S)-SH} \xrightarrow{HgCl_2} \textit{No Reaction}$$

$$\underset{\text{1° amine}}{R-NH-H} \xrightarrow[OH^-]{HOCl} \underset{\textit{N-chloroamine}}{R-NCl-H} \xrightarrow[OH^-]{HOCl} \underset{\textit{N, N-dichloroamin}}{R-NCl-Cl}$$

Secondary amines give mono-*N*-chloroamines

$$\underset{\text{2° amine}}{R-NR-H} \xrightarrow[OH]{HOCl} \underset{\textit{N-chloroamine}}{R-NR-Cl}$$

The chlorination of amines is carried. more conveniently with t-butyl hypochlorite, $(CH_3)_3C{-}OCl$.

(9) Reaction.with Grigoard Reagents. Primary and secondary amines react with Grignard reagents to form alkanes. Tertiary amines do not react since they do not contain a replaceable hydrogen atom.

$$\underset{1^\circ\text{ amine}}{R\text{—}\overset{H}{\overset{|}{N}}\text{—}H} + R'MgX \xrightarrow{\text{Ether}} \underset{\text{alkane}}{R'H} + RNHMgX$$

$$\underset{2^\circ\text{ amine}}{R\text{—}\overset{R}{\overset{|}{N}}\text{—}H} + R'MgX \xrightarrow{\text{Ether}} \underset{\text{alkane}}{R'H} + R_2NMgX$$

(10) Carbylamine Reaction. Primary amines form isocyanides or carbylamines when warmed with chloroform ($CHCl_3$) and an ethanolic solution of KOH.

$$\underset{1^\circ\text{ amine}}{R-NH_2} + CHCl_3 + 3KOH \xrightarrow{\Delta} \underset{\substack{\text{alkyl isocyanide,}\\ \text{or carbylamine}}}{R-\overset{+}{N}\equiv\overset{-}{C}} + 3KCl + 3H_2O$$

The isocyanide or carbylamine so produced has an extremely unpleasant odour and hence this reaction, frequently called *carbylamine reaction*, is used .as a qualitative diagnostic test (Carbylamine Test) for primary amines. Secondary and tertiary amines do not respond to this test.

MECHANISM. The formation of carbylamine takes the following pathway.

(*i*) Chloroform in the presence of base (KOH) produces dicnlorocarbene ($:CCl_2$).

$$CHCl_3 + \overset{-}{O}H \longrightarrow H_2O + \overset{-}{C}Cl_3 \longrightarrow :CCl_2 + \overset{-}{Cl}$$

(*ii*) Nucleophilic addition of primary amine to dicnlorocarbene and subsequent elimination of two HCl molecules yields the isocyanide.

$$R\text{—}\underset{H}{\overset{H}{N}}: + :CCl_2 \longrightarrow R\text{—}\underset{H}{\overset{H}{\overset{+}{N}}}\text{—}\overset{-}{C}Cl_2 \longrightarrow R\text{—}\ddot{N}H\text{—}CHCl_2$$

$$\xrightarrow{-HCl} R\text{—}\ddot{N}\text{=}CHCl \xrightarrow{-HCl} R\text{—}\ddot{N}\text{=}C: \longleftrightarrow \underset{\substack{\text{Alkyl isocyanide}\\ \text{or Carbyl amine}}}{R\text{—}\overset{+}{N}\equiv\overset{-}{C}}$$

(11) Oxidation of Amines. All types of amines are easily oxidized The course reaction is very variable and depends very much on the structure of the amine and the specific oxidant used.

(*a*) Primary amines on oxidation with acidified $KMnO_4$ give aldehydes and ammonia

$$R\text{—}CH_2\text{—}NH_2 \xrightarrow{O} \underset{\text{Aldimine}}{R\text{—}CH\text{=}NH} \xrightarrow[H^+]{+H_2O} \underset{\text{Aldehyde}}{R\text{—}CH\text{=}O} + NH_3$$

$$\underset{R}{\underset{|}{R\text{—}CH}}\text{—}NH_2 \xrightarrow{O} \underset{\text{Ketimine}}{\underset{R}{\underset{|}{R\text{—}C}}\text{=}NH} \xrightarrow[H^+]{+H_2O} \underset{\text{Ketone}}{\underset{R}{\underset{|}{R\text{—}C}}\text{=}O} + NH_3$$

$$R\text{—}\underset{R}{\overset{R}{C}}\text{—}NH_2 \xrightarrow{O} \underset{\text{A nitroalkane}}{R\text{—}\underset{R}{\overset{R}{C}}\text{—}NO_2}$$

(*b*) Secondary amines on oxidation with $KMnO_4$ give tetralkylhydrazine while with H_2O_2 or peracid they yield dialkylhydroxylamine.

$$2\ R\text{—}\underset{\displaystyle R}{\underset{|}{N}}\text{—}H + [O] \xrightarrow{KMnO_4} R\text{—}\underset{\displaystyle R}{\underset{|}{N}}\text{—}\underset{\displaystyle R}{\underset{|}{N}}\text{—}R + H_2O$$

Tetraalkylhydrazine

$$R\text{—}\underset{\displaystyle R}{\underset{|}{N}}\text{—}H + [O] \xrightarrow[\text{or peracid}]{H_2O_2} R\text{—}\underset{\displaystyle R}{\underset{|}{N}}\text{—}OH$$

Dialkylhydroxylamine

(*c*) *Tertiary amines* from amine oxides when oxidised with H_2O_2 or peracids.

$$R\text{—}\underset{\displaystyle R}{\overset{\displaystyle R}{N}}: + \ddot{\underset{..}{O}}: \xrightarrow[\text{or peracid}]{H_2O_2} R\text{—}\underset{\displaystyle R}{\overset{\displaystyle R}{\overset{+}{N}}}\text{—}\bar{O}$$

Amine oxide

e.g.

$$\underset{\text{Trimethylamine}}{(CH_3)_3N} + [O] \xrightarrow{H_2O_2} \underset{\text{Trimethylamine-}N\text{-oxide}}{(CH_3)_3\overset{+}{N}\text{—}\bar{O}}$$

IR and NMR Spectra of Amines

A characteristic feature of infrared spectra of 1° and 2° amines in dilute solutions is the absorption at 3500- 3300 cm^{-1}, corresponding to N-H stretching. 1° amines have two such bonds, 2° amines only

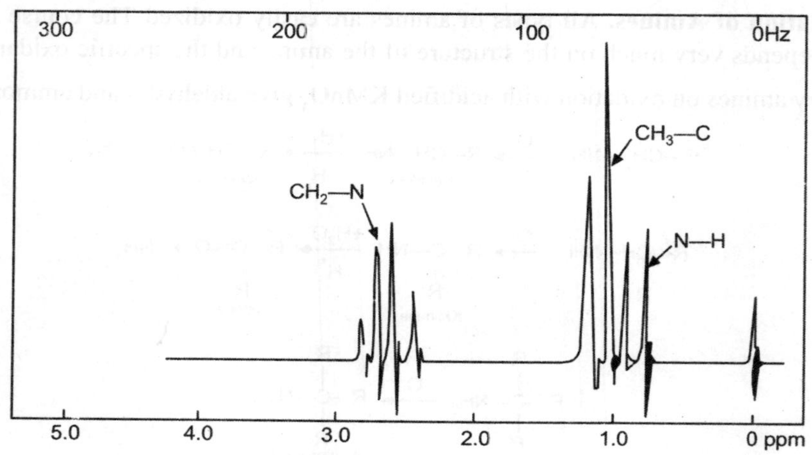

one while 3° amines (which contain no N-H bond) do not' absorb this region. Furthermore the area generated by one or two amino protons is generally quite small as compared with the rest of the spectrum. Hence the amino protons are sometimes observed by absorptions arising from other parts of the molecular structure. This complicates interpretation of the spectrum.

METHYLAMINE

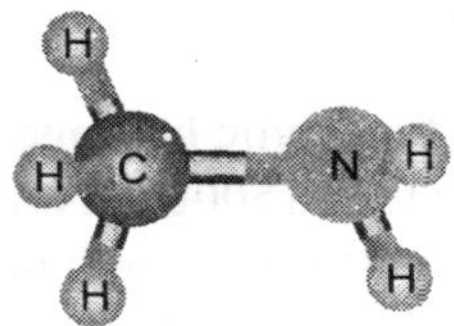

Preparation

Preparation of a 10% Methylamine solution in Methanol

By Sunlight

This is a convenient way of making a 10% methanolic solution of methylamine freebase from the commercially available 40% aqueous solution.

Put in a 2-neck flask 700 grams of commercial dry NaOH, and using a compensating pressure separatory funnel attached to one of the necks, add slowly over one hour 350 cc of aq. methylamine to the NaOH. In the other neck attach a rubber stopper with a teflon tube and to it a glass tube with a gas diffusor (a tiny glasfunnel with a glasfilter glued in it with 6 seconds glue or two component glue or just melt it tight with a torch), and bubble the MA-gas in 1000 cc of methanol in a flask in a methanol- or water/ice bath. Reaction is endothermic, finally you can heat a bit the NaOH containing flask to force the liberation of the last gas, but it's not really necessary, just shake it to see a homogeneus paste. The compensation of pressure can be done using a regular separatory funel if you attach a 1-hole rubber stopper with a teflon tube connecting to the NaOH flask, if you use a 3-neck flask you can use the other neck, and if you use a two neck flask, you can use in one of the necks a 2-hole rubber stopper, 1 hole connecting to the stopper in the neck of the separatory funel and one to the ... dripping hole of it (excuse my english...)

You can refill your sep. funnel with aq. methylamine if it's not enough big to complete the whole liberation of all the gas, close the valve and close the other tube, open it and put the aq. solution in, then stop it and free the other tube. The absortion of methylamine in methanol at 0-5 C is perfect, you don't even need to redirect possible fumes from the methanol, there aren't, but stop the flask with some papertissue to prevent the entry of humidity. The gas diffusor can be done with a regular glass tube, try to enlarge a bit one of the ends, and then try to weld inside it a piece of fine filterglass for the gas bubbles feed, it will act as a diffusor, not very good, but enough, and it won't heighten the pressure too much (one stopper could jump and you would have a toxic ambient atmosphere of methylamine in your viginity). Probably using a pipette will work, absortion is very good.

You can weight finally the 10% methanolic solution of methylamine to know the exact weight of gas you have in it (you weighted it before). You will have a bit of water, but don't worry, mix it with the ketone (350 grams), add silica gel (160 gr) and shake and let stand, shake and so on... during an hour, filter and wash the silica with 200-300 cc of methanol, add the NaBH4 at +7 C (an ice/salt bath works fine) to get from 75-87% of the desired product depending on the purity of your ketone. You can dehydrate the methylamine gas by passing it through a flask with NaOH, but it's not necessary due to the addition of silica gel.

Addendum by KrZ:

The mix {$MeNH_2$ + $2H_2O$ (The product of catalytic hydrogenation of methylamine) in methanol with Na_2SO_4} was split in half and each half sat with 1500g of Na_2SO_4 for an hour, with occasional stirring. The Na_2SO_4 (sodium sulphate) was filtered off for a long time, until it was quite dry in appearance, the mix halves were combined and washed 2x with 1L of CH_3OH (methanol), which was filtered off and added back to the solution.

Addendum by Sunlight

We did something similar with $MgSO_4$, trying to dry a mixture of 40% CH_3NH_2 and methanol, and the end product was titrated and it was about 3 % instead of 10 % w/w (weight/weight), the $MgSO_4$ had a strong smell of methylamine, so we thought most of the gas was ended in the water and in the magnesium sulphate, and we tried the above procedure, that works fine.

From Hoffman Degragation

The well-known Hofmann reaction by which, an amide, in the presence of chlorine or bromine and an alkali, is converted into a primary amine with one less carbon atom, is of considerable interest in organic chemistry. The reaction may be used in going "down series," and it serves as a laboratory method for the preparation of a primary aliphatic amine. The reaction is of interest also because it involves a molecular rearrangement.

The Hofmann reaction is often illustrated in the laboratory by the preparation of methylamine hydrochloride from acetamide. It has been our experience, however, that when this experiment is carried out with bromine and alkali according to the directions given in many laboratory manuals, a considerable portion of the product consists of ammonium chloride.

In this paper directions are given for the preparation of methylamine hydrochloride from acetamide using commercial calcium hypochlorite and sodium hydroxide. The product obtained with these reagents is contaminated with only a small amount of ammonium chloride, and this can be removed by treatment with aqueous alkali as described below.

The Hoffmann Rearrangement

$$2\,CH_3CONH_2 + Ca(OCl)_2 \rightarrow 2\,CH_3CONHCl + Ca(OH)_2$$

$$2\,CH_3CONHCl + Ca(OH)_2 \rightarrow (CH_3CONCl)_2Ca + 2\,H_2O$$

$$(CH_3CONCl)_2Ca + 2\,NaOH \rightarrow 2\,(CH_3CONCl)^{-}Na^{+} + Ca(OH)_2$$

$$(CH_3CONCl)^-Na^+ \text{—}Heat \rightarrow CH_3CON + NaCl \text{—}Rearr. \rightarrow CH_3N{=}C{=}O$$

$$CH_3N{=}C{=}O + H_2O + 2\ NaOH \rightarrow CH_3NH_2 + Na_2CO_3$$

$$CH_3NH_2 + HCl \rightarrow CH_3NH_3^+\ Cl^-$$

In this experiment, acetamide is first converted into N-chloroacetamide, which, in the presence of alkali, eliminates hydrogen chloride and undergoes rearrangement to form methylisocyanate. Hydrolysis of the latter gives methylamine which is isolated as the hydrochloride. It should be pointed out that the elimination of hydrogen chloride from N-chloroacetamide in the presence of alkali, actually consists of the removal of the hydrogen as a proton to form the alkali salt of N-chloroacetamide; the chloride ion is apparently released from the anion of this salt. The complete transformation may be represented by the following equations:

Experimental

Preparation of Methylamine Hydrochloride

The apparatus for this experiment should be set up before any materials are mixed.

A 500-mL distilling flask is fitted with a two-hole stopper carrying a thermometer and an inlet tube for air. The bulb of the thermometer should dip below the surface of the liquid, and the air tube should reach almost to the bottom of the flask. The distilling flask is attached to a condenser bearing an adapter. Two 250-mL Erlenmeyer flasks containing 35 mL each of 6 N HCl are placed in series as receivers. The first is fitted with a two-hole stopper. A glass tube attached to the adapter should dip below the surface of the HCl in the first receiver. The second receiver is connected to the first by means of another glass tube leading from above the acid solution in the first receiver to below the acid in the second.

To 16.4 g. of "H.T.H." in an Erlenmeyer flask is added 50 mL of water, and the mixture shaken until practically homogeneous. A small amount of material will remain undissolved. The solution is cooled to about 0°C in an ice-bath and 10 g. of crushed ice added.

The distilling flask is disconnected, and a cold solution of 10 g. of acetamide in 20 mL of water poured into it With the distilling flask immersed in an ice-bath, 100 g. of crushed ice is first added to the solution, followed by the addition of the cold "H.T.H." solution in three or four small portions with shaking after each addition. The temperature of the mixture should not rise above 0°C. More ice may be added to the reaction mixture if necessary. If the mixture should warm to more than 10°C (due to insufficient cooling), it should be discarded and the experiment repeated.

The flask is allowed to remain in the ice-bath for five to ten minutes. At the expiration of this time the flask is fitted to the condenser and a solution of 24 g. of NaOH in 40 mL of water at room temperature is added. The stopper is immediately replaced, and a current of air passed through the mixture during the remainder of the manipulations. The current of air should be sufficient to produce thorough mixing of the contents of the flask and to prevent bumping, yet not strong enough to cause much loss of the methylamine from the receivers.

By heating the flask the temperature of the mixture is raised rapidly to about 60°C. Between 65°C and 75°C the N-chloroacetamide decomposes with liberation of considerable heat, which, if not controlled, is likely to cause the contents of the distilling flask to bump over into the condenser. Consequently, the temperature is increased carefully from 60°C to 65°C and the flame removed. From this point the temperature will increase without the application of external heat. The current of air passing through the mixture is regulated so that it is just sufficient to prevent bumping. If the temperature rises to 80°C the flask is cooled in a bath of crushed ice and water. The temperature is held between 70°C and 80°C for five to ten minutes, or until heat is no longer spontaneously generated.

The contents of the distilling flask are then heated to boiling and distilled until 75 to 100 mL of distillate has been collected. The contents of the two receivers are combined, placed in a large evaporating dish, and evaporated over a wire gauze until the volume of the solution is about 15 mL The dish is then transferred to a water bath and the contents evaporated to dryness.

The solid residue consists of methylamine hydrochloride, together with a small amount of ammonium chloride. This product may be crystallized directly from absolute alcohol, in which case, the yield should be about seventy per cent. of the theoretical amount. It is recommended, however, that the crude product be purified according to the following procedure. The crude product is transferred to a casserole and dissolved in 15 mL of water. 10 mL of 7.5% sodium hydroxide is added and the solution heated to gentle boiling over a small flame. Since methylamine is a stronger base than ammonia, the latter, which is preferentially liberated from its hydrochloride, will be driven off with only a slight loss of methylamine. The casserole is kept in constant motion and the heating continued until all the water has evaporated and dense white fumes appear. A portion of the methylamine hydrochloride will be in the fused state. The flame is removed and the mass stirred while cooling. If the residue does not completely solidify when cold it is heated further. The residue is finely pulverized, transferred to a 250-mL Erlenmeyer flask, and refluxed a few minutes with 125 mL of absolute alcohol. A small amount of solid (NaCl) will remain undissolved. The solution is filtered through a steam funnel into a 250 mL Erlenmeyer flask. The latter is connected by means of a bent glass tube to a condenser and the alcohol distilled off until crystals begin to form in the hot liquid. The solution is then allowed to cool to room temperature and is finally cooled in an ice-bath. The crystals are filtered with suction, washed with 10 mL of absolute alcohol, and dried on a watch glass set on a hot steam coil. The yield of methylamine hydrochloride should be about 55% of the theoretical amount, melting at 228-230°C.

The use of air in this experiment serves to agitate the mixture and prevent it from bumping. If desired, a mercury-sealed mechanical stirrer may be used for this purpose and the air omitted. In this case one receiver will be sufficient. A short-stem funnel is fitted to the adapter, the larger end of the funnel dipping about 0.5 cm below hydrochloric acid contained in a beaker. By using a mechanical stirrer the reaction may be carried out on a larger scale with no essential modification in procedure.

Methylamine by catalytic hydrogenation of nitromethane

Procedure

Into a Hydrogenation vessel was placed the following:

- 3500ml Nitromethane
- 12500ml methyl alcohol

The vessel was stirred with a 3.75" stirbar, 1200rpm. The atmosphere was evacuated with vacuum, and the vessel was pressurized to 80psi with H_2. The vessel was left unattended with the self-adjusting pressure gage attached for 120 hours. At the end of this time, the vessel was shaken to remove any Pd/C that might have managed to kling to the side and dried there. The H2 was then vented outside, with the pressure release valve. The remaining H_2O was vacuumed from the pump with the output hose of the pump draped out the window. Atmosphere was then allowed to enter the inside of the vessel. 1L of the reaction mixture was filtered through celite twice to remove the suspended Pd/C. To this was added 300g of anhydrous $MgSO_4$, the mix was stirred for 15 minutes and then filtered to remove the $MgSO_4$ (Wanted to use $NaSO_4$ but someone said it wouldn't work as well, true?). The 1L was then placed in a 2L RBF and stirred, when vacuum was applied considerable bubbling was noted, once the bubbling subsided the mixture was then distilled, when the methyl alcohol fraction was complete there was nothing in the flask, nothing!!! The rest of the reaction mix was then filtered through celite, dried with $MgSO_4$, and used as it was, still with a bit of a charcoal tint, but nothing to interfer with the unholy scale aminations that came next.

And another one...

- 2500g CH_3NO_2
- 3000g CH_3OH
- 20g Pd/C 5%

They were stirred and hydrogenated at 80psi, after 24 hours the heat subsided, and at 48 hours H_2 uptake had long since ceased and the reaction was shut down. On depressurization, the ammonia-like odor was definitely evident. The catalyst was vacuum filtered off and collected, and the remainder still had a definite charcoal look to it. Filtering this through celite took care of most of that. The mix was split in half and each half sat with 1500g of Na_2SO_4 for an hour, with occasional stirring. The Na_2SO_4 was filtered off for a long time, until it was quite dry in appearance, the halves were combined and washed 2x with 1L of CH_3OH, which was filtered off and added back to the solution. What was left was about 7.5L of good loving. When 1L of this was converted to the HCl salt, 250g of salt was collected.

Methylamine HCl from Formaldehyde and Ammonium Chloride

This is the least desirable of all three processes. The yields are lower than the two rearrangements, and it requires substantial labour to get a decently pure product. Not "labor" as in difficult but "labor" as in a lot of it. I would suggest this only for those who have a large supply of Formaldehyde available to them (note - N. Coffey found formaldehyde at Home Depot - look for "Mildewcide" and dissolve it in enough water to make a 37% solution to depolymerize the paraformaldehyde).

Place 250g of Ammonium Chloride and 500g of technical Formaldehyde (37%, Formalin). Rig the flask for simple distillation such that a thermometer extends into the reaction mixture, and a Liebig or

West condenser. Heat the mixture on the steam bath until no more distillate comes over, then turn up the heat and hold the reaction temperature at 104C until, once again, nothing else comes over. This should take from 4 to 5 hours. The distillate may contain interesting things, so check out[1] for details on what to do with it. Next, the reaction flask should be cooled rapidly to room temperature by immersion into first a warm water bath (60C) swirled, and then an ice bath. Filter the solution on the vacuum Buchner funnel to recover ~62g of Ammonium Chloride crystals. Concentrate the filtrate using moderate vacuum and gentle heat until the volume is reduced to half. Filter the mother liquor once again after cooling quickly to yield a second batch of Ammonium Chloride, 19g.

Transfer the filtrate to a ceramic evaporating dish and heat on a water bath until a crystalline scum forms on the top. Cool the dish quickly then filter the mess on the vacuum Buchner to yield ~96g of Methylamine Hydrochloride. Concentrate the filtrate once again to obtain a second crop of crystals, ~18g.[2] Concentrate the filtrate a third time as far as possible using the water bath, then store the dish in a vacuum dessicator loaded with Sodium Hydroxide in the bottom for 24 hours. Add Chloroform to the residue left in the crucible to dissolve out Dimethylamine Hydrochloride (distill off the Chloroform to recover - good stuff) then filter on the venerable old vacuum Buchner funnel to yield an additional ~20g of Methylamine Hydrochloride, washing the crystals in the funnel with a small portion of Chloroform (~10mL).

Purification of the Methylamine HCl is in order now, so transfer all of the crude product to a 500mL flask and add either 250mL of absolute Ethanol (see for preparing this) or, ideally, n-Butyl Alcohol[4]. Heat at reflux with a Calcium Chloride guard tube for 30 minutes. Allow the undissolved solids to settle (Ammonium Chloride) then decant the clear solution and cool quickly to precipitate out Methylamine HCl. Filter rapidly on the vacuum Buchner funnel and transfer crystals to a dessicator[3]. Repeat the reflux-settle-cool-filter process four more times if using absolute Ethanol, or two more times if using n-Butyl Alcohol. The yield of Methylamine HCl should be 100g.

Physical Properties

Melting Point	= 0°C
Boiling Point	=185 – 189°C
Molecular Formula	= CH_5N

1. The byproducts of the first step are Dimethoxymethane and Sodium Formate.
2. The Methylamine solutions in all steps should be cooled rapidly to promote smaller crystal formation.
3. According to the original document, centrifuging is the most satisfactory method of drying products because of their hygroscopic nature. I suggest warming in an oven on a glass dish then transfering to a vacuum dessicator loaded with either concentrated Sulfuric Acid or Sodium Hydroxide in the bottom. It is not normally necessary to have absolutely dry Methylamine HCl anyway.
4. The solubility of Ammonium Chloride in absolute Ethanol is 0.6g/100g at 15C. The solubility in n-Butyl Alcohol is neglible, even at its boiling point. If you use n-Butyl Alcohol, you will only need to perform 3 reflux/filter operations to obtain sufficiently pure Methylamine Hydrochloride.

Formula Weight	= 31.0571
Composition	= C(38.67%) H(16.23%) N(45.10%)
Molar Refractivity	= 10.21 ± 0.3 cm^3
Molar Volume	= 48.7 ± 3.0 cm^3
Parachor	= 100.9 ± 4.0 cm^3
Index of Refraction	= 1.340 ± 0.02
Surface Tension	= 18.4 ± 3.0 dyne/cm
Density	= 0.637 ± 0.06 g/cm^3
Polarizability	= 4.05 ± 0.5 $10^{-24}cm^3$
Monoisotopic Mass	= 31.042199 Da
Nominal Mass	= 31 Da
Average Mass	= 31.05764 Da

Chemical Properties

It behaves absolutely like primary amine, and produces methyl ammonium chloride with hydrochloric acid.

Uses

Methylamines are used to make intermediates for a wide range of agricultural chemicals, including herbicides, fungicides, insecticides, biocides, and miticides. Agricultural chemicals are one of the widest uses for methylamines. One use for monomethylamine is to produce metam-sodium (n-methyldithiocarbamate), which is a soil fumigant and sterilant. It is made from the reaction of MMA, carbon disulfide, and caustic. Its applications are increasing, and it is currently used for potatoes, tomatoes, and home and garden applications. It is replacing methyl bromide as a soil fumigant, due to concerns of methyl bromide being a possible ozone depleter.

MMA is also phosgenated to make methyl isocyanate, which is used to make carbaryl, aldicarb, and trimethocarb. The insecticides carbofuran and carbosulfan are produced from these. Monomethylformamide (MMF), made from MMA and carbon monoxide, is being used to make methyl isocyanate in situ, due to concerns of storing and transporting this highly toxic material. MMF, sold by DuPont, can also be used as a solvent for other applications.

Dimethylamine is used to produce the herbicide 2,4-D (2,4-dichlorophenoxyacetic acid). Its DMA salt improves the water solubility of pesticide formulations. Related agricultural products include the ferric reaction with N,N-dimethyl dithiocarbamate to make ferbam, a fungicide for apples and tobacco;

the zinc reaction to make ziram, a fungicide for vegetables; and thiram (tetramethylthiuram disulfide), a disinfectant for grains, bulbs, tubers, and corn.

Other products from DMA include the herbicides diuron, linuron, hexazinone, and fluometuron. Quaternary biocides are produced from alkyldimethylamine (from DMA) and dialkylmethylamine (from MMA). Salts of DMA reacted with substituted naphthaquinones are used to make miticides and aphicides. DMA is also used to produce systemic insecticides such as octamethylpyrophosphoramide (OMPA) (schradan) that are absorbed into the plant rendering it toxic to sapsucking insects. Trimethylamine reacted with sulfur dioxide has been used to produce a volatile insecticide.

Methylamines are used directly as catalysts or as raw materials to produce other compounds with catalytic activity. MDEA (see Gas and Oil Treatment) made from MMA is used as a urethane catalyst. DMAE (see Water Treatment) made from DMA is also used as a urethane catalyst to promote foam rise and gel strength for insulation applications, such as refrigerators. DMAPA (see Personal Care) is also used as a urethane catalyst. It is made from the reaction of dimethylamine with acrylonitrile, followed by hydrogenation. Hydrochlorides of MMA and DMA in the presence of cupric chloride are used for the polymerization of acetylene to vinyl and divinyl acetylene. MMA and DMA are used as catalysts for the dealdolization of diacetone alcohol. TMA is a catalyst for ethyl alcohol to ethylene oxide and for Diels-Alder diene association reactions. DMA and calcium sulfate are used to prepare dinitroalkanes and nitroolefins. MMA and DMA can be used for the condensation of phenol and formaldehyde. DMA is a catalyst to condense acetone with formaldehyde to produce synthetic resins used in paper bonding and coating.

Electronics Dimethylamine and benzyl chloride are used to make benzyldimethylamine, which can be used as an accelerator for epoxy resins in laminates for electrical equipment. Tetramethylammonium hydroxide from TMA is a stripper for resists in printed circuit boards. TMA-hydrochloride is used to make specialty chemicals for the electronics industry.

Explosives Water gel explosives have been used in the mining industry due to their ease and safety in handling. The prime ingredient is monomethylamine nitrate made from MMA, formaldehyde, and ammonium nitrate. Perchlorates of mono-, di-, and trimethylamine have also been used in the mining industry. There have been some military uses for 2,4,6- trinitrophenylmethyl nitramine made from the nitration of dinitromethylaniline with nitric acid. Dinitromethylaniline is the product of MMA plus dinitrochlorobenzene in the presence of caustic. Fuses have been made from trimethylamine with nitric acid and nitrous perchlorate-paraffin mixtures.

Fuel additives are used to improve engine performance in a variety of ways. DMCHA and DMAPA (see Personal Care) are used for gasoline and motor oil additives. MMA and chlorobenzene are reacted with copper as a catalyst to make n-methylaniline for aviation fuel. DMA hydrochloride is also used as an additive in aviation fuel as an antiknock compound. MMA nitrate is a freezing point depressant in jet engines using ammonium nitrate-urea fuels. Mixtures of MMA, DMA, and TMA in water have been tested as coolants in some high-speed engines.

Gas and Oil Treatment Alkylalkanolamines have a large use in the treatment of natural gas and oil. Two of the largest volume alkylalkanolamines, methyldiethanolamine (MDEA) and monomethylethanolamine (NMEA), are made from the reaction of monomethylamine and ethylene oxide. MDEA is especially important in removing acidic components in natural gas such as hydrogen sulfide, mercaptans, carbonyl sulfide, and carbon monoxide. Dimethylammonium dimethylcarbamate, made from the reaction of dimethylamine and carbon dioxide, is used to remove sulfur compounds, olefins, and aromatic hydrocarbons from raw oil.

Paper Chemicals. Trimethylamine is used to make paper chemicals such as its reaction with epichlorohydrin to make 3-chloro-2-hydroxypropyltrimethylammonium chloride and the epoxy product glyciditrimethylammonium chloride. These are both used as cationizing starches. DMDAC (see Surfactants) from dimethylamine is also used in the paper industry.

Personal Care. The personal care industry is an area with growing use for methylamines as raw materials. N,N-dimethylaminopropylamine (DMAPA) and DMDAC (see Water Treatment) are used to make cosmetics.DMAPA is also an intermediate for betaines and fatty amide/amine oxides for liquid soaps. DMAE (see Water Treatment) is reacted in polymers for hair products and cosmetics, as well as other personal care items such as hand lotions, shaving creams, and shampoos.

Pharmaceuticals

The manufacture of intermediates to make pharmaceuticals is one of the most diverse uses of methylamines. The reaction of monomethylamine and ethylene oxide creates two key intermediates: methyldiethanolamine (MDEA) and monomethylethanolamine (NMEA) (see Gas and Oil Treatment). MDEA is used to make analgesics with sedative and antispasmodic effects, such as meperidine. MMA reacted with phosgene or carbon dioxide yields N,N'dimethylurea. This is used to make xanthine derivatives, such as caffeine and theophylline. The reaction of MMA and phenylsuccinic acid yields phensuximide, an anticonvulsant. N-benzyl-n-methylamine is used in pharmaceuticals. MMA is also used to make intermediates for chlordiazepoxide and lorazepam tranquilizers, nortryptylene antidepressant, isometheptene veterinary smooth muscle relaxant, iothalamic acid x-ray contrast medium, and levothyroxine.

Sympathomimetic drugs are made from MMA, such as adrenaline (amination of 3,4-dihydroxyphenyl chloromethyl ketone from pyrocatechol), ephidrine (reductive amination of phenylmethyl diketone), neosynephrine, vanedrine, and veritol. Dimethylamine is used to make DMAE (see Water Treatment), which has many uses to produce pharmaceuticals. It is used to make antihistamines, such as pheniramine, chlorpheniramine, and diphenhydramine, for the symptomatic relief of allergies, such as hay fever. DMAE is also used in the synthesis of local anesthetics, such as amethocaine, procaine, and pontocaine. Propoxyphene analgesic and doxepin antidepressant are made from DMA. Salicylates of both MMA and DMA are used as ointments for skin disorders. DMA is used to make intermediates for tranquilizing agents.

Photographic Chemicals P-methylaminophenol sulfate, used as a photographic developer, is made by using monomethylamine to replace an hydroxyl group on hydroquinone. N-methyl acetoacetamide, used as a coupling agent for pigment manufacture, is produced from MMA and diketene. MDEA (see Gas and Oil Treatment) has a growing use in making other photographic chemicals.

Resin Manufacturing. Methylamines are used to make resins for water demineralization and deionization. Cationic exchange resins are made from trimethylamine reacted with chloromethylated polystyrene-divinyl benzene copolymers. Anionic resins are made from adding dimethylamine and hydrochloric acid to melamine formaldehyde condensation products.

Rocket Fuels. 1,1-dimethylhydrazine is used as a propellant for booster rockets. It has replaced hydrazine in some applications, because it is more soluble in petroleum fuels and has a lower freezing point. It is made from dimethylamine salt and sodium nitrite to make dimethylnitrosamine, which is reduced using zinc, or it can be directly oxidized with chloramine.

Rubber Chemicals. Dimethylamine is used in several applications in the rubber industry. Dimethyldithiocarbamate salts are used for vulcanization accelerators. Tetramethylthiuram mono- and disulfide (TMTD) is a rubber accelerator, especially for butyl rubbers. DMA oleate is an emulsifier for use in synthetic rubber production. DMA is used directly as a modifier in the vapor phase polymerization of butadiene and as a stabilizer for natural rubber latex (instead of using ammonia).

Solvents. Solvents are the biggest use for dimethylamine (DMA) and a growing use for monomethylamine (MMA). They are used to make the polar, aprotic solvents dimethylformamide (DMF), dimethylacetamide (DMAc), and n-methylpyrrolidone (NMP). DMF is made from the reaction of DMA and carbon monoxide or methyl formate. Its uses include urethane coatings, spinning solvent (primarily for acrylics), reaction solvent, extraction solvent (such as butadiene extraction), and processing solvent (including solvent for dicyandiamide for epoxy-laminated printed circuit boards).DMAC is made from the reaction of DMA and acetic acid or acetic anhydride. It is also used as a spinning solvent and reaction solvent as well as for the manufacture of dyes, pharmaceuticals, and vinyl polymers. NMP uses MMA reacted with gammabutyrolactone. Its varied uses include reaction solvent (engineering resins), extraction solvent (lube oil refining), spinning solvent, paint stripper (replacing methylene chloride), electronics (for strippers and cleaners), and enamel coatings (magnet wire with polyamide/imide resins).MMA is also used for extractive distillation of butylene and butadiene by forming an azeotrope with butylene. DMA can be reacted with phosphorous trichloride in the presence of trimethylamine to make tris-(dimethylamido)-phospate. This has been used to dissolve polyacrylonitrile, as a solvent for polyvinylidene chloride, and as a selective solvent for acetylene.

Surfactants. Surfactants are used in the manufacture of soaps, shampoos, detergents, and other specialty wetting applications. Dimethylamine is reacted with fatty tertiary amines with C_{10}–C_{18} alkyl groups from alpha-olefins or fatty alcohols to make nonionic alkyldimethylamines. Salts of n-acyl-n-methyltaurates and n-acyl sarcosinates are used in many specialty surfactants. DMDAC (see Water Treatment), made from DMA, is used in shampoos.

Water Treatment. Dimethylamine is used as a raw material to produce agents that remove suspended solids for water clarification in industrial and municipal raw, potable, and wastewater treatment plants. They are used to improve solid/liquid separation in sedimentation, filtration, and flotation operations. Polyelectrolytic cationic flocculants made from DMA, such as dimethyldiallylammonium chloride (DMDAC) or epichlorohydrindimethylamine, are homopolymerized or copolymerized with acrylamide. Also, DMA can be reacted with formaldehyde and polyacrylamide to make Mannich polymers. These polymers can be quarternized with dimethyl sulfate or methyl chloride to improve their performance at high pHs. Dimethylaminoethanol (DMAE) methacrylate-acrylamide copolymers are also used as cationic flocculants.

Other Uses. Methylamines are used to make many other intermediates covering a myriad of applications. DMAE (see Water Treatment) is used for water-based surface coatings as a dispersant and pH control agent. It is also used as a viscosity index improver for lube oils. MMA is used to make cyclic polyimides. DMAE and DMAPA (see Personal Care) are used for corrosion inhibitors in boiler feed water and return condensate steam systems.

ETHYL AMINE

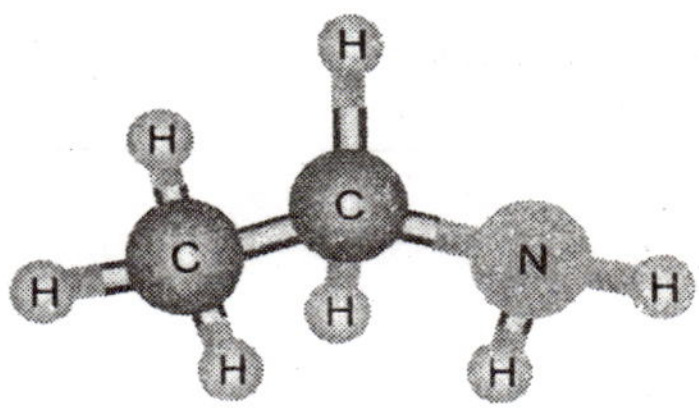

Preparation

180 ml 70% ethylamine was poured into 1qt cold isopropyl alcohol(IPA) + methyl alcohol (75%, 25%). Magnesium sulphate + sodium sulphate was added to the solution to remove all water. The drying agents were then decant off after several hours of falling to the bottom.

50g aluminium was put in a 2 litre flask with 1.5 litre clean IPA and allowed to amalgamate for 25 minutes with 400mg $HgCl_2$. After which the 1.5 litre was discarded and replaced with the 700ml ethanol and 300ml methyl alcohol along with the 1 litre ethyl amine mix. It was heated for 5 minutes until warm and 95ml of undistilled ketone was dripped in while the reaction was heated by the hotplate at an increasing rate over 20 minutes. There was quite a few problems getting the stirring going; during the first 50 minutes at least 1/2 of that time was spent trying to get the stirring going again. It got so bad that the solution had to be dumped out in a large beaker in a better attempt to use a more suitable container for stirring. The reaction was run for a total of 3.5 hours from start of drip till the end of stirring. Dry Sodium hydroxide was intermitenly added to the stirring reaction to speed it up. The unreacted aluminium was filtered off and washed with methyl alcohol, basified the mix, no sludge since tons of unreacted aluminum. Extracted with 1.75 litre toluene and washed several times.

Physical Properties

Boiling point:	16.6°C
Melting point:	–81°C
Relative density (water = 1):	0.7 (liquid)
Solubility in water:	miscible
Vapour pressure, kPa at 20°C:	121
Relative vapour density (air = 1):	1.55
Relative density of the vapour/air-mixture at 20°C (air = 1):	1.66
Flash point:	17°C c.c.
Auto-ignition temperature:	385°C
Explosive limits, vol% in air:	3.5-14
Octanol/water partition coefficient as log Pow:	–0.27/-0.08 (calculated)

Chemical Properties

$C_2H_5-NH_2$ (ethylamine)

$$+\overset{+}{H}\overset{-}{Cl} \longrightarrow C_2H_5N\overset{-}{H}_3\overset{-}{Cl}$$ (ethylammonium chloride)

$$+CH_3Br \xrightarrow[OH^-]{-HBr} C_2H_6NHCH_3 \xrightarrow{CH_3Br} C_2H_5N(CH_3)_2 \xrightarrow{CH_3Br} C_3H_5\overset{+}{N}(CH_3)_3\overset{-}{Br}$$ (ethylmethylamine; ethyldimethylamine)

$$+CH_3COCl \longrightarrow C_2H_5-NH-COCH_3+HCl$$ (N-ethylacetamide)

$$+C_6H_5SO_2Cl \longrightarrow C_2H_5-NH-SO_2-C_6H_5 \xrightarrow{NaOH} C_2H_5-\overset{-}{N}-SO_2-C_6H_5$$ (N-ethylbenzenesulphonamide; soluble sod. salt)

$$+HONO \longrightarrow C_2H_5OH+N_2-+H_2O$$ (ethyl alcohol)

$$+RCHO \longrightarrow C_2H_5-N=CH-R$$ (aldimine)

$$+CS_2 \longrightarrow C_2H_2-NH-CS-SH \xrightarrow{HgCl_2} C_2H_5-N=C-S+HgS+2HCl$$ (dithiocarbamic acid; ethyl isothiocyanate)

$$+HOCl \xrightarrow{OH^-} C_2H_5-NHCl \xrightarrow[OH^-]{HOCl} C_2H_5-NCl_2$$ (N-chloroethylamine; N,N-dichloroethylamine)

$$+RMgX \xrightarrow{ether} RH+C_2H_5NHMgX$$ (alkane)

$$+CHCl_3+3KOH \xrightarrow{D} C_2H_5-\overset{+}{N}\equiv\overset{-}{C}+3KCl+3H_2O$$ (ethyl isocyanide)

$$+O \xrightarrow{KMnO_4} CH_3-CH=NH \xrightarrow[H^+]{+H_2O} CH_2-CHO+NH_3$$ (aldimine; acetaldehyde)

Uses

Pesticides, rubber accelerant, antioxidant, ion exchange resin, fuel for planes, solvent, surfactant, etc.

DIMETHYLAMINE

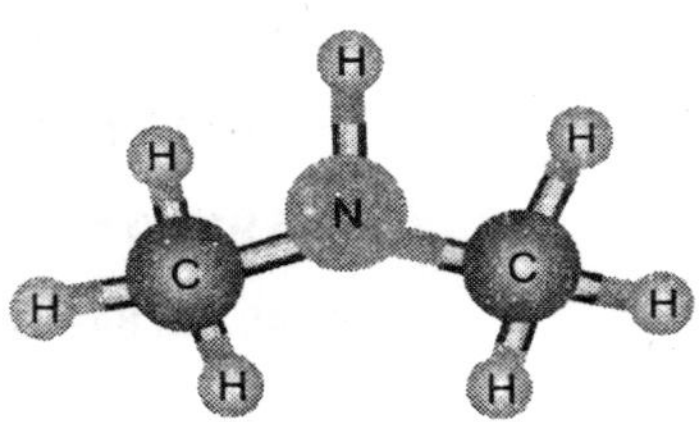

It occurs in small in Peruvian Guano and pyrolignous acid

Preparation

(*i*) Dimethylamine cane be prepared in a pure state by the hydrolysis of p-nitroso-N, N-dimethylaniline with sodium hydroxide.

N,N-dimethyl-4-nitrosoaniline + H_2O $\xrightarrow{NaOH}$ (4-nitrosophenolato-κO)sodium + N-methylmethanamine + H_2O

(*ii*) By heating ammonium chloride with four equivalents of formaldehyde (formalin)

$$NH_4Cl + 4HCHO \rightarrow (CH_3)_2NH_2Cl + 2HCOOH$$

(Dimethylammonium Chloride)

$$(CH_3)_2NH_2Cl + NaOH \rightarrow (CH_3)_2NH + NaCl + H_2O$$

Properties

Melting Point	= –93ºC
Boiling Point	= 7ºC
Molecular Formula	= C_2H_7N
Formula Weight	= 45.08368
Composition	= C(53.28%) H(15.65%) N(31.07%)

Molar Refractivity	= 14.90 ± 0.3 cm^3
Molar Volume	= 70.3 ± 3.0 cm^3
Parachor	= 139.6 ± 4.0 cm^3
Index of Refraction	= 1.344 ± 0.02
Surface Tension	= 15.5 ± 3.0 dyne/cm
Density	= 0.640 ± 0.06 g/cm^3
Polarizability	= 5.91 ± 0.5 $10^{-24}cm^3$
Monoisotopic Mass	= 45.057849 Da
Nominal Mass	= 45 Da
Average Mass	= 45.084629 Da

Chemical Properties

Dimethylamine ($(H_3C)_2NH$)

$+HCl \longrightarrow (CH_3)_2\overset{+}{N}H_2\,\overset{-}{Cl}$ dimethylammonium chloride

$+CH_3Br \xrightarrow[OH^-]{-HBr} (CH_3)_3N \xrightarrow{CH_3Br} (CH_3)_4NBr$ trimethylamine; tetramethylammonium bromide

$+CH_3COCl \longrightarrow (CH_3)_2N-COCH_3 + HCl$ N,N-dimethylacetamide

$+C_6H_5SO_2Cl \longrightarrow (CH_3)_2N-SO_2-C_6H_5 \xrightarrow{NaOH}$ No Reaction — N-dimethylbenzene sulphonamide (insoluble in NaOH soln.)

$+HONO \xrightarrow{-H_2O} (CH_3)_2N-NO \xrightarrow{H_2SO_4}$ green $\xrightarrow{NaOH}$ deep blue or violet — N-nitroso-dimethylamine (yellow oil)

$+R_2CH-CO-R \longrightarrow (CH_3)_2N-CR=CR_2$ enaquine

$+CS_2 \longrightarrow (CH_3)_2N-CS-SH \xrightarrow{HgCl_2}$ No Reaction — dithiocarbamic acid

$+HOCl \xrightarrow{OH^-} (CH_3)_2N-Cl$ N-chlorodimethylamine

$+R'MgX \xrightarrow{ether} R'H + (CH_3)_2NMgX$

$+[O] \xrightarrow{KMnO_4} (CH_3)_2N-N(CH_3)_2 + H_2O$ tetramethylhydrazine

Uses

Dimethylamine is used as a raw material of solvent (dimethylformamide, dimethylacetamide), and alkyl alkanolamines and dimethyl hydrazine (ingredients of rocket fuel). It is used to make other organic chemicals including rubber vulacanization accelerators (tetra methyl thiuram disulphide, zinc dimethyldithiocarbamate, sodium dimethyldithiocarbamate and potassuim dimethyldithiocarbamate), pesticides (isoproturon, ziram, thiram, diuron, monuron, chloroxuron, fenuron), surfactants (alkyldimethylamine oxides, dimethylaminoethanol, quaternary ammonium compounds), photographic chemicals, corrosion inhibitors, explosives, dyes and pharmaceuticals. Dimethylamine is used in rayon and nylon industry to improve the tensile strength.

TRIMETHYLAMINE

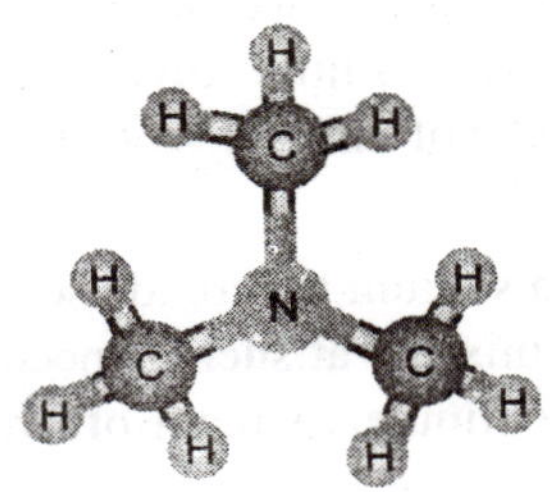

Introduction

Trimethylamine, also known as **N, N- Dimethylamine** and **trimethylamine**, is a colourless, hygroscopic, and flammable simple amine with a typical fishy odor in low concentrations and an ammonia-like odor in higher concentrations. Trimethylamine has a boiling point of 2.9 °C and is a gas at room temperature. Trimethylamine usually comes in pressurized gas cylinders or as a 40% solution in water. Trimethylamine is a nitrogenous base and its positively charged cation is called **trimethylammonium** cation. A common salt of trimethylamine is **trimethylammonium chloride**, a hygroscopic colourless solid.

Trimethylamine is a product of decomposition of plants and animals. It is the substance mainly responsible for the fishy odor often associated with fouling fish, bacterial vagina infections, and bad breath. It is also associated with taking large doses of choline.

Preparation

Five hundred grams (9.35 moles) of technical ammonium chloride and 1330 g. (14.8 moles) of paraformaldehyde are mixed and allowed to react as in the procedure described for the preparation of trimethylamine hydrochloride. To obtain from this crude reaction mixture a water or alcohol solution of trimethylamine, it is not satisfactory merely to treat with alkali and lead the trimethylamine through the solvent, since the gas is inefficiently absorbed; in fact, it is almost impossible to get a concentrated solution in this way without loss of material. The best way to prepare a solution is to liquefy the trimethylamine (b.p. 3.5°) and in this form run it into the solvent. For this purpose, the 5-l. flask in which the initial reaction is carried out is fitted with a rubber stopper holding a separatory funnel and a glass tube for leading off the trimethylamine (Note 1). This tube is run into one opening of a large U-tube or Wolff bottle holding soda-lime for drying the gas. The exit tube from this drying bottle has a glass tube leading to the top of an upright condenser which consists of a spiral tube cooled with a freezing mixture (Note 2).

$$3(CH_2=O)_n \xrightarrow[\Delta]{2NH_4Cl} 2(CH_3)_3N.HCl$$

$$2(CH_3)_3N.HCl \xrightarrow{NaOH} 2(CH_3)_3N$$

The bottom of this condenser leads through a rubber stopper fitting tightly in a 2- or 3-l. flask, holding the solvent and immersed in an ice-salt mixture. An exit tube from this flask leads to a Wolff bottle kept at a low temperature and holding a little solvent, and through this solvent any unabsorbed trimethylamine must bubble. The exit tube of this last Wolff bottle is closed with a soda-lime tube if an anhydrous solvent is being used (Note 3).

The separatory funnel is filled with sodium hydroxide solution (1100 g. in 2 l. of water) which is then allowed to flow into the reaction mixture at such a speed that, after the air in the apparatus has been replaced by trimethylamine, a continuous stream of drops of trimethylamine flows from the condenser into the solvent and practically no gas goes through the end Wolff bottle. At the beginning, the evolution of trimethylamine is rapid, and hence the addition of the alkali must be very slow (Note 4); as soon as the reaction mixture is partly decomposed, however, the alkali may be added more rapidly. After it has all been added, the reaction mixture is heated ten to fifteen minutes to make sure that all the amine is evolved. The time for this alkali treatment is about three to four hours.

With such an experiment as is described above, using 1000 g. of absolute alcohol (p. 249) in the solvent flask and about 150 g. in the end Wolff bottle, 1469 g. of solution is obtained in the main flask and 174 g. in the Wolff bottle, thus indicating that a total of 491 g. (89 per cent of the theoretical amount) of trimethylamine is present. There is thus obtained a 32 per cent solution in the main flask, which has a specific gravity of about 0.760 at +5° (Note 5). By titration with standard hydrochloric acid (Note 6), the same amounts of trimethylamine are indicated as are found by weighing. In several experiments, the yields were consistently 85–90 per cent of the theoretical amount (Note 7).

Notes

1. As trimethylamine is very volatile and easily lost, particular care must be taken to have all connections in the apparatus tight.

2. Special attention should be drawn to the condenser used since it is valuable in many other experiments wherever substances having a boiling point between "10° and +30° are being handled (see, also, p. 117). Any gallon (4-l.) can, having an opening in the top in which a stopper may be inserted, is satisfactory for a jacket. The bottom of the can is cut out, and the can is then inverted. A

spiral tube such as that which may be obtained from the inside of a spiral condenser is placed in the can with the lower end fitting tightly through the stopper in the neck. By filling the can with an ice and salt mixture a condensing temperature of –10° or below can easily be obtained.

3. If water is to be used as a solvent, no soda-lime tube is necessary on the exit of the end Wolff bottle. Moreover, with water as a solvent it is not absolutely necessary to have the soda-lime tube between the generating flask and the solvent. It is advisable to have it there, however, since it tends to catch any small amount of coloured liquid coming from the reaction flask and thus prevents contamination of the water solution obtained.

4. During the evolution of the trimethylamine, it is necessary occasionally to shake the flask containing the solvent. It happens that, after about a 10–15 per cent solution is formed, the trimethylamine tends to form a top layer which is not dissolved until agitated.

5. In order to avoid loss of material, the specific gravity of any trimethylamine solution must be taken at a low temperature. In this experiment, +5° was chosen.

6. Titration of the strength of the trimethylamine in the solvent is best done by pipetting out 1 cc. of the solution and allowing it to run immediately into 50 cc. of water. This is to prevent volatilization during titration. The amine is then titrated in the usual way with standard hydrochloric acid, using methyl orange as an indicator.

7. Since tests in the trimethylamine hydrochloride preparation (p. 531) indicate that only trimethylamine is formed in this reaction, it is unnecessary to purify the trimethylamine solution.

Discussion

Trimethylamine can be prepared by the action of formaldehyde on ammonium chloride under pressure;[1] by the action of formaldehyde and formic acid on ammonia,[2] a method which has been checked by one of the editors and is highly recommended; from paraformaldehyde and ammonium chloride;[3] and from methyl alcohol, ammonia, and a catalyst. The numerous commercial processes which lead to mixtures of methylamines have not been reviewed since they are not convenient for the laboratory.

Properties

Melting Point	= –117.1 °C
Boiling Point	= 2.9 °C
Molecular Formula	= C_3H_9N
Formula Weight	= 59.11026
Composition	= C(60.96%) H(15.35%) N(23.70%)
Molar Refractivity	= 19.66 ± 0.3 cm^3
Molar Volume	= 85.3 ± 3.0 cm^3

Parachor	= 177.1 ± 4.0 cm^3
Index of Refraction	= 1.378 ± 0.02
Surface Tension	= 18.6 ± 3.0 dyne/cm
Density	= 0.692 ± 0.06 g/cm^3
Dielectric Constant	= Not available
Polarizability	= 7.79 ± 0.5 $10^{-24}cm^3$
Monoisotopic Mass	= 59.073499 Da
Nominal Mass	= 59 Da
Average Mass	= 59.111618 Da

Chemical Properties

Chemically, it gives all the reactions of a typical tertiary amine. When oxidized with H_2O_2 or peracid, it forms N – oxide H_2O_2

$$(CH_3)_3N + [O] \xrightarrow{H_2O_2} (CH_3)_3N^+ - O$$

trimethylamine-N-Oxide

On heating under pressure with hydrochloric acid, it decomposes to form methyl chloride.

$$N(CH_3)_3 + 4\,HCl \xrightarrow[\text{pressure}]{\Delta} 3\,H_3C{-}Cl + NH_4Cl$$

N,N-dimethylmethanamine — Methyl chloride

Thus trimethylamine is used as a source of industrial methyl chloride

Tetraalkylammonium Salts

They are regarded as derived from ammonium salts in which all the four hydrogen atoms of the ammonium ion have been substituted by alkyl groups. They are commonly referred to as **Quanterernary Ammonium Salts.**

$$H_4N^+\,\bar{Cl} \xrightarrow[+4R]{-4H} R_4N^+\,\bar{Cl}$$

Ammonium chloride — tertraalkylammonium chloride or quaternary ammonium chloride

Quaternary (4°) ammonium salts are completely ionic substances. Tetraalkylammonium ion (R_4N^+) is a stable cation like Na^+ or K^+ and exists only in combination with amines. Thus quaternary ammonium salts behave like inorganic salts. The reactions of Quaternary salts have no relation whatever to those of amines.

Tetraalkylammonium Halides $R_4N^+X^-$ They arc prepared by treating an ethanonic, solution of ammonia with excess of alkyl halide (RX) in a sealed tube at 100°. Alternatively tetraalkylammonium salts can be obtained by direct reaction of a tertiary amine with an alkyl halide.

$$R_3N: + R-X \longrightarrow [R_4N^+]\bar{X}$$

3° Amine — Tetraalkylammonium Halide

Quaternary ammonium halides are white, crystalline solids, soluble in water in which solution they are completely dissociated. Thus,

$$(CH_3)_4N^+\bar{I} \longrightarrow (CH_3)_4N^+ + \bar{I}$$

Tetramethylammonium cation

When a quaternary ammonium halide is heated in vacuum, it gives the tertiary amine and alkyl halide.

Quaternary ammonium salts in which one of the four alkyl groups contains a "long carbon chain, are used as detergents and germicides.

Tetraalkylammonium Hydroxides, $R_4{}^+NOH^-$. These are prepared by treatment of quaternary ammonium halides with moist silver oxide {AgOH) in water.

$$R_4N^+\bar{X} + AgOH \longrightarrow R_4N^+\bar{O}H + AgX\downarrow$$

4° ammonium halide — Tetraalkylammonium hydroxide (4° ammnium hydroxide)

For example

(*a*)

$$(CH_3)_4N^+\bar{I} + AgOH \longrightarrow (CH_3)_4N^+\bar{O}H + AgI$$

Tetramethylammonium iodide — Tetramethylammonium cation

(*b*)

$$CH_3CH_2CH_2CH_2-N^+(CH_3)_3\bar{I} + AgOH \longrightarrow CH_3CH_2CH_2CH_2-N^+(CH_3)_3\bar{O}H + AgI$$

n-butyl-trimethylammonium iodide — *n*-butyl-trimethylammonium hydroxide

Quaternary ammonium hydroxides are white deliquescent crystalline solids extremely soluble in water. They are completely ionized in solution.

$$R_4\overset{+}{N}\,\overset{-}{O}H \longrightarrow R_4\overset{+}{N} + \overset{-}{O}H$$

4° ammonium hydroxide 4° ammonium cation

They are as strong bases as NaOH or KOH. Their aqueous; solutions precipitate hydroxides of heavy metals from the solutions of their salts. They absorb carbon dioxide.

Pyrolysis of Quaternary Ammonium Hydroxides. Tetramethylammonium hydroxide on thermal decomposition or pyrolysis yields trimethylamine and methyl alcohol.

$$(CH_3)_4\overset{+}{N}\,\overset{-}{O}H \xrightarrow{\Delta\ \Delta} (CH_3)_3N: + CH_3OH$$

Tetramethyl ammonium hydroxide Trimethyl amine

All other tetraalkylammonium hydroxides, in which one of the substituent groups on the nitrogen atom bears at least one β hydrogen, on pyrolysis form a tertiary amine and an alkene by E2 mechanism.

1° Amine

$$R_3\overset{+}{N}-CH_2-CH_2-H\ \overset{-}{O}H \xrightarrow{\Delta\ \Delta} R_3N: + CH_2{=}CH_2 + H_2O$$

4° Ammonium Hydroxide 3° Amine Ethylene

This reaction which has been known since 1851, is Hofmann Elimination. Thus tetraethylamine on pyrolysis gives triaethylamine and ethylene

$$(C_2H_5)_3\overset{+}{N}-CH_2-\underset{\beta}{C}H_3\ \overset{-}{O}H \xrightarrow{\Delta\Delta} (C_2H_5)_3N + CH_2{=}CH_3 + H_2O$$

Tetracthylammnium hydroxide Triethylamine Ethylene

When there are four different alkyl groups attached to the nitrogen atom of the ammonium ion, the pyrolysis of the tetraalkylammonium hydroxide may theoretically lead to the formation of more than one type of alkene. In such cases, the elimination takes place in a fashion that the major product is the alkene containing the least alkylated double bond *i.e.,* the least substituted alkene *(Hofmann Rule)*. For example,

$$H_3C-\overset{+}{N}(CH_3)_2-CH_2-CH_2-CH_2-OH \xrightarrow{\Delta\Delta} H_3C-N(CH_3)-CH_2-CH_2-CH_3 + CH_2{=}CH_2 + H_2O$$

3-hydroxy-*N,N,N*-trimethylpropan-1-aminium → *N,N*-dimethylpropan-1-amine

$$\xrightarrow{\Delta\Delta} H_3C-N(CH_3)-CH_2-CH_3 + H_3C-CH{=}CH_2 + H_2O$$

N,N-dimethylethanamine + prop-1-ene

Comparison of Properties of 1°, 2°, and 3° Amines

The primary, secondary and tertiary amines can be distinguished from each other by a study of the following properties.

	Primary (RNH_2)	Secondary (R_2NH)	Tertiary (R_3N)
(1)	**Basic Chracter:**		
	More basic than amonia	More basic than 1° amines	Less basic than 2° amines
(2)	**Formation of Salts with acids :**		
	Form alkylammonium salts,	Form dialkylanmonium salts,	Form trialkylammonium salts,
	$RNH_2 + \overset{+}{H}\overset{-}{Cl} \rightarrow R\overset{+}{N}H_3\overset{-}{Cl}$	$R_2NH + \overset{+}{H}\overset{-}{Cl} \rightarrow R_2\overset{+}{N}H_2\overset{-}{Cl}$	$R_3N + \overset{+}{H}\overset{-}{Cl} \rightarrow R_3\overset{+}{N}H\overset{-}{Cl}$
(3)	**Reaction with Alkyl halides (Alkylation):**		
	React with three molecules of RC to form 4° salt:	React with two molecules of RX to form 3° salt.	React with one molecule of RX to form 4° salt.
	$RNH_2 \xrightarrow{RX} R_2NH \xrightarrow{RX} R_3N \xrightarrow{RX} R_4\overset{+}{N}\overset{-}{X}$	$R_2NH \xrightarrow{RX} R_3N \xrightarrow{RX} R_1\overset{+}{N}\overset{-}{X}$	$R_3NH + RX \rightarrow R_3\overset{+}{N}\overset{-}{X}$
(4)	**Reaction with Acid chlorides (RCOCl) Acylation :**		
	Form N-substituted amides,	Form N, N-disubstituted amides,	No Reaction
	$RNH_2 + R'COC \rightarrow RNHCOR' + HCl$	$R_2NH + R'COCl \rightarrow R_2NCOR' + HCl$	
(5)	**Reaction with Benzene sulphonyl chloride ($C_4H_5SO_2Cl$):**		
	Form N-alkylbenzene sulphonamides soluble in NaOH solution.	Form N, N-dialkylbenzene sulphonamides, insolube in NaOH solution	No Reaction
	$RNH_2 + C_6H_5SO_2Cl \rightarrow RNHSO_2C_6H_5$	$R_2NH + C_6H_5SO_2Cl \rightarrow R_2NSO_2C_6H_5$	
(6)	**Reaction with Nitrous acid (HONO):**		
	Form mixtures of alcohols, alkenes, alkyl halides etc., with evolution of N_2. For exmaple,	Form oily nitrosoamines which give Libermann Reaction.	Form salts
	$RNH_2 + HONO \rightarrow ROH + N_2 \uparrow$	$R_2NH + HONO \rightarrow R_2NNO + H_2O$ $R_2NNO + phenol + conc\ H_2SO_4 \rightarrow green\ soln. \xrightarrow{NaOH} deep\ blue$	$R_3N + HNO_2 \rightarrow R_3\overset{+}{N}H\overset{-}{NO_2}$

	Primary (RNH_2)	Secondary (R_2NH)	Tertiary (R_3N)
(7)	**Reaction with Aldehydes and Ketones:**		
	Form aldimines with aldehydes and ketimines with ketones. $RNH_2 + R'CHC \rightarrow RN{=}CHR' + H_2O$ $RNH_2 + R_2'CO \rightarrow RN{=}CR'_2 + H_2O$	Form enamines, $R_2NH + R'_2CHCOR' \rightarrow R_2NCR'' = Cr'$	No reaction
(8)	**Reaction with CS_2 :**		
	Form dithiocarbamic acid which gives alkylithiocyanate with $HgCl_2$. $RNH_2 + CS_2 \rightarrow RNH\overset{S}{\overset{\|}{C}}SH \xrightarrow{HgCl_2} RNCS + HgS + 2HCl$	Form dithiocarbamic acid, not decomposed by $HgCl_2$ $R_2NH + CS_2 \rightarrow R_2N\overset{S}{\overset{\|}{C}}SH \xrightarrow{HgCl_2} X$	No reaction
(9)	**Halogenation with Hypochlorous acid in presence of Alkali:**		
	Form N, N-dichloroamine, $RNH_2 \xrightarrow[OH^-]{HOCl} RNCl_2$	Form N-chlorcamine, $R_2NH \xrightarrow[OH^-]{HOCl} R_2NCl$	No reaction
(10)	**Reaction with Grignard Reagents:**		
	Form alkanes, $RNH_2 + R'MgX \rightarrow R'H + RNHMgX$	Form alkanes, $R_2NH + R'MgX \rightarrow R'H + R_2NMgX$	No reaction
(11)	**Carbylamine Reaction:**		
	When warmed with CHCl3 and ethanolic KOH, 1° amines produce carbylamine having obnoxious odour, $RNH_2 + CHCl_3 + 3KOH \rightarrow RNC + 3KCl + 3H_2O$	No Reaction	No reaction
(12)	**Oxidation:**		
	With $KMnO_4$ form aldehydes and ammonia, $RCH_2NH_3 \xrightarrow{O} RCHO + NH_3$	With $KMnO_4$ form tetraalkylhydrazine, $R_2NH \xrightarrow{O} R_2N-NR_2$	With H_2O_2 form N-oxio $R_2N + O \longrightarrow R_3\overset{+}{N}-\overset{-}{O}$

CHAPTER

16

Cyan Compounds

The cyan compounds are characterized by the presence of a cyan group (CN). They are considered to be derivatives of hydrogen cyanide (HCN). Since the cyanide ion (:C≡N:) has unshared electron pair on both the carbon and nitrogen atoms, the proton (H+) can attach itself to either of these atoms. Thus hydrogen cyanide exists as a tautomeric equilibrium mixture of the cyanide form (I) and the isocyanide form (II). As shown by spectroscopic examination the cyanide form dominates.

$$H{-}C{\equiv}N: \rightleftharpoons H^{+} :\bar{C}{\equiv}N: \rightleftharpoons :\bar{C}{\equiv}N:H \text{ or } H{-}\overset{+}{N}{\equiv}C$$

The organic cyan compounds derived from the cyanide form are called Alkyl Cyanides or Nitriles, while those derived from isocyanide form are called **Alkyllsocyanides** or **Isonitriles**.

$$H{-}C{\equiv}N \longrightarrow \underset{\text{alkyl cyanides}}{R{-}C{\equiv}N} \quad \text{or} \quad RCN$$

$$H{-}\overset{+}{N}{\equiv}\bar{C} \longrightarrow \underset{\text{alkyl isocyanides}}{R{-}\overset{+}{N}{\equiv}\bar{C}} \quad \text{or} \quad RNC$$

The methods of formation and reactions of the two series of cyan compounds leave no doubt that in alkyl cyanides the alkyl group (R) is attached to nitrogen and in alkyl iso-cyanides it is attached to carbon.

The cyan compounds have proved to be important synthetic reagents.

Alkyl Cyanides or Nitriles

They are alkyl derivatives of hydrogen cyanide (or hydrocyanic acid) in which the substituent R is linked to the carbon of the cyan group –CN. They are frequently referred to LS Nitriles (containing nitrogen). Alkyl Cyanides or Nitriles have the general formula R-C=N where -C°N is the functional group.

Structure

There is a close similarity between the structure of a -CN group of alkyl cyanides and lat of acetylene. The carbon and the nitrogen atoms-of the cyan group are both in *sp* state f hybridization. The valence orbitals of the carbon atom in *sp* state of hybridization are sown in Fig. 16.1(a). The two *sp* hybrid orbitals are linear while'"tile remaining pure *p* orbitals e prependicular to each other and also to the hybrid orbitals. All these four orbitals have one electron each.

The electronic configuration of N-atom ($1s^2$; $2s^2$, $2p_x^1$. $2p_y^1$. $2p_z^1$) indicates that there are five electrons in four orbitals in the valence shell. The 26' orbital and one of the three orbitals (say $2p_z$) hybridize to give two *sp* hybrid orbitals. Two *p* orbitals ($2p_x$and $2p_z$) are left ft as such. One of the hybrid Orbitals will have only one electron while the other accommodates e lone pair originally belonging to 2s orbital. The two *p* orbitals lie perpendicular to each other and also to the *sp* hybrid as shown Fig. 16.1*(b)* as in case of carbon.

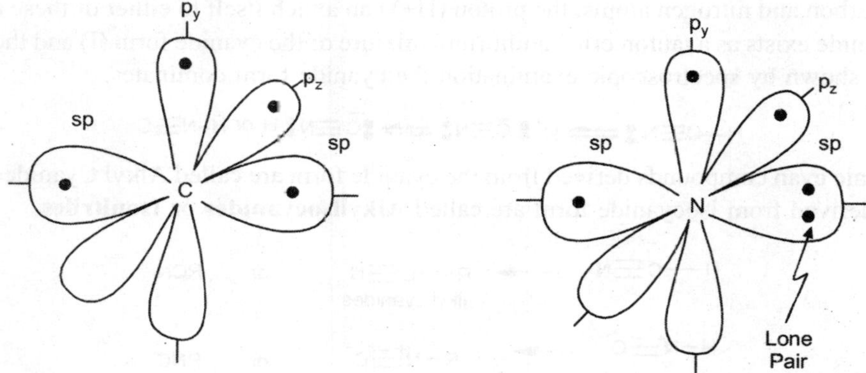

Fig. 16.1(a). Valence orbitals of Carbon in *sp* state of hybridization

Fig. 16.1(b). Valence orbitals of Nitrogen in *sp* state of hybridization

In the formation of –CN group, one half-filled *sp* orbital on carbon atom overlaps with half-filled *sp* orbital on nitrogen atom, forming a δ bond. The second *sp* orbital of carbon overlaps with the *Sp*³ orbital of an alkyl R forming another δ bond in line with the first one. Now there are two *p* orbitals *(2pv* and *2p.)* on carbon and nitrogen atoms both. These affect sidewise overlaps to form two 7t bonds (7tr and 7*t*,) as shown in Fig. 16.2. It may be noted that the second *sp* orbital on N atom contain~ a lone pair of electrons.

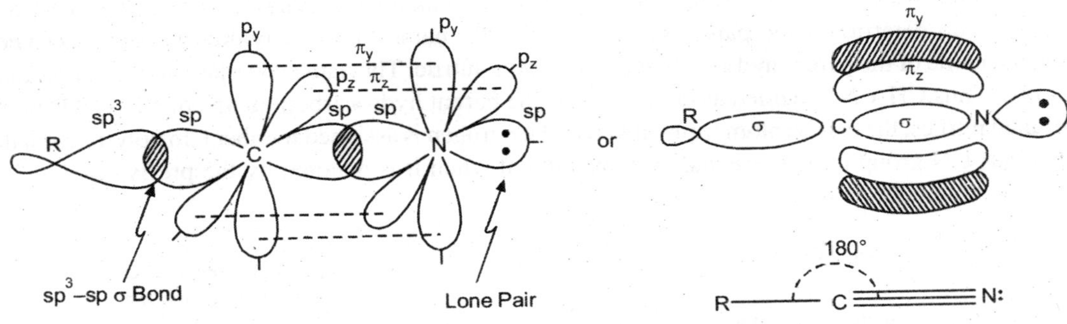

Fig. 16.2. Showing the Orbital structure of Nitriles

Nomenclature

The compounds having the formula R–C≡N are named in three ways.

(1) Cyanide System. In this system, they are named as if they were salts of hydrogen cyanide. Just as salts of hydrogen chloride are the *chlorides,* those of hydrogen cyanide would be named as *cyanides.* Thus the cyanide name of an individual member is derived by putting down the name of the alkyl group and adding the word 'cyanide'. For example; CH.CN named as methyl cyanide and C_2H_5CN as ethyl cyanide. Their class name is therefore, *Alkyl Cyanides.*

(2) Common System. Their common names are derived from the *common names* of acids to which they are hydrolysed. The ending *-ic acid* of the corresponding acid is dropped and the suffix -*onitrile* is added. For example, CH, CN upon hydrolysis yields acetic acid,

$$CH_3 - C \equiv N + H_2O \xrightarrow{H^+/OH^-} \underset{\text{acetic acid}}{CH_2 - COOH} + NH_3$$

and is, therefore, named as acetonitrile.

$$\text{Acetic Acid} - \text{IC Acid} + \text{Onitrile} \rightarrow \text{Acetonitrile}$$

Table 16.1 : Names of some Alkyl Cyanides or Nitriles

Structure	Cyan Name	Common Name	IUPAC Name
CH_3—CN	methyl cyanide	acetonitrile	ethanenitrile
CH_3CH_2–CN	ethyl cyanide	propiononitrile	propanenitrile
$CH_3CH_3CH_2$–CN	propyl cyanide	butyronitrile	butanenitrile
CH_3–CH(CH_3)–CN	isopropyl cyanide	isobutyronitrile	2-methyl prepaneuitrile

(3) IUPAC System. In this system, the longest carbon chain containing the cyan group (CN) as the terminal group determines the parent name. The IUPAC name of an individual nitrile is derived by adding the name of the parent hydrocarbon on the suffix -nitrile. The name is written as single continuous one word. Thus CH_3CN is named as ethanenitrile, the parent hydrocarbon being ethane. While naming substituted nitriles, the carbon atom triply linked to the nitrogen is assigned number l. In polyfunctionalized compounds CN group may be treated as a substituent when it is denoted by the prefix *-cyano.*

$$\overset{4}{C}H_3-\overset{3}{C}H_2-\overset{2}{C}(CH_3)_2-\overset{1}{C}N$$
2, 2-dimethylbutanenitrile

$$\overset{4}{C}H_3-\overset{3}{C}H(Cl)-\overset{2}{C}H_2-\overset{1}{C}N$$
3-chlorobutanenitrile

$$NC-\overset{3}{C}H_2-\overset{2}{C}H_2-\overset{1}{C}OOH$$
3-cyanopropanoic acid

Methods of Preparation

Alkyl cyanides or nitriles can be obtained by many methods of which the more important ones are listed below.

(I) Action of Metal Cyanides on Alkyl halides. Nitrites can be prepared conveniently by heating an alkyl halide with sodium or potassium cyanide in aqueous-ethanolic solution.

$$\underset{\text{alkyl halide}}{R-X} + \overset{+}{Na}CN \longrightarrow \underset{\text{nitrile}}{R-CN} + \overset{+}{Na}\overset{-}{X}$$

e.g. $$\underset{\text{ethyl bromide}}{CH_3CH_2-Br} + \overset{+}{Na}\overset{-}{CN} \longrightarrow \underset{\text{propanenitrile}}{CH_3CH_2-CN} + \overset{+}{Na}\overset{-}{Br}$$

Small amounts of alkyl isocyanide are also formed which can be removed by partial hydrolysis, because it is hydrolysed more rapidly than the alkyl cyanide.

When nitriles are prepared from alkyl halides by this method, an extra carbon atom is introduced. This is the key step for ascending the series of carboxylic acids (or 1° alcohols)

$$RCOOH \xrightarrow[H^+]{CH_3OH} RCOOCH_3 \xrightarrow{LiAlH_4} RCH_2OH \xrightarrow{PBr_3}$$
$$RCH_2Br \xrightarrow{KCN} RCH_2CN \xrightarrow[H^+]{H_2O} RCH_2COOH$$

(2) Dehydration of Amides or Ammonium salts of Carboxylic acids. Simple amides on drastic dehydration by heating with phosphorus pentoxide yield nitriles.

$$R-\overset{O}{\overset{\|}{C}}-NH_2 \xrightarrow{P_2O_5} R-C\equiv N + H_2O$$

$$\underset{\text{Propionamide}}{H_3C-CH_2-\overset{O}{\overset{\|}{C}}-NH_2} \xrightarrow{P_2O_5} \underset{\text{Propiononitrile}}{H_2\overset{CH_3}{\overset{|}{C}}-C\equiv N} + H_2O$$

Higher-molecular-weight amides may be dehydrated by heating alone.

Ammonium salts of carboxylic acids dehydrate first to form amide which on loss of a molecule of water yields nitrile. Thus when carboxylic acid vapours mixed with ammonia are passed over heated alumina at 500°, a nitrile results *(Industrial).*

$$\underset{\text{carboxylic acid}}{R-\overset{O}{\overset{\|}{C}}-OH}+NH_3 \longrightarrow \underset{\text{amm salt}}{R-\overset{O}{\overset{\|}{C}}-O^-NH_4^+} \xrightarrow[-H_2O]{Al_2O_3, 500^\circ} \underset{\text{amide}}{R-\overset{O}{\overset{\|}{C}}-NH_2} \longrightarrow \underset{\text{nitrile}}{R-C\equiv N}$$

(3) Dehydration of Aldoximes upon dehydration with acetic anhydride or thionyl chloride yield nitriles.

$$\underset{\text{aldoxime}}{R-\underset{H}{\underset{|}{C}}=N-OH} \xrightarrow[-H_2O]{(CH_3CO)_2O} \underset{\text{nitrile}}{R-C\equiv N}$$

e.g.,

$$\underset{\text{acetaldoxime}}{CH_3-\underset{H}{\underset{|}{C}}=N-OH} \xrightarrow[-H_2O]{(CH_3CO)_2O} \underset{\text{ethanenitrile, acetonitrile}}{CH_3-C\equiv N}$$

(4) Dehydration of Higher Amines. Higher amines dehydrated by passing their vapours a Cu or Ni catalyst at high temperature

$$\underset{\text{Amine}}{R-\underset{H}{\underset{|}{\overset{H}{\overset{|}{C}}}}-\underset{H}{\underset{|}{\overset{H}{\overset{|}{N}}}}} \xrightarrow[300^\circ]{Cu} \underset{\text{Nitrile}}{R-C\equiv N} + 2H_2$$

(5) Ammonoxidation of Alkanes or Aldehydes. An alkane having a terminal methyl group is oxidized at elevated temperature (500 – 600) over a catalyst in presence of ammonia to form nitrile

$$\underset{\text{alkane}}{R-CH_3}+NH_3+3/2O_2 \xrightarrow{550^\circ} \underset{\text{nitrile}}{R-C\equiv N}+3H_2O$$

e.g.,

$$\underset{\text{propane}}{CH_3CH_2CH_3}+NH_3+3/2O_2 \xrightarrow{550^\circ} \underset{\text{propanenitrile}}{CH_3CH_2-C\equiv N}+3H_2O$$

Aldehydes give a similar reaction in presence of sodium methoxide ($NaOCH_3$) and Cu^{2+} ion at 30°C

$$CH_3-CHO+NH_3+O_2 \xrightarrow[Cu^{2+},\ 30^\circ]{NaOCH_3} \underset{\text{ethanenitrile}}{CH_3-C\equiv N}+2H_2O$$

(6) Action of Grignard Reagents with Cyanogen Chloride. Grignard reagents react with cyanogen chloride (Cl – C=N) to form nitriles

$$\underset{\text{Grignard reagent}}{RMgX}+Cl-C\equiv N \xrightarrow{\text{ether}} \underset{\text{nitrile}}{R-C\equiv N}+Mg^{2+}X^-Cl^-$$

e.g.,

$$\underset{\text{ethylmagnesium bromide}}{C_2H_5MgBr}+Cl-C\equiv N \xrightarrow{\text{ether}} \underset{\text{propanenitrile}}{C_2H_5-C\equiv N}+Mg^{2+}Br^-Cl^-$$

(7) Addition of HCN to Alkanes. Hydrogen cyanide adds to terminal alkanes in the gaseous phase at 350° in presence of alumina to give nitriles

$$\underset{\text{2-methylpropane}}{CH_3-\underset{\underset{CH_3}{|}}{C}=CH_2} + HCN \xrightarrow[300°]{Al_2O_3} \underset{\text{2, 2-dimethylpropane-nitrile}}{CH_3-\overset{\overset{CH_3}{|}}{\underset{\underset{CH_3}{|}}{C}}-CN}$$

Physical Properties

(1) Lower molecular weight (up to Cu) aliphatic nitriles are colourless liquids at room temperature. The higher members are crystalline solids.

(2) They have fairly pleasant smells, resembling that of bitter almonds.

(3) Nitriles are dipolar compounds. Their dipole moments being pretty high ($CH_3CN = 4.0D$), they are highly associated molecules: Their boiling points are, therefore, abnormally high.

(4) The lower members are soluble in water with which they form hydrogen bonds. The solubility decreases as the hydrocarbon group gets larger. Thus acetonitrile is miscible with water and propiononitrile is fairly soluble, while higher members are oily and slightly soluble. All cyanides are soluble in organic solvents.

(5) They are moderately poisonous in contrast to hydrogen cyanide which is a deadly poison.

(6) Nitrites absorb in their infrared spectra in the region 2000 to 2300cm^{-1} owing to the stretching vibrations of the carbon-nitrogen triple bond (C≡N).

Table 16.2 : Physical properties of some Nitriles

Name	Formula	mp°C	bp°C	Density g/ml, 20°
Acetonitrile	CH_3-CN	–45	82	0.78
Propiononitrile	CH_3-CH_2-CN	–92	97	0.78
n-Butyronitrile	$CH_2-CH_2CH_2-CN$	–112	118	0.80
Isobutyronitrile	$CH_3-\underset{\underset{CH_3}{|}}{CH}-CN$	–72	104	0.78

Chemical Properties

The cyan group (C≡N) in nitriles is strongly polarized as indicated by the high dipole moment of 40D. It is polar in the same way as a carbonyl group (C=O). Thus nitriles can be represented as a resonance hybrid of two canonical forms, in which there is a high contribu-tion from the dipolar form.

Because of the dipole, a nitrile is subject to electrophilic attack on nitrogen and to nucleophilic attack on carbon. The cyan group being electron attracting activates the ex hydrogen which. can be easily removed as a proton by strong bases. In fact the cyan group could be compared to the carbonyl group in its chemical behaviour. Thus nitriles give the addition reactions just like aldehydes and ketones.

(1) Weak Basic Character Despite the presence of an unshared pair of electrons at the nitrogen atom, the nitriles are not basic enough to form salts with -aqueous acids they are therefore less basic than amines and ammonia. However they form addition compounds with strong acids in the absence of water.

$$\left[R{-}C{\equiv}\ddot{N}{:} \longleftrightarrow R{-}\overset{+}{C}{=}\overset{-}{\ddot{N}}{:} \right] \quad \text{or} \quad R{-}\overset{\delta+}{C}{\equiv}\overset{\delta-}{N}{:}$$

Resonance Hybrid

The weaker basic character of nitriles has been associated with a greater *s* character in the orbital containing the unshared electron pair. Unlike amines where tie lone pair of electrons is in sp^3 orbital, the lone pair of N atom in nitriles belongs to a *sp* orbital. This orbital (sp) evidently has greater *s* character than sp^3 orbital. Therefore, the lone pair of nitrogen in nitriles is less readily available and they are much weaker bases than amines.

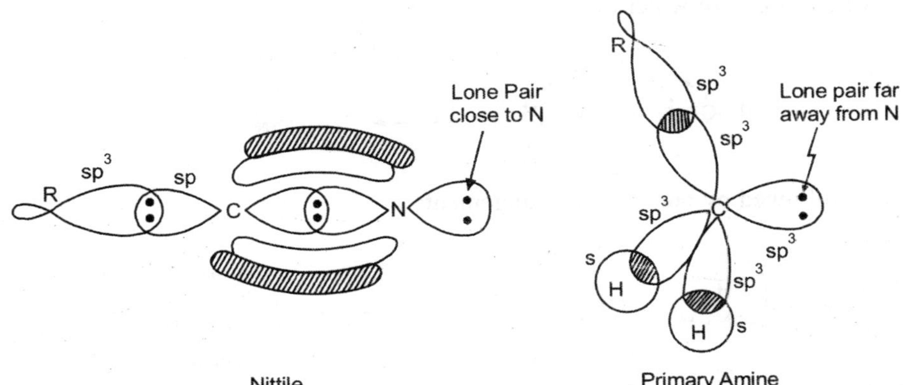

Nittile

Lone pair held *closer* to the N atom is *sp* orbital is *less* readily available; hence nitriles are LESS BASIC.

Primary Amine

Lone pair lying relatively *away* from the N atom in sp^2 orbital is *more* readily available; hence amines are MORE BASIC.

(2) Hydrolysis. On boiling with aqueous mineral acid or aqueous alkali, nitriles are hydrolysed to a carboxylic acid and ammonia. This reaction fakes place through the formation of an amide which is then hydrolysed to a carboxylic acid.

$$\underset{\text{nitrile}}{R-C\equiv N} \xrightarrow[+H_2O]{H^+ \text{ or } OH^-} \underset{\text{amide}}{R-\overset{\overset{\large O}{\|}}{C}-NH_2} \xrightarrow[+H_2O]{H^+ \text{ or } OH^-} \underset{\text{carboxylic acid}}{R-\overset{\overset{\large O}{\|}}{C}-OH} + NH_3$$

e.g.

$$\underset{\text{acetonitrile}}{CH_3-C\equiv N}\xrightarrow[+H_2O]{H^+ \text{ or } OH^-}\underset{\text{acetamide}}{CH_3-\overset{O}{\overset{\|}{C}}-NH_2}\xrightarrow[+H_2O]{H^+ \text{ or } OH^-}\underset{\text{acetic acid}}{CH_3-\overset{O}{\overset{\|}{C}}-OH}+NH_3$$

When a mineral acid (say HCl) is used for hydrolysis, the product is a carboxylic acid and ammonia. On the other hand, if an .alkali (say NaOH) is used, the hydrolysis results in formation of sodium salt of the carboxylic acid.

$$R-C\equiv N+2H_2O+HCl\longrightarrow RCOOH+NH_4^+Cl^-$$

$$R-C\equiv N+H_2O+NaOH\longrightarrow RCOO^-Na^++NH_3$$

Mechanism of Acid Catalysed Hydrolysis

(*a*) Protonation of nitrile

$$\underset{\text{Alkyl Cyanide}}{R-C\equiv N:}+H^+\longrightarrow \underset{I}{R-C\equiv \overset{+}{N}-H}\longleftrightarrow \underset{II}{R-\overset{+}{C}=\ddot{N}-H}$$

(*b*) Nucleophilic addition of water to II:

$$R-\overset{+}{C}=\ddot{N}-H+H-\ddot{O}-H\longrightarrow R-\overset{+OH_2}{C}=\ddot{N}-H$$

(*c*) Proton loss followed by tautomeric arrangement:

$$R-\overset{+OH_2}{C}=N-H\xrightarrow{-H^+}R-\overset{O-H}{C}=\ddot{N}-H\rightleftharpoons \underset{\text{Amide}}{R-\overset{O}{\overset{\|}{C}}-\overset{H}{\ddot{N}}-H}$$

The amide thus produced undergoes the same sequence of steps leading to the formation of carboxylic acid.

$$\underset{\text{Amide}}{R-\overset{O}{\overset{\|}{C}}-\ddot{N}H_2}\xrightarrow{H^+}R-\overset{O}{\overset{\|}{C}}-\overset{+}{N}H_3\longrightarrow R-\overset{\bar{O}}{\overset{|}{\overset{+}{C}}}-\overset{+}{N}H_3\xrightarrow{H_2\ddot{O}:}R-\overset{\bar{O}}{\underset{+OH_2}{C}}-\overset{+}{N}H_3\longrightarrow \underset{\text{Carboxylic acid}}{R-\overset{O}{\overset{\|}{C}}-OH}+\overset{+}{N}H_4$$

Mechanism of Base Catalysed Hydrolysis

(*a*) Nucleophilic attack at the carbon atom by OH ions.

$$HO^- + R{-}C{\equiv}N: \longrightarrow R{-}\underset{}{\overset{O-H}{C}}{=}\ddot{N}:^-$$

(*b*) Nucleophilic N atom takes up a proton from water regenerting OH ion. The product rearranges to produce stable amide.

$$R{-}\overset{O-H}{C}{=}\ddot{N}:^- + H{-}O{-}H \xrightarrow{-\bar{O}H} R{-}\overset{O-H}{C}{=}\ddot{N}H \rightleftharpoons R{-}\overset{O}{\overset{\|}{C}}{-}\ddot{N}H_2$$

The amide thus formed gets further hydrolysed to give carboxylic acid and ammonia.

$$\underset{\text{Amide}}{R{-}\overset{}{\underset{O}{\underset{\|}{C}}}{-}\ddot{N}H_2} + \bar{O}H \longrightarrow R{-}\overset{O-H}{\underset{O^-}{C}}{-}\ddot{N}H_2 \xrightarrow[H-O-H]{-\bar{O}H} R{-}\overset{O-H}{C}{-}\ddot{N}H_2 \rightleftharpoons \underset{\text{Carboxylic acid}}{R{-}\overset{O}{\overset{\|}{C}}{-}OH} + NH_3$$

(2) Alcoholysis. Nitriles on boiling with excess of alcohol in the presence of concentrated Sulphuric acid or hydrogen chloride, form esters

$$\underset{\text{Nitrile}}{R{-}C{\equiv}N} + \underset{\text{Alcohol}}{R'{-}OH} + H_2O \xrightarrow{H^+} \underset{\text{Ester}}{R{-}C{-}OR'} + NH_4^+$$

$$\underset{\text{Propanenitrile}}{CH_3CH_2{-}C{\equiv}N} + \underset{\text{Methanol}}{CH_3OH} \xrightarrow{H^+} \underset{\text{Methyl propionate}}{CH_2CH_2{-}\overset{O}{\overset{\|}{C}}{-}OCH_3} + NH_4^+$$

(3) Addition of Ammonia and Hydroxylmaine. Nitriles when reacted with ammonia or hydroxylamine in the presence of $NaNH_2$ as catalyst, form amidines and amidoximes

$$\underset{\text{Nitrile}}{R{-}C{\equiv}N:} + \ddot{N}H_3 \xrightarrow{NaNH_3} \underset{\text{Amidine}}{R{-}\overset{NH}{\overset{\|}{C}}{-}NH_2}$$

$$R{-}C{\equiv}N: + \underset{\text{Hydroxylamine}}{H_2\ddot{N}OH} \xrightarrow{NaNH_3} \underset{\text{Amidoxlme}}{R{-}\overset{NH_2}{C}{=}N{-}OH}$$

MECHANISM Both these reagents add by nucleophilic addition mechanism similar to those in case of carboxyl compounds. For illustration

$$R{-}C{\equiv}N: + \ddot{N}H_3 \longrightarrow R{-}\overset{H{-}\overset{H}{\overset{|+}{N}}{-}H}{C}{=}\ddot{N}:^- \longrightarrow R{-}\overset{NH_2}{C}{=}NH$$

(4) Addition of Gig nards Reagent Nitriles react with Grignard reagents to form ketimines which on hydrolysis yields ketones

$$\underset{\text{Nitrile}}{R{-}C{\equiv}N} + \overset{\delta-\ \ \delta+}{R'\ MgX} \longrightarrow \underset{\text{Ketimine}}{R{-}\overset{R'}{\overset{|}{C}}{=}NMgX} \xrightarrow[H^+]{H_2O} \underset{\text{Ketone}}{R{-}\overset{R'}{\overset{|}{C}}{=}O} + \overset{2+}{Mg}\ \bar{X}\ \bar{O}H$$

e.g.,

$$\underset{\text{Acetonitrile}}{CH_3{-}C{\equiv}N} + \underset{\text{Methylmagnesium bromide}}{CH_3MgBr} \longrightarrow CH_3{-}\overset{CH_3}{\overset{|}{C}}{-}NMgBr \xrightarrow[H^+]{H_2O} \underset{\text{Acetone}}{CH_3{-}\overset{CH_3}{\overset{|}{C}}{=}O} + \overset{2+}{Mg}\bar{Br}\bar{O}H$$

(5) Reduction (a) Partial Reduction When nitriles are reduced with lithium tri ethooxyaluminium hydride $LiAlH(OCH_3)_3$, in ether, aldimine is first produced. This subsequent hydrolysis yields aldehydes.

$$\underset{\text{nitrile}}{R-C\equiv N} + LiAlH(OCH_2CH_3)_3 \xrightarrow{\text{ether}} \underset{\text{aldimine}}{R-CH-NH_2} \xrightarrow[H^+]{H_2O} \underset{\text{aldehyde}}{RCHO}$$

(b) Complete reduction to primary amines. On treatment with lithiumaluminium hydride $LiAlH_4$ nitriles undergo complete reduction to yield primary amines

$$\underset{\text{nitrile}}{R-C\equiv N} + 4H \xrightarrow[\text{ether}]{LiAlH_4} \underset{\text{primary amine}}{R-CH_2-NH_2}$$

This reduction may also be carried with sodium and ethanol *(Mendius Reaction),* or with hydrogen and metal catalyst (Raney nickel).

$$\underset{\text{acetonitrile}}{CH_3-C\equiv N} + 2H_2 \xrightarrow{\text{Raney Ni}} \underset{\text{ethylamine}}{CH_3-CH_2-NH_2}$$

(6) Alkylation Nitriles having α-hydrogen atoms are alkylated readily when heated with an alkyl halide in presence of finely divided sodium amide ($Na^+ NH^-{}_2$)

$$\underset{\text{Alkyl cyanide}}{R{-}\overset{H}{\underset{H}{\overset{|\alpha}{\underset{|}{C}}}}{-}C{\equiv}N} + \underset{\text{Alkyl halide}}{R'{-}X} \xrightarrow{\overset{+}{Na}\overset{-}{N}H_2} \underset{\text{2° cyanide}}{R{-}\overset{H}{\underset{R'}{\overset{|\alpha}{\underset{|}{C}}}}{-}C{\equiv}N} + HX$$

$$R{-}\overset{H}{\underset{R}{\overset{|\alpha}{\underset{|}{C}}}}{-}C{\equiv}N + R'{-}X \xrightarrow{\overset{+}{Na}\overset{-}{N}H_2} \underset{\text{3° cyanide}}{R{-}\overset{R'}{\underset{R}{\overset{|}{\underset{|}{C}}}}{-}C{\equiv}N} + HX$$

This reaction provides a procedure for preparing tertiary amines

MECHANISM The base: NH_2 furnished 'by sodium amide initiates the reaction producing carbanion which is then alkylated by alkyl halide as follows:

(*a*)

$$R{-}\overset{H}{\underset{H}{\overset{|}{\underset{|}{C}}}}{-}C{\equiv}N + :\bar{N}H_2 \xrightarrow[-NH_3]{\text{Fast}} \underset{\substack{\text{2° Carbanion}\\ \text{Stable}}}{R{-}\overset{H}{\overset{|}{\underset{\bullet\bullet}{\bar{C}}}}{-}C{\equiv}N}$$

(b)

$$R-\underset{\cdot\cdot}{\overset{H}{\overset{|}{C}}}-C\equiv N + R'-X \longrightarrow R-\underset{R'}{\underset{|}{\overset{H}{\overset{|}{C}}}}-C\equiv N + \bar{X}$$

2° Cyanide

*(7) **Thorpe Nitrile Condensation**.* Nitriles undergo condensation in the presence of sodium in ether, when only the ex hydrogen atoms are involved. Two molecules of the nitrile undergo Claisen-type condensation as follows.

$$\underset{\text{(2 moles of nitrile)}}{\underset{\text{Propanenitrile}}{CH_3-CH_2-\overset{N}{\overset{|||}{C}}}} + \underset{CH_3}{\underset{|}{\overset{H}{\overset{|}{C}}}}H-CN \longrightarrow \underset{\text{3-imino-2-methylpentanenitrile}}{CH_3-CH_2-\overset{NH}{\overset{||}{C}}-\overset{CH_3}{\overset{|}{C}}H-CN}$$

This reaction is known as *Thorpe Nitrile Condensation.*

Alkyl Isocyanides or Isonitriles

As the name implies, the isocyanides or isonitriles are structural isomers of the cyanides, or nitriles. They are regarded as alkyl derivatives of the isocyanide form of - hydrogen cyanide ($H-N^+\equiv C:^-$). They have the general formula

$$\underset{\text{I}}{R-\overset{+}{N}\equiv\bar{C}:} \longleftrightarrow \underset{\text{II}}{R-\ddot{N}=C:}$$

Formula	Common name	IUPAC Name
CH_3-NC	methyl isocyanide	methylcarbylamine
CH_3CH_2-NC	ethyl isocyanide	ethylcarbylamine
$CH_3CH_2CH_2-NC$	propyl isocyanide	propylcarbylamine

Methods of Preparation

(1) By heating an lkyl iodide with silver cynide (AG–C≡N:) in aqueous ethanolic solution.

$$\underset{\text{alkyl iodide}}{R-I} + Ag-C\equiv N: \xrightarrow{\Delta} \underset{\text{alkyl isocyanide}}{R-N^+\equiv C^-:} + AgI$$

A small amount of the isomeric cyanide is also produced.

(2) By heating a primary amine and chloroform ($CHCl_3$) with ethanolic potassium hydroxide solution. (*Hofmann Carbylmine Reaction*).

$$\underset{\text{1° amine}}{R-NH_2} + \underset{\text{chloroform}}{CHCl_3} + 3KOH \xrightarrow{\Delta} \underset{\text{alkylisocyanide}}{R-N^+\equiv C^-:} + 3KCl + 3H_2O$$

(3) By heating isocyanates (RNCO) with alkyl phosphites.

$$\underset{\text{alkyl isocyanate}}{R-N=C=O} + \underset{\text{triethyl phosphite}}{(C_2H_5O)_3P} \xrightarrow{150-190^\circ} \underset{\text{alkyl isocyanide}}{R-N^+\equiv C^-:} + (C_2H_5O)_3PO$$

(4) By heating *N*-alkylformamides with phosphoryl chloride in pyridine solution.

$$\underset{\textit{N}\text{-alkyl formamide}}{R-NH-CHO} \xrightarrow[\text{pyridine}]{POCl_3} \underset{\text{alkyl isocyanide}}{R-\overset{+}{N}\equiv\overset{-}{C}:} + H_2O$$

Physical Properties

Isonitriles are highly unpleasant smelting liquids. They are almost insoluble in water the nitrogen atom not having a lone pair of electrons for hydrogen bonding. They have lower boiling, points than the respective isomerio nitriles (CH_3CN bp 81°C and CH_3NC bp 68°C)

Chemical Properties

This is attributed to their lower dipole moments (3D) compared to that of nit riles (4D).

The N: ≡ C: stretching absorption is in the infrared region 2185-2120 cm^{-1} (S).

(Chemical) It is clear from the structure of isocyanides that they have highly polar molecules having an unshared pair of electrons on the triply bonded carbon. This is why they are a very reactive class of compounds. Their reactions usually involve nucleophilic attack on the reagent (A-B) followed by the nucleophilic attack by the fragment B:- on the carbon atom of the isocyanide. Thus both the fragments of the attacking reagent are added to the isocyanide carbon.

$$R-\overset{+}{N}\equiv\overset{-}{C}: + A-B \longrightarrow \underset{\text{First addition product}}{R-\overset{+}{N}\equiv C-A} + :\overset{-}{B} \longrightarrow \underset{\text{Second addition product}}{R-\ddot{N}=\underset{\substack{|\\B}}{C}-A}$$

Some of the common reactions of alkyl isocyanides are listed below.

(1) Isomerisation. Isocyanies on heating with NaCN or KCN rearrange to more stable cyanides.

$$\underset{\text{Alkyl isocyanide}}{R-\overset{+}{N}\equiv\overset{-}{C}} \xrightarrow[100^\circ]{KCN} \underset{\text{Alkyl cyanide}}{R-C\equiv N}$$

MECHANISM. Here the cyanide ion brings about a nucleophilic displacement reaction.

$$\underset{\text{C-Alkideion}}{:N\equiv\overset{-}{C}:} + \underset{\text{Alkyl Isocyanide}}{R-\overset{+}{N}\equiv\overset{-}{C}:} \longrightarrow \underset{\text{Alkyl cyanide}}{:N\equiv C-R} + \ddot{N}\equiv\overset{-}{C}:$$

(2) Addition Reactions. Isocyanides react with hydrogens, sulphur, mercuric oxide, Grignard reagents, amines, alcohols etc., to form addition compounds. For example,

$$R-\overset{+}{N}\equiv\overset{-}{C}: + Cl_2 \longrightarrow R-N=CCl_2$$

Alkyl isocyanide → Alkyliminocarbonyl chloride

$$R-\overset{+}{N}\equiv\overset{-}{C}: + S \longrightarrow R-N=C=S$$

Alkyl isothiocyanate

$$R-\overset{+}{N}\equiv\overset{-}{C}: + HgO \longrightarrow R-N=C=O$$

Alkyl isocyanate

(3) Hydrolysis. They are hydrolysed by dilute acids to form a primary amine and formic acid, but are not hydrolysed by alkalis.

$$\underset{\text{Alkyl isocyanide}}{R-\overset{+}{N}\equiv\overset{-}{C}:} + 2H_2O \xrightarrow{\text{acid}} \underset{1^\circ\text{ amine}}{R-NH_2} + \underset{\text{formic acid}}{HCOOH}$$

$$R-\overset{+}{N}=\overset{-}{C}: + H_2O \xrightarrow{\text{alkail}} \text{No reaction}$$

In contrast, nitriles (RCN) undergo both acid and base hydrolysis to form the corresponding carboxylic acids (ROOH). This reaction brings out clearly that in isocyanides the alkyl group R is attached to nitrogen, while in nitriles it is attached to the terminal carbon. Thus,

$$\underset{\text{Methyl isocyanide}}{CH_3-\overset{+}{N}\equiv\overset{-}{C}:} + 2H_2O \xrightarrow{H^+} \underset{\text{Methylamine}}{CH_3-NH_2} + HCOOH \quad \text{.........}I$$

$$\underset{\text{Acetonitrile}}{CH_3-C\equiv N:} + 2H_2O \xrightarrow{H^+} \underset{\text{Acetic acid}}{CH_3-\overset{O}{\overset{\|}{C}}-OH} + NH_3 \quad \text{.........}II$$

In reaction (*i*) the original $H_2C - N$ bond remains intact in reaction (*ii*) the original $H_2C - C$ bond remains intact.

MECHANISM Isocyanide undergoes acid hydrolysis by the following mechanism.

$$R-N=C: + H^+ \longrightarrow R-N=C-H \ (+\ OH_2) \xrightarrow[\text{Transfer}]{\text{Proton}} \underset{\text{Substituted Formamide}}{R-\underset{H}{N}-\overset{}{\underset{O}{\overset{}{C}}}-H} \xrightarrow{\text{Hydrolysis}} \underset{1^\circ\text{ Amine}}{RNH_2} + HCOOH$$

($:\ddot{O}H_2$ attacks the carbon of $R-N=C:$)

(4) Reduction. Isocyanides are reduced to secondary amines either by sodium and ethanol, or by catalytic hydrogenation (Pt, H_2)

$$\underset{\text{Alkyl isocyanide}}{R-\overset{+}{N}\equiv\overset{-}{C}:} + 2H_2 \xrightarrow{Na,\ C_2H_5OH} \underset{2^\circ\text{ amine}}{R-NH-CH_3}$$

e.g., $CH_3CH_2{-}\overset{+}{N}{\equiv}\overset{-}{C}{:} + 2H_2 \longrightarrow CH_3CH_2{-}NH{-}CH_3$
Propanenitrile → Athylmethylamine

It may be recapitulated that nitriles 00 reduction with sodium and ethanol form primary amines. Thus,

$$\underset{\text{Alkyl isocyanide}}{CH_3{-}\overset{+}{N}{\equiv}\overset{-}{C}{:}} + 2H_2 \xrightarrow{Na,\ C_2H_5OH} \underset{\text{Dimethylamine}}{CH_3{-}NH{-}CH_3}$$

$$\underset{\text{Alkyl isocyanide}}{CH_3{-}C{\equiv}N{:}} + 2H_2 \xrightarrow{Na,\ C_2H_5OH} \underset{\text{Ethylamine}}{CH_3{-}CH_2{-}CH_3}$$

This reaction also proves the structural difference between isocyanides

VINYL CYANIDE

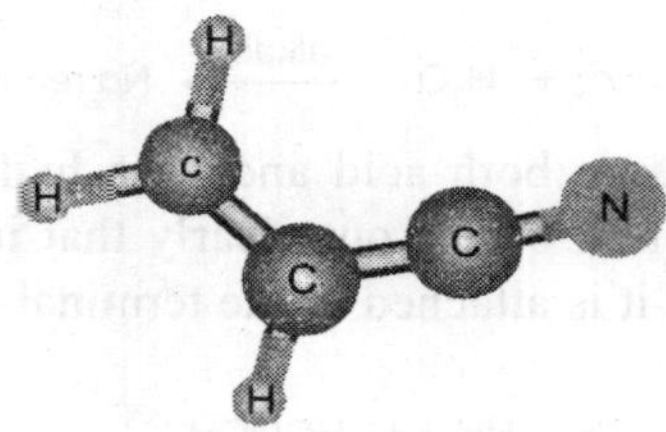

It is also known as Acrylonitrile

Introduction

> The Environmental Protection Agency (EPA) and the German MAK commission has classified acrylonitrile to be a human carcinogen.

Acrylonitrile, vinyl cyanide, CH_2=CHCN, is a colorless man-made liquid with a slightly pungent odor. It was first prepared in 1893 by a **French Chemist, Ch. Moreau**, who dehydrated ethylene cyanohydrin with phosphorus pentoxide. Its prominent use came shortly before World War II in connection with the development of copolymers for use in oil-resistant rubber. Since that time the demand for acrylonitrile has dramatically increased due largely to acrylic fibers, first introduced commercially in 1950 by DuPont under the trademark Orlon. Acrylonitrile has a boiling point of 77.3°C, a freezing point of -83.55°C, and a vapor pressure of 100 mm Hg at 23.6°C.In 1977, the U.S. Federal Drug Administration declared acrylonitrile to be an indirect food additive and banned its use in beverage containers and other food-packaging applications.

Production

In the Sohio process, propylene, oxygen, ammonia, and water are fed to a fluidized-bed reactor containing a complex metal-oxide catalyst. Small amounts of water are fed to the reactor to increase the selectivity of acrylonitrile and decrease the temperature within the reactor. The reactor temperature is in the range of 400°C to 500°C and the residence time is a few seconds. The pressure of the reactor is 0.3-2 bar and the yield is 80-90 percent depending on the catalyst used. This process is highly selective and requires no recycling. Each of the reactants are fed in their respective stoichiometric ratios. The by-products formed during the ammoxidation reaction are hydrogen cyanide and acetonitrile, which are separated during the distillation process.

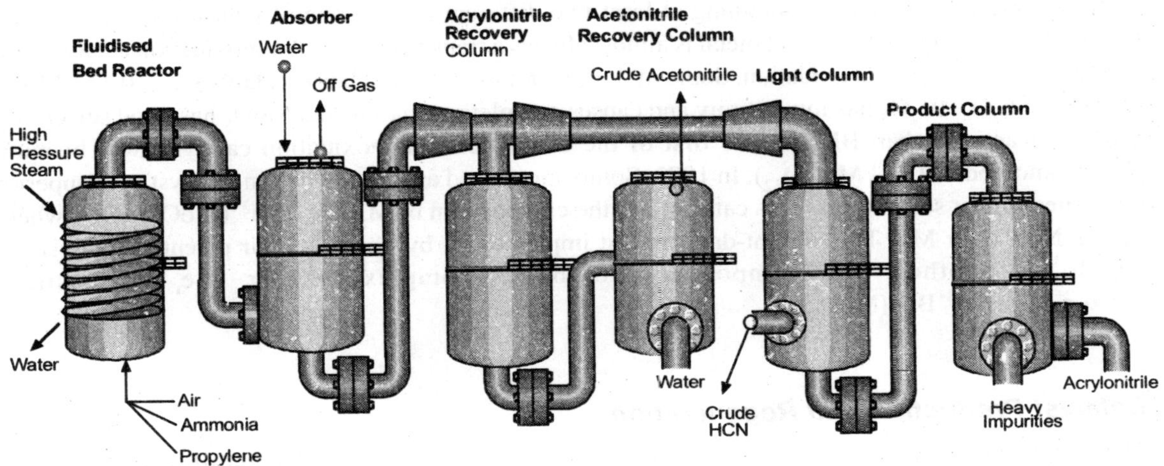

Fig. 16.3 : Showing the Industrial Preparation of Acrylonitrile

The DuPont process for the production of acrylonitrile involved the catalytic addition of hydrocyanic acid to acetylene.

$$HC=CH + HCN \rightarrow CH_2=CHCN$$

This commercial reaction was carried out at 80°C in dilute hydrochloric acid containing cuprous chloride. Although the yield from this reaction was good, the raw materials were relatively expensive and some undesirable impurities were difficult to remove. The cuprous chloride catalyst used in this process required frequent regeneration and expense. DuPont eventually abandoned this procedure in 1970.

In 1947, Allied Chemical and Dye Corporation patented the manufacture of acrylonitrile via an ammoxidation process of propylene but it was not until 1960 when Standard Oil of Ohio(Sohio) developed the first commercially viable catalyst for this process. Today, over 90% of acrylonitrile production in the world is based on the Sohio process. In this process, propylene, oxygen, and ammonia are catalytically converted directly to acrylonitrile using a fluidized-bed reactor operated at temperatures between 400°C and 500°C and gauge pressures between 0.3 and 2 bar.

$$2\ CH_2=CHCH_3 + 3\ O_2 + 2\ NH_3 \rightarrow 2\ CH_2=CHCN + 6\ H_2O$$

Acrylonitrile can be used to produce a wide variety of products. Such products are adiponitrile (used in production of Nylon 66), ABS and SAN resins (automobile plastics), acrylic fibers, and acrylamide and nitrile elastomers (oil-resistant rubber). Nearly 60% of the acrylonitrile produced is used to make acrylic fibers. These acrylic fibers find use primarily in wearing apparel and in home furnishings such as carpets and draperies. By 2001, the demand for acrylonitrile is expected to be 3.8 billion pounds.

Catalysts

The Sohio process utilizes a series of complex metal-oxide catalysts. These catalysts are oxide combinations or compounds containing at least two different metals, one of which coming from the later row 5a elements. The second metal is almost always a transition metal Ammoxidation catalysts are highly selective for partial oxidation. Some of the best-known ammoxidation catalysts are BI_2O_3+MoO_3 and UO_3+Sb_2O_4. BI_2O_3 has low activity and causes complete oxidation and MoO_3 has good selectivity and lower activity than BI_2O_3. The first of the commercial ammoxidation catalysts was bismuth phosphomolybdate ($Bi_9PMo_{12}O_{52}$). In 1972, Sohio announced an ammoxidation catalyst with superior properties(higher selectivity). This catalyst has the composition of $M_8^{2+}Fe_3^{3+}Bi^{3+}(MoO_4)_{12}O_2^{2-}$ where M^{2+} is Ni, Co, or Mg. The present-day catalyst implemented by Sohio is their patented catalyst 41, which is another multicomponent molybdate complex having the composition of $Co_6^{2+}Ni_2^{2+}Fe_3^{3+}Bi^{3+}(MoO_4)_2$.

Catalyst Production and Regeneration

The metal oxide catalyst used in the ammoxidation process for the production of acrylonitrile are usually prepared by precipitation from solution. In this process, an aqueous metal salt solution, consisting of bismuth nitrate and ammonium molybdate, is rapidly mixed to form product precipitates. Depending on Bi/Mo ratio, pH, and the time of interaction, these precipitates are filtered, washed, dried, and heated to temperatures of 500°C-800°C to form the final catalyst. The catalyst is then sent to a grinder to obtain catalyst particles ranging from 10-100 microns. During the ammoxidation reaction process, small particles tend to fuse together to form much larger ones. This process steadily deactivates the catalyst and reduces its activity. Catalyst activity can also be reduced during the reaction process by losses of oxygen attached to the bismuth molecule. Reoxidation of the catalyst occurs by adsorption and dissociation of oxygen on anion vacancies associated with molybdenum. Anion vacancies on the bismuth are reoxidized by transfer of oxygen from the oxidized molybdenum (Satterfield 276). Catalyst lifetime in the fluidized-bed reactor is approximately 1 year.

Physical Properties

Boiling Point	= 77°C
Melting Point	= –84 °C
Molecular Formula	= C_3H_3N
Formula Weight	= 53.06262

Composition	= C(67.90%) H(5.70%) N(26.40%)
Molar Refractivity	= 15.58 ± 0.3 cm^3
Molar Volume	= 66.5 ± 3.0 cm^3
Parachor	= 148.8 ± 4.0 cm^3
Index of Refraction	= 1.384 ± 0.02
Surface Tension	= 25.0 ± 3.0 dyne/cm
Density	= 0.797 ± 0.06 g/cm^3
Polarizability	= 6.17 ± 0.5 $10^{-24}cm^3$
Monoisotopic Mass	= 53.026549 Da
Nominal Mass	= 53 Da
Average Mass	= 53.063762 Da

Chemical Properties

It polymerizes readily in the presence of catalyst (organic peroxides and other oxidizing agents) to form polyacrylonitriles

$$\underset{\text{acrylonitrile}}{x\ CH_2 = CH - CN} \xrightarrow{\text{cat.}} \underset{\text{polyacrylonitrile}}{\left[-CH_2 - \underset{\substack{| \\ CN}}{CH} - \right]_x}$$

Polyacrylonitriles is used for making synthetic fibers as Orion, Acrion, Cresion and Zefran

Owing to the electron – withdrawing effect of the nitrile group

$$H_2C = CH - C \equiv N \longleftrightarrow \overset{+}{H_2}C = CH = C = \overset{-}{N}:$$

the double bond in acrylonitrile exhibits high reactivity towards nucleophilic reagents. Thus in the presence of basic catalysts, the double bond in acrylonitrile undergoes addition of a number of reagents such as water, alcohols, hydrogen sulphide, amines and also with active methylene compounds (R-CH_2-CO-R '). In these reactions the active hydrogen of the reagent (H-A) is replaced by cyanoethyl group (-CH_2CH_2CN).

$$\underset{\text{acrylonitrile}}{CH_2 = CH - CN} + H - A \xrightarrow{\text{base}} \underset{\text{cynoethyl derivative}}{A - CH_2 - CH_2 - CN}$$

This procedure by which active hydrogen of a compound is substituted by cyanoethyl group is referred to as Cyanothylation. Thus,

$$CH_2 = CH - CN \left| \begin{array}{l} +H_2O \xrightarrow{base} HO - CH_2CH_2CN \\ +ROH \xrightarrow{base} RO - CH_2CH_2CN \\ +H_2S \xrightarrow{base} HS - CH_2CH_2CN \\ +RNH_2 \xrightarrow{base} RNH - CH_3CH_3CN \\ +RCH_2COR' \xrightarrow{base} R'CH(CO-R) - CH_2CH_2CN \end{array} \right.$$

acrylondrille

MECHANSIM The cynoethylation in case of a ketone containing active methylene group may be interrupted to take place as follows

Since the product still has one active hydrogen the process is still repeated

$$R-CH(H)-CO-R' + \bar{O}H \longrightarrow R-\ddot{C}H-CO-R' + H_2O$$

$$R-\ddot{C}H-CO-R' + CH_2=CH-C\equiv N \xrightarrow{H_2O} R-CH(CH_2-CH=C=NH)-CO-R \longrightarrow R-CH(CH_2-CH_2-C\equiv N)-CO-R'$$

Alkyl Cyanates and Isocyanates

These are the derivatives of cyanic acid (HO – C ≡ N), an oxidation product of hydrogen cyanide (H – C ≡ N). Since hydrogen cyanide exists into two tautomeric forms so does its oxidation product.

Alkyl cyanates can be obtained by treating sodium alkoxide with cyanogen chloride.

$$\underset{\text{Sodium Alkoxide}}{R-O-Na} + \underset{\text{cyanogen chloride}}{Cl-C\equiv N} \rightarrow \underset{\text{alkyl cyanate}}{R-O-C\equiv N} + NaCl$$

The cyanate ion is an ambident nucleophile in nucleophilic substitution because it can react to form an alkyl cyanate R-OCN (exception) or an alkyl isocyanate R-NCO (rule). Aryl cyanates (C_6H_5OCN) can be formed by a reaction of phenol with cyanogen chloride (ClCN) in the presence of a base.

Cyanates are salts of cyanic acid for example potassium cyanate (KNCO).

ALKYL IOSOCYANATES AND ISOCYANIDES
METHYL ISOCYANATE

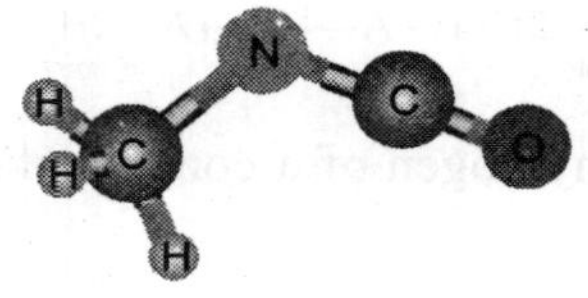

Introduction

Methyl isocyanate (MIC) is an organic compound with the molecular formula C_2H_3NO, arranged as $H_3C\text{-}N\text{=}C\text{=}O$. Synonyms are isocyanatomethane, methyl carbylamine, and MIC. It was discovered in 1888 as an ester of isocyanic acid.

Manufacture

Methyl isocyanate is usually manufactured from monomethylamine and phosgene. These substances react at a range of temperatures, but for large scale production it is advantageous to combine these reactants at higher temperature in the gas phase. A mixture of methyl isocyanate and two moles of hydrogen chloride is formed, but N-methylcarbamoyl chloride (MCC) forms as the mixture is condensed and leaves one mole of hydrogen chloride as a gas.

$$COCl_2 + H_3C\text{—}NH_2 \longrightarrow H_3C\text{-}NH\text{-}C(=O)\text{-}Cl \; (\text{MCC}) + HCl$$

The methyl isocyanate is obtained by treating the MCC with a tertiary amine (e. g.: dimethylaniline, pyridine(Slocombe & Hardy 1949) or by separating it by using distillation techniques (Merz 1965).

$$H_3C\text{-}NH\text{-}C(=O)\text{-}Cl \; (\text{MCC}) \rightleftharpoons H_3C\text{—}N\text{=}C\text{=}O + HCl$$

Methyl isocyanate is also manufactured from N-methyformamide and air. In the latter process it is immediately consumed in a closed-loop process to make methomyl (Chem.Week 1985a). Other manufacturing methods have been reported (Giesselmann et al. 1980; Chem. Week 1985b).

Physical Properties

Boiling point:	39°C
Melting point:	–45°C
Relative density (water = 1):	0.96
Solubility in water:	reaction
Vapour pressure, kPa at 20°C:	46.4
Relative vapour density (air = 1):	1.97
Relative density of the vapour/air-mixture at 20°C (air = 1):	1.4
Flash point:	–7°C c.c.
Auto-ignition temperature:	535°C
Explosive limits, vol% in air:	5.3-26

Chemical Properties

Methyl isocyanate reacts readily with many substances that contain N-H or O-H groups and some other compounds. It also reacts with itself to form a trimer or higher molecular weight polymers.

Methyl isocyanate reacts with water to form 1,3-dimethylurea and carbon dioxide with the evolution of heat (325 calories per gram of MIC that reacts).

$H_3C{-}N{=}C{=}O + H_2O \longrightarrow CO_2 +$

With xs water: 1, 3-dimethylurea, m.p. 106°C
or with xs MIC: 1, 3, 5-trimethylbiuret, m.p. 125°C

At 25°C, in excess water, one-half of the MIC is consumed in 9 minutes (Castro et al. 1985); if the heat is not efficiently removed from the mixture the rate of the reaction will increase and rapidly cause the MIC to boil. If MIC is in excess, 1,3,5-trimethylbiuret is formed along with carbon dioxide (Union Carbide 1967).Compounds that contain hydrogen attached to nitrogen, such as ammonia or primary or secondary amines, will rapidly react with MIC to form substituted ureas. Other N-H compounds, such as amides and ureas, react much more slowly with MIC (March 1985).Alcohols and phenols, which contain an O-H group, react slowly with MIC, but the reaction can be catalyzed by trialkylamines or dialkyltin dicarboxylate.

1-naphthol + $H_3C{-}N{=}C{=}O$ (MIC) $\xrightarrow{\text{Catalyst}}$ Carbaryl

Oximes, hydroxylamines, and enols also react with MIC to form methylcarbamates (Union Carbide 1967). These reactions produce the products described below (Uses). When treated with catalysts, MIC reacts with itself to form a solid trimer, trimethyl isocyanurate, or a higher molecular weight polymer.

$3H_3C{-}N{=}C{=}O \xrightarrow{\text{Catalyst}}$ Trimethyl isocyanurate, m.p. 178°C

Sodium methoxide, triethyl phosphine, ferric chloride, and certain other metal compounds catalyze the formation of the MIC-trimer, while the higher molecular weight polymer formation is catalyzed by certain trialkylamines. Since the formation of the MIC trimer is exothermic (298 calories per gram of MIC), the reaction can lead to violent boiling of the MIC. The high-molecular-weight-polymer hydrolyzes in hot water to form the the trimethyl isocyanurate. Since catalytic iron and tin salts can be formed from impurities in commercial grade MIC, this product must not be stored in steel drums or tanks (Union Carbide 1967).

Uses

Methyl isocyanate is an intermediate chemical in the production of carbamate pesticides (such as carbaryl, carbofuran, methomyl, and aldicarb). It has also been used in the production of rubbers and adhesives.

Hazards

Methyl isocyanate is extremely toxic and can damage by inhalation, ingestion and contact in quantities as low as 0.4 ppm. Damage includes coughing, chest pain, dyspnea, asthma, irritation of the eyes, nose and throat as well as skin damage. Higher levels of exposure, over 21 ppm, can result in pulmonary or lung edema, emphysema and hemorrhages, bronchial pneumonia and death. Although the detectable odor of methyl isocyanate is a concentration triple the permissible exposure, its potent lachrymal properties provide an excellent warning of its presence (at a concentration of 2-4 parts per million (ppm) subject's eyes are irritated, while at 21 ppm subjects could not tolerate the presence of methyl isocyanate in air)(Kimmerle and Eben 1964).

Proper care must be taken to store methyl isocyanate because of its ease of exothermically polymerizing (see Reactions) and its similar sensitivity to water. Only stainless steel or glass containers may be safely used; the MIC must be stored at temperatures below 40ºC.

The toxic effect of the compound was apparent in the Bhopal accident, when around 40,000 kilograms of methyl isocyanate were released over a populated area on December 3, 1984, killing thousands of people.

METHYL ISOCYANIDE

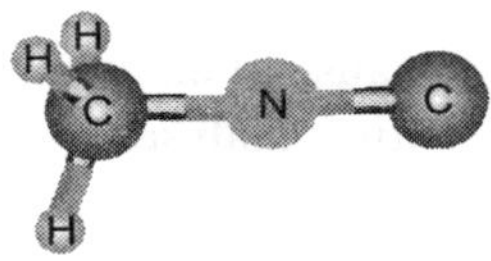

In a 2-l. four-necked flask (Note 2) equipped with a 250-ml. pressure-equalizing dropping funnel, a sealed mechanical stirrer, a thermometer, and a receiver trap (Note 3) are placed 1034 g. (8.0 moles)

of quinoline and 572 g. (3.0 moles) of *p*-toluenesulfonyl chloride (Note 4). The solution is heated to 75° by an oil bath and the system evacuated to a pressure of 15 mm. The receiver is cooled in a bath of liquid nitrogen (Note 5). While the solution is vigorously stirred and maintained at this temperature, 118 g. (2.0 moles) of N-methylformamide (Note 4) is added dropwise to maintain a smooth distillation rate. The addition is complete in 45–60 minutes.

The material which collects in the receiver is distilled through a 15-cm. Vigreux column at atmospheric pressure. Methyl isocyanide, a colorless, vile-smelling liquid, is collected at 59–60°; weight 57–61 g. (69–74%) (Note 6). Analysis by gas liquid chromatography indicates that the purity exceeds 99% (Note 7).

$$CH_3NHCHO \xrightarrow[75°,\ vacuum]{p\text{-TsCl, quinoline}} CH_3-\overset{+}{N}\equiv\overset{-}{C}$$

2. Notes

1. An explosion involving methyl isocyanide has been reported.[2] For this reason, prudence dictates the use of adequate shielding in all heating operations.

2. A three-necked flask may be used by employing a suitable adapter on one of the necks. The checkers used a standard wide-bore, 75-degree side-arm adapter fitted with a long-stemmed thermometer extending into the reaction solution. A Trubore® stirrer equipped with a semicircular Teflon® paddle was also used.

3. A vapor trap having a wide-bore inlet tube and the appropriate condensate capacity is used. The checkers used a 4.8 cm. × 30.0 cm. trap having a 2.0 cm. × 18.0 cm. inlet tube. Best results are obtained when the trap is connected directly to the flask or the adapter (see (Note 2)) by a wide-bore tube. Ground-glass joints should be used throughout the apparatus.

4. All materials were obtained from Eastman Kodak Company. Quinoline, practical grade, b.p. 72–74° (0.2 mm.), was freshly distilled from zinc dust. If undistilled quinoline is employed, a major contaminant, which appears to be methyl isocyanate, will be formed. "White label" *p*-toluenesulfonyl chloride and N-methylformamide were used without further purification.

5. The checkers found that a bath of dry ice and acetone worked equally well.

6. A single transfer under high vacuum afforded a product of identical purity. The major contaminants appear to be small amounts of high-boiling starting materials.

7. When a 2-m. polypropylene glycol on firebrick column at 75° is used, the retention volume of methyl isocyanide is 55 cc. of helium. Because of an unknown factor in conditioning the column, it is advisable to perform at least two consecutive analyses.

The checkers employed a 5-ft. 20% Carbowax 20 M (terminated with terephthalic acid) on Chromosorb W (acid washed) column at 60°. Methyl isocyanide showed a retention volume of 300 cc. of helium. Only traces of lower-boiling impurities were observed.

3. Discussion

Methyl isocyanide has been prepared chiefly by minor modifications of the original method of Gautier,[3] which is the alkylation of silver cyanide by an alkyl halide.

Physical Properties

Melting Point	= -45°C
Boiling Point	= 59.6°C
Density	= 0.756
Molecular Formula	= C_2H_3N
Formula Weight	= 41.05192
Composition	= C(58.52%) H(7.37%) N(34.12%)
Monoisotopic Mass	= 41.026549 Da
Nominal Mass	= 41 Da
Average Mass	= 41.052725 Da

Chemical Properties and Uses

Methyl isocyanide is a typical isocyanide and behaves like other alkyl isocyanides.Methyl isocyanide is used as a chemical intermediate for the production of carbamate insecticides and herbicides

PHENYL ISOCYANIDE

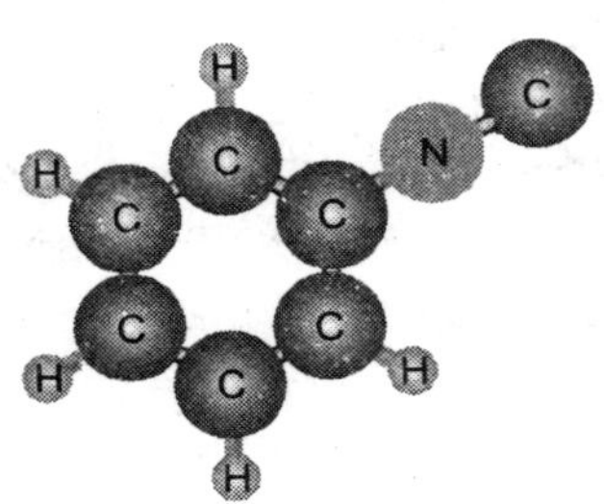

Preparation

Phenyl isocyanide can be prepared by heating aniline and chloroform with potassium hydroxide in alcoholic mixture at 55°C or with dry powdered potassium hydroxide. It is also formed by the reaction of phenyl hydrazine with chloroform, carbon tetrachloride hexachloroethane, and bromoform and iodoform or chloral.

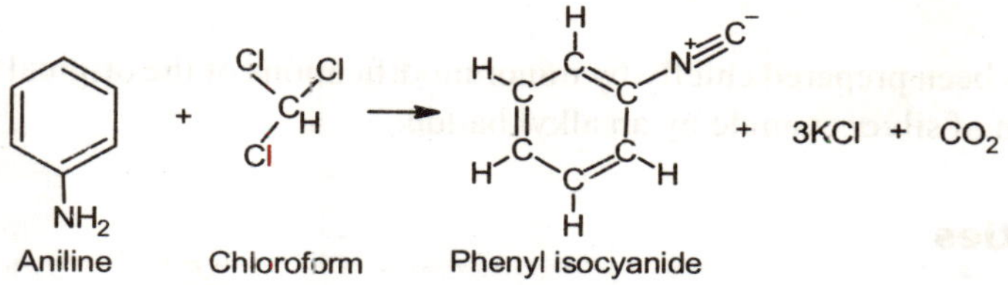

Aniline + Chloroform → Phenyl isocyanide + 3KCl + CO_2

Physcial Properties

Boiling point	= 78°C
Molecular Formula	= C_7H_5N
Formula Weight	= 103.1213
Composition	= C(81.53%) H(4.89%) N(13.58%)
Density	= 0.975
Monoisotopic Mass	= 103.042199 Da
Nominal Mass	= 103 Da
Average Mass	= 103.123861 Da

Chemical Properties

When heated phenyl isocyanide at 200°C it rearranges to benzonitrile

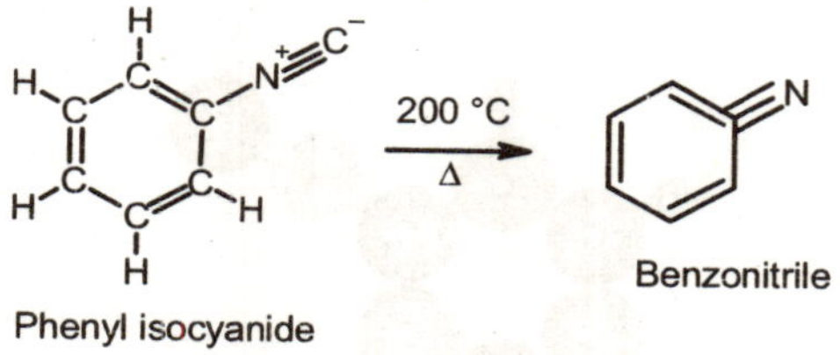

Phenyl isocyanide —(200 °C, Δ)→ Benzonitrile

Phenyl isocyanide reacts with dry hydrogen chloride and gets easily chlorinated as below

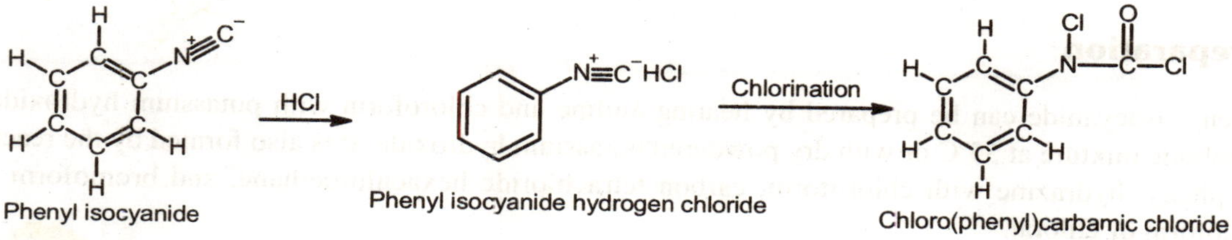

Phenyl isocyanide —(HCl)→ Phenyl isocyanide hydrogen chloride —(Chlorination)→ Chloro(phenyl)carbamic chloride

Dilute acids reacts with phenyl isocyanide to form formanilide and sulphur yields phenyl isocyanate (phenyl mustard oil)

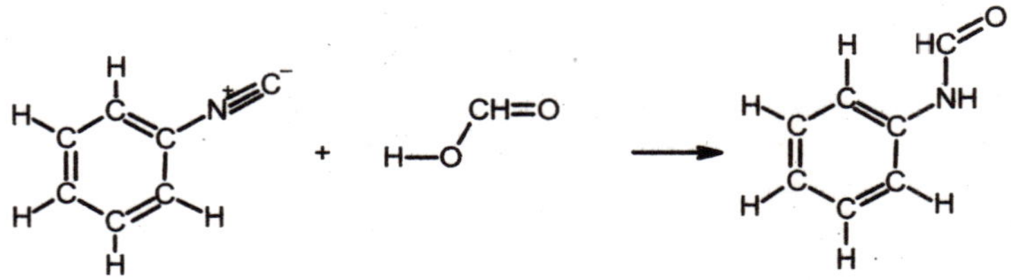

phenyl isocyanide + formic acid → fomanilide

When reacts with sulphur phenyl isothiocyante is produced

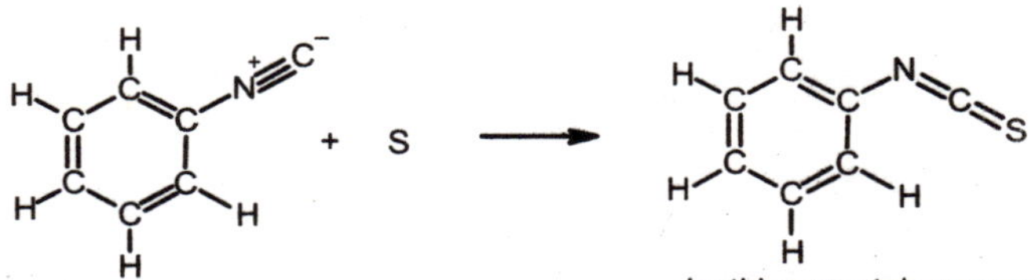

Isocyanobenzene
phenyl isocyanide
+ S →
Isothiocyanatobenzene
phenyl isothiocyanate

phenyl isocyanide is used for the preparation of some illegal adulteration for edible oils and it is also used for the organic synthesis in labs.